LIBERATION ECOLOGIES

Driven by momentous economic and political changes and by apocalyptic visions of impending global ecological doom, the environmental question and the future of development have taken center stage. Yet no discussion of environment or development can begin without interrogating these keywords and the various discourses and practices in which they are situated.

Liberation Ecologies explores contemporary debates over the real definitions of "development" and "environment." Bringing together some of the very best theorists in the Development field, the book discusses the theory, growth, and impact of political ecology, and casts new light on the effects of development on environment in the Third World. Presenting a wide range of case studies from, amongst others, Ecuador, Madagascar, Gambia, China, Zimbabwe, and Indonesia, the book questions the realities and impact of sustainability, and confirms that eradicating poverty through enhancing and protecting livelihood strategies is as much an environmental sustainability issue as a "simple" asset or resource question.

Liberation Ecologies highlights new theoretical engagements between political ecology and poststructuralism, and challenges many conventional notions of development, politics, democracy, and sustainability.

Richard Peet is Professor of Geography, Clark University, Massachusetts; **Michael Watts** is Professor of Geography and Director of the Institute of International Studies, University of California, Berkeley.

LIBERATION ECOLOGIES

Environment, development, social movements

*Edited by Richard Peet and
Michael Watts*

London and New York

First published 1996
by Routledge
11 New Fetter Lane, London EC4P 4EE

Simultaneously published in the USA and Canada
by Routledge
29 West 35th Street, New York, NY 10001

Reprinted 1998

Typeset in Garamond by
Keystroke, Jacaranda Lodge, Wolverhampton
Printed and bound in Great Britain by
T.J. International, Padstow, Cornwall

British Library Cataloguing in Publication Data
A catalogue record for this book is available from the British Library

Library of Congress Cataloging in Publication Data
A catalog record for this book is available from the Library of Congress

ISBN 0–415–13361–0 (hbk)
ISBN 0–415–13362–9 (pbk)

CONTENTS

CONTENTS

ILLUSTRATIONS

TABLES

FIGURES

CONTRIBUTORS

Anthony Bebbington is Professor of Geography, University of Colorado, Boulder, Colorado.

Judith A. Carney is Professor of Geography, University of California, Los Angeles, California.

Arturo Escobar is Professor of Anthropology, University of California, Berkeley, California.

Lucy Jarosz is Professor of Geography, University of Washington, Seattle, Washington.

Donald S. Moore is Wantrup Fellow, Institute of International Studies, University of California, Berkeley, California.

Joshua S.S. Muldavin is Professor of Geography, University of California, Los Angeles, California.

Richard Peet is Professor of Geography, Graduate School of Geography, Clark University, Worcester, Massachusetts.

Haripriya Rangan is Visiting Professor of Geography, University of Kentucky, Lexington, Kentucky.

Richard A. Schroeder is Professor of Geography, Rutgers University, New Brunswick, New Jersey.

Krisnawati Suryanata is Professor of Geography, University of Denver, Denver, Colorado.

Michael Watts is Director, Institute of International Studies, University of California, Berkeley, California.

Lakshman Yapa is Professor of Geography, Pennsylvania State University, Pennsylvania.

Karl S. Zimmerer is Professor of Geography, University of Wisconsin, Madison, Wisconsin.

PREFACE

Most of the contributors to *Liberation Ecologies* have some connection with the Geography departments at Clark University in Worcester, Massachusetts, and the University of California at Berkeley. It is perhaps appropriate that two institutions most closely associated with research on the relation between culture, material life, and environment should form the crucible in which an effort is made to push forward the frontiers of thinking about nature–society relations. There is of course little in this volume of a theoretical or conceptual nature reminiscent of the work of George Perkins Marsh, or for that matter Carl Sauer. But the subject matter of the book – wetlands in Africa, local farmer knowledges in Latin America, soil degradation in China – and the moral commitment to the health of the environment are surely in keeping with the Berkeley and Clark traditions.

A number of chapters in this volume were initially written for two Special Issues of the journal *Economic Geography* focusing on the environment–development–social movements nexus. All of these chapters have been substantially rewritten and several other chapters commissioned to ensure greater geographical and thematic breadth. The starting point for the project was the very success of recent geographic work on the environment – the emergence of "political ecology" in which geographers such as Piers Blaikie, Susanna Hecht, and Harold Brookfield played central roles. The theoretical heart of this body of scholarship was the linking of political economy (typically of a Marxist or neo-Marxist variety) with ecology, or earlier forms of cultural ecology. In attempting to chart what Blaikie and Brookfield called the shifting dialectic of nature–society relations, important new avenues were opened for research and activism. These included analyses of how the capacity to manage resources could be constrained by the relations of production in which peasants were enmeshed, how particular forms of state subsidy stimulated the mining of the soil, or how local forms of knowledge could be harnessed in ecologically adaptive ways. *Liberation Ecologies* begins from these insights, attempts to explore the absences, silences, and weaknesses of political ecology, and presents the work of a new generation of scholars socialized and intellectually formed by the body of work they now seek to extend, deepen, even criticize.

Not surprisingly, this project has been shaped in complicated ways by larger intellectual currents in North Atlantic and Third World academia: the skepticism about so-called meta-narratives, the proliferation of poststructural critiques of rationality, science, and development, and wide-ranging debates over the character and potential of environmentally based social movements (typically counterposed against the supposed exhaustion of class-based politics). All of these larger debates appear in various guises in the chapters in this volume, and the synergy between political ecology and poststructural social theory, in our view, lends the case studies their particular power and originality. It is a sign of the historical moment that no discussion of "environment" or "development" can begin without interrogating the meanings of these keywords and the various discourses and practices in which they are situated. The engagement between knowledge–power institutions and political ecology courses through virtually all the contributions to *Liberation Ecologies* and lends it some originality. Lest it be thought that we are invoking thematic unity or political consensus around a new, established form of analysis ("beyond" political ecology), it needs to be said that there are important tensions and differences among the contributors (including the editors!) – as well as important confluences – yet in our minds this lends to the project an additional *frisson*.

The editors have been centrally engaged with the project since its initiation three years ago and have subsequently worked closely with the authors to ensure an engagement with issues raised in the introduction (Chapter 1). There is no effort in our introduction to lay out a theoretical agenda, or to outline a prospectus of research. Rather, we attempt to locate political ecology and studies of development–environment on a larger intellectual canvas in such a way that new sorts of questions might be asked, new avenues opened up, new synergies created. It is driven naturally in our case by a normative and political commitment to the liberatory potential of environmental concerns and to the community movements rising around natural themes in the South.

Michael Watts would like to acknowledge the support of the Social Science Research Council/MacArthur Foundation International Peace and Security Program which supported his research during the period in which this book was written. Many of the ideas in the introduction and conclusion (Chapters 1 and 12) were first aired in the course of teaching graduate seminars on "Nature, Culture, and Social Theory" at the University of California, Berkeley. I am especially indebted to the following individuals who were central intellectual figures in those seminars: Priya Rangan, Peter Taylor, Iain Boal, Sharad Chari, William Boyd, Carolyn Trist, James McCarthy, Susanne Friedberg, Ravi Rajan, the Bay Area CNS Collective, Dick Norgaard, and Donald Moore. In the final stages of preparing the book, my son Ethan Louis Watts was born. He has red hair and a touch of green in his eyes. His commitment to liberation ecology seems not to be in question.

Richard Peet would like to thank Bruce Bratley, Jody Emel, Tom Estabrook, Elaine Hartwick, Ann Oberhauser, Kathy Olsen, Phil Steinberg, Elliot Tretter, and Danny Weiner for providing a discursive environment in, and especially around, Worcester, Massachusetts, in which the ideas contributed to this book could evolve, and a Clark University Faculty Development grant for financial support during the summer of 1995, when the project was completed.

1

LIBERATION ECOLOGY

Development, sustainability,
and environment in an age of
market triumphalism

Richard Peet and Michael Watts

> Even society as a whole, a nation, or all existing societies put together,
> are not owners of the Earth. They are merely its occupants, its users; and
> like good caretakers, they must hand it down improved to subsequent
> generations.
>
> (Marx, *Capital*, Vol. 1)

> In this world which is so respectful of economic necessities, no one really
> knows the real cost of anything which is produced. In fact the major part
> of the real cost is never calculated; and the rest is kept secret.
>
> (Debord, *The Society of the Spectacle*)

Driven by momentous political and economic changes and by apocalyptic
visions of impending global ecological doom, the environmental question has
returned to center stage, and with a vengeance (Turner *et al.* 1990; World Bank
1992). In the return of the repressed we frequently hear the language of "sustainability" and "sustainable development." The meanings of these terms are hotly
contested (O'Connor 1994). But the new lexicon is so endemic it appears with
as much frequency in the frothy promotional literature of the World Bank as in
the rhetoric of the Sierra Club, the US military, or the myriads of Third World
grassroots environmental and community movements. Whatever its semantic
ambiguity, sustainability has the effect of linking three hitherto relatively disconnected discourses. It is now taken for granted that the *global environmental
crisis*, and a renewed concern with *global demography* (the return of the
Malthusian specter) are inseparable from the terrifying map of *global economic
inequality* (Adams 1991; Lipietz 1988), from the devastating portrait of our
times painted in the 1995 World Health Organization report in which poverty
wields its destructive influence at every stage of human life. In sharp contrast to
the 1960s, even conventional views confirm that eradicating poverty through
enhancing and protecting livelihood strategies is as much an environmental
sustainability issue and a fertility question (in which women's employment and

1

education figures centrally) as a "simple" asset or resource endowment question (World Bank 1992).

This new emphasis on nature–society relations in the context of concerns over the growing polarity of world income (UNDP 1992) has its genesis in a distinctive *fin de siècle* intellectual and political-economic atmosphere. First, the *collapse of many actually existing socialisms* and the rise of a neo-liberal hegemony in policy circles signals, for many, the exhaustion, if not the extinction, of socialist and, in many cases, import substitution or welfarist models of development. Second, there is a *resurgence of environmentalist concerns articulated explicitly in global terms* (e.g. global climate change, ozone depletion) as the 1992 UN Conference on Environment and Development in Rio made clear. Global ecology and the discourse of global environmental management and governance, however, is attached to a renewal of the old debate over the specter of Malthusian over-population (World Bank 1992), though the UN Cairo Population Conference in 1994, unlike earlier conferences, revealed a new sensitivity to the plight of women in population planning. And, third, *the rise of political ecology*, which offered a powerful Marxist-influenced analysis of resource use and environmental conservation during the 1970s and early 1980s (Blaikie and Brookfield 1987) is increasingly shaped, prodded, and challenged by wide-ranging debates within social and development theory. In the late 1980s and early 1990s, a number of loosely related theoretical ideas embracing post-Marxist, action-research oriented, and poststructuralist ideas contributed to "a renewed interest in the diversity of development experience" (Booth 1995: xiii). This vibrant body of work challenges the ostensibly modernist and Eurocentric character of development itself, positing in its place various "alternatives to development" (Escobar 1995; Sachs 1993) and a kind of postmodern discourse on development (Slater 1992). Indeed, poststructural concerns with knowledge–power, institutions and regimes of truth, and cultural difference have proven compelling in the rethinking of both development theory and political ecology, as this book testifies.

Located on this expansive canvas of intellectual and political-economic ferment, *Liberation Ecologies* explores, through a series of rich case studies drawn from Latin America, Africa, East, Southeast, and South Asia, the current debates over development and the environment. In choosing this title we seek to emphasize a number of concerns. Obviously we wish to mark the potential liberatory or emancipatory potential of current political activity around environment and resources. However, we also wish to signal the fact that the proliferation of environmental concerns linked to questions of development has other profound theoretical and practical consequences. One is that the politics of the environment seem to embrace a wide terrain including not just new social movements, but transnational environmental alliances and networks, multilateral governance through, for example, the Global Environmental Facility of the World Bank, and a sensitivity to a panoply of local conflicts and resistances that may not warrant the term "movement." Another is that theories about environment and development – political ecology in its various guises – have been pushed and extended both by the realities of the new social movements themselves, and by intellectual developments associated with discourse theory and poststructuralism. These exciting new developments – many of which appear in the chapters which follow – represent for us the possibility of a more robust political ecology which integrates politics more centrally, draws upon aspects of discourse theory which demand that the politics of meaning and the construction of knowledge be taken seriously, and engages with the wide-ranging critique of development and modernity particularly associated with Third World intellectuals and activists such as Vandana Shiva, Arturo Escobar, and Victor Toledo. *Liberation Ecologies* highlights, in other words, new theoretical engagements between political ecology and poststructuralism on the one hand, and a practical political engagement with new movements, organizations, and institutions of civil society challenging conventional notions of development, politics, democracy, and sustainability on the other.

Our introduction is structured around three broad themes which link environment–development theory and the particular conditions of the 1980s and 1990s outlined so far. The first is political ecology itself – that is to say the efforts begun in the 1980s to "combine the concerns of ecology and a broadly defined political economy . . . [which] encompasses the constantly shifting dialectic between society and land-based resources" (Blaikie and Brookfield 1987: 17). This leads to a discussion of how political ecology may be extended through poststructural critiques of Western reason and discourse theory. Second, we examine the lineaments of this turn to poststructuralism and discourse in general, but with particular reference to development theory (what we call "mapping" development discourse). Finally, we turn to environmental politics, and specifically ideas on social movements and other political forms which are struggles for livelihood but nonetheless are ecological "insofar as they express objectives in terms of ecological requirements for life" (Martinez-Alier 1990: 7).

POLITICAL ECOLOGY: A DEVELOPMENT DISCOURSE FOR THE 1990s?

Twenty-five years after the first stirrings of Earth Day, Spaceship Earth, and a worldwide environmental awareness, it is clear that environmentalism – now couched in the language of sustainability – is back on the political agenda. Some nineteen green parties are active in a dozen West European states, environmental movements dot the landscape of the former socialist bloc, and the link between ecology and Third World poverty has been sealed in such unthreatening and centrist documents as the 1987 Brundtland Commission Report and the Rio Declaration of 1992. It is tempting to see this proliferation of green politics as history repeating itself, but the current conjuncture is quite different from that of the 1960s and early 1970s. First, the restructuring of capitalism in the North Atlantic economies has radically transformed the regulatory environment, while

3

new institutional forms of globalization and market integration (WTO, NAFTA), coupled with new and more destructive technologies and substances in a climate of aggressive deregulation in privatized economies, suggests a quite different world from 1969. Second, the growth of peripheral Fordism and high rates of industrial growth in some of the new industrializing states (Brazil, Korea, Taiwan) has exacted a heavy environmental toll, while the terrifying environmental record in the former "socialist" bloc is now slowly becoming public knowledge (Feshbach and Friendly 1992). Indeed, there is a profound sense in which the very crisis of socialism itself was precipitated by serious environmental and resource problems generated by the economics of shortage. And, third, the recognition of new long-term catastrophic *global* tendencies (global warming, ozone depletion, biogenetic hazards) has spawned new efforts at multilateral and transnational institutional regulation and governance: witness UNCED in Rio, the Montreal protocols on climate change, and the efforts to green GATT (Esty 1994; Sand 1995).

The intellectual firmament of the last fifteen years is also markedly different from that of the first environmental wave of the 1960s, which was dominated by Darwinian or Malthusian thinking and simple organic analogies. Perhaps the most important line of recent social scientific thinking about environment and development is "political ecology." The term can be traced with some certainty to the 1970s when it emerged as a response to the theoretical need to integrate land-use practice with local–global political economy (Wolf 1972) and as a reaction to the growing politicization of the environment (Cockburn and Ridgeway 1979). Subsequently taken up by geographers, anthropologists, and historians (Bryant 1991), it is perhaps most closely associated with Blaikie (1985) and Blaikie and Brookfield (1987). In their view, political ecology combines the concerns of ecology with "a broadly defined political economy" (1987: 17): accordingly environmental problems in the Third World, for example, are less a problem of poor management, overpopulation, or ignorance, as of social action and political-economic constraints. Standing at the center of Blaikie and Brookfield's political ecology is the "land manager" whose relationship to nature must be considered in "a historical, political and economic context" (1987: 239).

Political ecology is part of a larger body of work which had its origins in the critique of ecological anthropology and "cultural ecology" in the late 1970s (Watts 1983). This earlier theory gained currency during the first wave of the post-war environmental movement in the late 1960s, and drew attention to the adaptive capacities of indigenous societies both in the efficacy of their "cognized models" of the local environment (for example, farmers in the Ivory Coast possessed a sophisticated understanding of local soil conditions and botanical relations), and in their *structural similarities* to all biological populations and living systems. Rappaport's (1967) classic account of the role of ritual pig killing among the Tsembaga Maring in local environmental regulation of fragile tropical ecosystems was a model of this ethnographically rich "systems thinking" about human adaptation to the environment. Culture – for example ritual practices or social structure – was seen to function as a homeostat or regulator with respect

4

to environmental stability. These studies took concepts derived from ecological theory or cybernetics and applied them directly to the sphere of social life; peasant societies were adaptive systems just like any other biological population, and culture was posited as an ecologically functional attribute of the evolutionary demands of the environment. Societies were closed homeostatic systems populated, as Jonathan Friedmann caustically observed, by "cybernetic savages." Typically working in rural and agrarian Third World societies, cultural ecologists nonetheless unearthed important data on local ethnoscientific knowledges and the relations between cultural practices and resource management – something which has re-emerged in the current concern with indigenous technical knowledge and the activities of the Center for Indigenous Knowledge for Agriculture and Rural Development in The Hague – but typically placed these in an overarching regulatory structure derived from the cybernetic and self-correcting properties of closed living systems. Many societies studied were actually part of large, complex, open political economies and it was precisely this openness – in short, market and state involvements of various sorts – which in many cases seemed to undermine, or be in contradiction with, the ideas of equilibrium and homeostasis on which geographers and anthropologists had drawn (Nietschmann 1973).

By the late 1970s, propelled by the appeal of Marxism and political economy and the proliferation of a radical peasant studies literature which privileged production over biological relations, ecologically concerned social scientists attempted to weld together the compelling questions of how communities were being integrated into, and transformed by, a global economy ("economic change") with local resource management and environmental regulation and stability (Grossman 1984; Watts 1983). During the 1980s, this attempt at synthesis met a second phase of environmental activism (the rise of the green movements world-wide) and a recognition of the deepening *global* human-induced modifications of the environment in part driven by the rapid industrialization of parts of the South and a renewed institutional concern with the consequences of high rates of demographic growth especially in Africa, South Asia, and parts of the Middle East (Meyer and Turner 1992). Forged in the crucible of Marxian or neo-Marxian development theory, this new "political ecology" was not inspired by the isolated rural communities studied by Rappaport but by peasant and agrarian societies in the throes of complex forms of capitalist transition. Market integration, commercialization, and the dislocation of customary forms of resource management – rather than adaptation and homeostasis – became the lodestones of a critical alternative to the older cultural or human ecology.

Political ecology and its limits

Environmental degradation is created . . . by the rational response of the poor households to changes in the physical, economic and social circumstances in which they define their survival strategies.

(de Janvry and Garcia 1988: 3)

5

prob w/ theoretical coherence —

If political ecology reflects a confluence between ecologically rooted social science and the principles of political economy, its theoretical coherence nonetheless remains in question. A broad and wide-ranging approach, encompassing the work of such diverse scholars as Susanna Hecht, Harold Brookfield, Anna Bramwell, Susan Stonich, Michael Redclift, and Ram Guha, political ecology seems grounded less in a coherent theory as such than in similar areas of inquiry (cf. Bryant 1992 who specifically identifies contextual sources of ecological change, questions of access, and political ramifications of environmental alteration; see also Bramwell 1989). Some of the tensions and heterogeneities within the approach are reflected in Blaikie and Brookfield's (1987) key text *Land Degradation and Society*. The authors raise a number of important issues including the social origins of degradation, the plurality of perceptions and definitions of ecological problems, the need to focus on the land manager (and his/her opportunities and constraints), and the pressure of production on resources.

Land Degradation and Society contains three broad motifs which turn on the relation between poverty and degradation. The first is the concept of marginality and how political, economic, and ecological marginality can be self-reinforcing: "land degradation" as they say "is both a cause and a result of social marginalization" (1987: 23). Second, the pressure of production on resources is transmitted to the environment through social relations which compel the land manager to make excessive demands ("the pressure of deprivation" as they call it). And finally, in keeping with poststructuralism, they acknowledge that the facts of degradation are contested, and that there will always be multiple perceptions (and explanations) – one person's degradation is another's soil fertility. All of this amounts to a radical critique of the pressure-of-population-on-resources view of environment and points to the need for a rethinking of both conservation and development.

a critique o neo-Malthus

Blaikie and Brookfield's important intervention represents within geography, and social theory more generally, a sophisticated extension of previous efforts to integrate questions of access and control over resources – relations of production as realms of possibility and constraint – with human ecology. But there are also complementarities between the ideas in *Land Degradation and Society* and those of other social theorists working on questions of ecological crisis and rehabilitation. Like the work of Little and Horowitz (1987), regional political ecology focuses on the producer and ecological pressure points; it shares with Redclift (1987) an emphasis on the contradictions of development; and with Jane Collins (1987) a sensitivity to the social causes of degradation and the need for a rethinking of development itself. Like Bunker's (1985) Amazonian study, Blaikie and Brookfield employ a regional analysis sensitive to spatial variation and environmental heterogeneity; like Perrings (1987) they raise the suggestion that the market–price system as a means to regulate the environment is limited by the time perspectives of economic agents under capitalism and by the presence of uncertainty. And not least they share with Martinez-Alier and Schluepmann

6

(1987) a belief that value in land (what Blaikie and Brookfield call "landesque capital") is inconsistent with both neo-classical economic theory and Marx's labor theory of value.

Collectively this body of work has punched a huge hole in the pressure-of-population-on-resources view, and the market distortion or mismanagement explanation of degradation. In their place it has affirmed the centrality of *poverty* as a major cause of ecological deterioration (de Janvry and Garcia 1988; Martinez-Alier 1990; Mellor 1988; see Blaikie and Brookfield 1987: 48). This represents an important advance in our thinking about nature–society relations but it nonetheless requires a much greater refinement, and an explicit theorization which is typically lacking because of political ecology's frequent appeal to plurality. What then are some of the limits and weaknesses of the political ecology that emerged in the late 1970s and early 1980s?

First, those who place an undue emphasis on poverty and poor peasants must recognize that impoverishment is no more a cause of environmental deterioration than its obverse, namely affluence/capital. Hecht and Cockburn (1989) make this point with respect to the rates of deforestation in the Amazon basin. The danger is to neglect the obvious power of capital as a material force in degradation and, as a consequence, come close to blaming the victim albeit in terms of the situational rationality of the land manager who is compelled to mine the soil or fell the forest. Second, the focus on poverty is perhaps not unrelated to the bias toward rural, agrarian, and Third World matters in *Land Degradation and Society*, and indeed in political ecology more generally (a bias which this book reinforces in part!). How, for example, might poverty or political ecology help explain worker injuries in the maquila plants in northern Mexico, toxic dumping in Nigeria, or urban water pollution in Turin? And, third, Blaikie and Brookfield privilege land – with good reason in view of its special significance in largely agrarian Third World states – as opposed to other "resources." The point we emphasize is that a poverty-centered analysis is, as the authors concede, only part of the story: there are other stories to tell of worker health and safety, air pollution, the decay of Third World cities, and of the restructuring of capitalism, and so on. The extent to which this partiality is of any analytical consequence rests, of course, on the theory which grants to poverty its causal powers.

Poverty, then, is at best only a *proximate* cause of environmental deterioration. In other words, one has to have a theory capable of explaining how the poverty of specific land managers is reproduced through determinate structures and by specific relations of production. Blaikie and Brookfield move some way toward this goal by isolating production but in an extremely diffuse and inconsistent way. Specifically, they invoke marginalization (which is an awkward label for several complex and contradictory processes) and an absence of control over resources. In short they make the land manager, and occasionally the production unit, the fulcrum, trapped within complex webs of relations, all held together by a political economy which, in a rather unhelpful way, is lumped together as "exogenous" (1987: 70). These exogenous, and largely untheorized, inputs into

7

the political ecology decision-making model (i.e. how and why the land manager acts) purportedly explain declines in land quality, a process which, according to Blaikie and Brookfield, *only* elicits three responses: perception correction, "change the social data," and migration. In short, a very broad and untheorized exogenous cause seems deterministically to produce quite specific outcomes; which is hardly the sort of dialectical analysis that they themselves suggest should be on offer.

In spite of the fact that Blaikie and Brookfield talk of the selection of strategic factors which have causal power, we are not given a theory which helps us in the act of selection. Rather we are provided with "a chain of explanation" (one begins with the Nepalese farmers and ends with Nepal's relationship to India) in which there is no sense how or why some factors become causes. Coupled with their emphasis on plurality, the authors actually produce not "a theory which allows for . . . and identifies complexity" (1987: 239), but an extremely diluted, diffuse, and on occasion voluntarist series of explanations. Degradation can arise under falling, rising, or stable population pressures, under an upswing or downswing in the rural economy, under labor surplus and labor shortage; in sum, under virtually *any* set of conditions. The best that Blaikie and Brookfield provide is what they call a "conjunctural" explanation which seems to operate under all empirical circumstances.

Despite their claims for theory construction and the importance of social structural antecedents, Blaikie and Brookfield actually present an *ad hoc* and frequently voluntarist view of degradation. Political ecology is radically pluralist and largely without politics or an explicit sensitivity to class interest and social struggle. Yet any analysis of land-based resources must surely confront and incorporate politics inscribed in various social arenas: familial-patriarchal, production-labor process, and the state (Burawoy 1985). Politics must be central to political ecology in order to give the bare bones of "poverty" some sort of flesh if it is to be employed analytically. Political ecology comes closest to theory when it invokes surplus extraction and yet the authors on occasion seem more inclined to abandon theory altogether. Rather than outlining an explicit theory of production or political economy and an arsenal of middle-level concepts, Blaikie and Brookfield only provide a plurality of disconnected linkages and levels. Hence their discussion of degradation in socialist economies can only conclude that it exists, and cannot offer any insights into the question they pose, namely "is there a distinctive socialist environmental management?" (1987: 208), which presupposes a theorization of socialist political economy.

In short, political ecology's conception of political economy appears fuzzy ("almost every element in the world economy," 1987: 68) and diffuse. Their emphasis on plurality comes perilously close to voluntarism while their chains of explanation seem incapable of explaining how some factors become causes. Particularly striking is the fact that *political* ecology has very little *politics* in it (an issue which a number of chapters in this book take on – especially Chapters 4, 9 and 10 by Bebbington, Schroeder and Suryanata, and Rangan). There is no serious attempt at treating the means by which control and access of resources or property rights are defined, negotiated, and contested within the political arenas of the household, the workplace, and the state.

These lacunae in Blaikie and Brookfield's book, coupled with its broad inter-disciplinary focus, have pushed the field of political ecology in a number of important and interesting directions. In our view, these developments have been driven by a dialogue with a larger intellectual environment – ideas drawn from poststructuralism, gender theory, critical theories of science, environmental history, and Marxist political economy – and by the realities of a panoply of ecological movements and struggles throughout the Third World and in post-socialist transitional states. No attempt is made here to review this burgeoning field of political ecology – indeed many of these concerns are raised directly by chapters in this book – but rather we point to several fruitful avenues for debate and empirical exploration. It is striking, nonetheless, how political ecology has, from its inception, wrestled with the way management questions – whether in the form of regulatory apparatuses, local knowledge systems, new community or resource-user groups – must occupy an important space in civil society. As we suggest on our map of development discourse (see Table 1.1), political ecology, like much of development theory in the 1990s, also seems to be increasingly concerned with institutions and organizations in the context of shifting configurations of state and market roles.

New directions, new questions

A number of loosely configured areas of scholarship have extended the frontiers of political ecology and have elaborated and developed the important work of Blaikie, Brookfield and others. The first attempts to refine political economy within the ambit of political ecology: in other words to make the causal connections between the logics and dynamics of capitalist growth and specific environmental outcomes rigorous and explicit. Some of the most exciting new work centers on efforts at explicitly re-theorizing political economy and environ-ment at several different levels. At the philosophical level there are debates over Marxism and ecology (Benton 1989; Grundemann 1991; see also Leff 1995) and whether the labor process is compatible with eco-regulation and the notion of biological limits. The work of James O'Connor (1988) and the journal *Capitalism, Nature, Socialism* (*CNS*) starts from the "second contradiction of capitalism." In this view Marx identifies production conditions (nature, labor power, and communal conditions of production) which capital cannot produce for itself as commodities. The state mediates, and hence politicizes, conflicts around these conditions (environmental movements, feminism, and social move-ments) in an effort at maintaining capitalist accumulation. Many contributions to *CNS* explore these ideas in various parts of the Third World. Also there are attempts at harnessing specific concepts drawn from political economy as a way of linking the two structures of nature and society. For example how the simple

9

reproduction squeeze compels self-exploitation among peasants who mine the soil; or how functional dualism can facilitate labor migration which undermines local conservation or constrains sustainable herding practices (Faber 1992; Garcia Barrios and Garcia Barrios 1990; Little and Horowitz 1987; Stonich 1989; Toulmin 1992; Watts 1987). Much remains to be done, however, in theorizing the specific dynamics of actually existing socialisms and the environment (Herskovitz 1993). Here, of course, the devastating ecological consequences of socialist political economy must be located not with respect to markets and profit but in relation to what Janos Kornai calls "the economics of shortage," that is to say the complementary and contradictory rationalities of centralized state planning (and its attachments to industrial gigantism and heavy goods) on the one hand, and the reciprocities and networks at the enterprise level on the other.

A second broad thrust questions the absence of a serious treatment of *politics* in political ecology. Efforts at integrating political action – whether everyday resistance, civic movements, or organized party politics – into questions of resource access and control have proven especially fruitful (Broad 1993; Kirby 1990). At the household management level, several studies focus on gender and domestic politics and struggles around the environment (Agarwal 1992; Guha 1990; MacKenzie 1991; see also Chapters 8 and 9 by Carney and Schroeder and Suryanata in this volume) specifically focused on property rights, labor, and the micro-politics of access and control within the domestic sphere. At other levels of analysis – the state, interstate, and multilateral institutions, and local, i.e. community level resource control – important new work has forged analytical links between power relations, institutions, and environmental regulation and ecological outcomes. Rich's book on the World Bank (1994) and more generally the ecological establishment, Peter Hass's studies of transnational scientific communities (1993) – epistemic communities in his lexicon – and international environmental agreements, Peter Sand's (1995) analysis of post-UNCED legal frameworks, all illustrate David Harvey's suggestion that "control over resources of others in the name of planetary health [and] sustainability . . . is never too far from the surface of many western proposals for global environmental management" (1993: 25). Peluso's brilliant study (1993a) links the historiography of criminality with everyday resistance to show how state power and forest management institutions are contested by Indonesian peasants, and raises the larger issues of the colonial legacy and of coercive patterns of conservation. In subsequent work on Kenya, Peluso shows how the militarization of environmental and resource conservation can be legitimated by international conservation groups (Peluso 1993b). What is at stake here is the more general question of participation, community rights, and local needs in environmental protection and conservation strategies (Utting 1994). Finally, the emancipatory potential which unites nature with social justice is a key theme in the emerging body of work on the ecology of the poor (Broad 1993; Gadgil and Guha 1992; Hecht and Cockburn 1989; Martinez-Alier 1990) and in the large body of work on Indian environmental movements (see *IICQ* 1992). Contained within this work is a sensitivity to the panoply of political forms – movements, domestic struggles over property and rights, contestations within state bureaucracies – and the ways in which claims are made, negotiated, and contested.

A third focus is the complex analytical and practical association of political ecology and the institutions of civil society. The growth of environmental movements largely unregulated by, and distinct from, the state poses sharply the question of the relations between civil society and the environment. There are two obvious facets of these relations, both of which have received some attention. The first is the origins, development, and trajectories of the environmental associations and organizations (see Escobar 1995; Ghai 1992; *Socialist Review* 1992). What are the spaces within which these movements develop and how, if at all, do they articulate with other organizations and resist the predations of the state (see Bebbington and Moore, Chapters 4 and 6 in this volume)? The second draws on the substantial literature on local knowledges and ecological populisms (Richards 1985; Warren 1991). The concern is not simply a salvage operation – recovering disappearing knowledges and management practices – but rather a better understanding of both the regulatory systems in which they inhere (see the literature on common property, Ostrom 1990) and the conditions under which knowledges and practices become part of alternative development strategies. In this latter sense we return to the politics of political ecology but more directly to the institutional and regulatory spaces in which the knowledges and practices are encoded, negotiated, and contested (see Bebbington and Jarosz, Chapters 4 and 7 in this volume) and ultimately to the relation between democracy and environmentally sound livelihoods.

A fourth theme employs discursive approaches to tackle head on Blaikie and Brookfield's point about the plurality of perceptions and definitions of environmental and resource problems. Several new lines of thinking are important. One draws upon the critical studies of science as a way of exploring the politics of what one might call "regulatory knowledge"; why particular knowledges are privileged, how knowledge is institutionalized, and how the facts are contested. Beck's (1994) work on risk and reflexive modernization, and Shrader-Frechette's (1990) work on risk and rationality are important illustrations of this sort of research. Another line of thinking traces the history of particular institutions – say forestry – and how particular knowledges and practices are produced and reproduced over time (Sivaramakrishnan 1995; Rangan, Chapter 10 in this volume). The genesis and transmission of conservation ideas, and the institutions of national parks and their management, have been explored productively in this way (Beinart and Coates 1995; Grove 1993; Neumann 1992). Another line of work examines the globalization of environmental discourse and the new languages and institutional relations of global environmental governance and management. Taylor and Buttel (1992), for example, trace the moral and technocratic ways in which the new global discourse on the environment is privileged, and how in the formulation of environmental science some courses of action are facilitated over others.

11

The question of doing environmental history represents a fifth aspect of an invigorated political economy (in this regard see the new journal *Environmental History* edited by Richard Grove). In providing much-needed historical depth to political ecology, environmental historians raise important theoretical and methodological questions for the study of long-term environmental change. The obvious theoretical contrasts between Worster (1977), Merchant (1993), and Cronon (1992) point to an extraordinary heterogeneity in the field. Contained within each is the idea of writing alternative histories from the perspective of long-term ecosystemic changes which cannot be captured with the clumsy unilinear models of agricultural and environmental change. The relations between agrarian intensification and the environment are rarely so simple. Tiffen and Mortimore's (1994) study of Machakos District in Kenya shows how population increased five-fold between 1930 and 1990 but the environmental status actually improved over the same period. Soil structure improved and even woodfuel was sustained. Similarly Fairhead and Leach (1994), working in Sierra Leone, locate forest quality and biodiversity in the influence of past land-use practices. Like new work on Amazonia and South Africa, they show how habitation and cultivation can improve soil and support denser woodlands. As they put it, "vegetation patterns are the unique outcomes of particular histories not predictable divergences from characteristic climaxes" (1994: 483). In a sense the new environmental historians meet on the same ground as a quite different intellectual tradition, derived from the so-called agrarian question (cf. Kautsky 1906), which attempts to chart the ways in which the biological character of agriculture shapes the trajectories of capitalist development (Kloppenberg 1989). Opportunities for exploring the long-term capitalization of nature through "appropriation" and "substitution" (Goodman *et al.* 1990; O'Connor 1994), and their environmental ramifications, can, and should be, readily seized by political ecologists.

Finally, there is the much-needed interrogation of the term "*ecology*" in political ecology and the extent to which political ecology is harnessed to a rather outdated view of ecology rooted in stability, resilience, and systems theory (Zimmerer 1994). Botkin (1990) and Worster (1977), among others, describe the relatively new ecological concepts which pose problems for the theory and practice of political ecology. The shift from 1960s systems models to the ecology of chaos, that is to say chaotic fluctuations, disequilibria, and instability, suggests that many previous studies of range management or soil degradation resting on simple notions of stability, harmony, and resilience may have to be rethought (Zimmerer 1994). The new ecology is especially sensitive to rethinking space–time relations to understand the complex dynamics of local environmental relations in the same way that the so-called dialectical biologists (Levins and Lewontin 1985) rethink the evolutionary dynamics of biological systems. Notwithstanding Worster's (1977) warning that disequilibria can easily function as a cover for legitimating environmental destruction, some of the work on agro-ecology (Altieri and Hecht 1990; Gleissman 1990; see also Zimmerer, Chapter 5 in this volume) suggests that the rethinking of ecological science can be effectively deployed in understanding the complexities of local management (for example in intercropping and pest management).

All of these new directions are not necessarily of a theoretical piece, and it remains to be seen where the conceptual confluences and tensions will arise within the political ecology of the 1990s. What is striking, however, is the extent to which *these new directions attempt to engage political ecology with certain ideas and concepts derived from poststructuralism and discourse theory.* There is in other words an extraordinary vitality within the field reflecting the engagements within and between political economy, the power–knowledge field, and critical approaches to ecological science itself. As a shorthand we refer to these confluences and engagements as "liberation ecology." The implication in this notation is to recognize the emancipatory potential of what we will call the "environmental imaginary" and to begin to chart the ways in which natural as much as social agency can be harnessed to a sophisticated treatment of science, society, and environmental justice. Of course, a major site of such engagement is in the analysis of social and environmental movements, a field which draws together the explosive growth of organizations and civic movements around sustainability with an implicit critique (and an alternative vision) of "development." It is to the philosophical and social theoretical underpinnings of development and the environmental imaginary that we now turn.

DISCOURSE, RATIONALITY, AND DEVELOPMENT

Cogito ergo sum.
(Descartes)

Poststructural theory's fascination with discourse originates in its rejection of modern conceptions of truth. In modern philosophy truth resides in the exact correspondence between an externalized reality and internal mental representations of that reality. Enlightenment philosophy considered all minds to be structurally similar, truths to be universal, and knowledge potentially the same for everyone. By comparison, following the philosophers Wittgenstein, Heidegger, and Dewey, the postmodern theorist Rorty (1979: 171) argues that the notion of knowledge as representation should be abandoned in favor of knowledge without foundations: "knowledge as a matter of conversation and of social practice, rather than as an attempt to mirror nature." For Foucault (1972, 1973, 1980; Dreyfus and Rabinow 1982; Rabinow 1986), similarly, each society has a *regime of truth*, with control of the "political economy of truth" constituting part of the power of the great political and economic apparatuses: these diffuse "truth," particularly in the modern form of "scientific discourse," through societies, in a process infused with social struggles. In the poststructural view, then, truths are statements within socially produced discourses rather than objective "facts" about reality.

13

Discourse theory

A "discourse" is an area of language use expressing a particular standpoint and related to a certain set of institutions. Concerned with a limited range of objects, a discourse emphasizes some concepts at the expense of others. Significations and meanings are integral parts of discourses just as, for example, the meaning of words depends on where a statement containing them is made (Macdonell 1986: 1–4). Hence for Barnes and Duncan (1992: 8) discourses are "frameworks that embrace particular combinations of narratives, concepts, ideologies and signifying practices, each relevant to a particular realm of social action." Discourses vary among what are often competing, even conflicting, cultural, racial, gender, class, regional, and other differing interests, although they may uneasily coexist within relatively stable ("hegemonic") discursive formations.

Discourse theory came to prominence in the context of the critique of Western rationality. Horkheimer and Adorno (1991) found European rationality liberating at the cost of political alienation. Foucault (1980: 54) found Western rationality's claim to universal validity to be "a mirage associated with economic domination and political hegemony." But as Young (1990: 9) points out, the French poststructural philosophical tradition is concerned particularly with the relation between the universal truth claims of the Enlightenment and the universal power claims of European colonialism; the new critical stress on this relation has stimulated a "relentless anatomization of the collusive forms of European knowledge." Hence Derrida (1971: 213) says: "the white man takes his own mythology, Indo-European mythology, his own *logos*, that is, the *mythos* of his idiom, for the universal form of that he must still wish to call Reason." In this view, then, Enlightenment reason is a regional logic supporting, reflecting, and justifying a history of global supremacy rather than a universal path to absolute truth. Reason, in a word, is ideological.

This critique of truth and re-emphasis on discourses of power when projected into space produces a new approach to inter-regional relations, among other things focused on the discursive relations between hegemonic and dominated regions. Said (1979: 2) argues that "the Orient" helped define Europe as its contrasting image (i.e. as "its other"); "Orient*alism*" is a "mode of discourse with supporting institutions, vocabulary, scholarship, imagery, doctrine, even colonial bureaucracies and colonial styles" through which European culture was able to "produce" the Orient (politically, imaginatively, etc.) in the post-Enlightenment period. Because the Orientalist discourse limits thought, the Orient was not, and is not, a free subject of thought or action. Vico observed that humans make their own history based on what they know; extending this to geography, Said finds localities, regions, geographical sectors like Orient and Occident, to be humanly "made." Subsequent work extends this notion of "discourses on the other" to a whole history of the different European conceptions ("science fictions") of "alien cultures" (Hulme 1986; McGrane 1989; Todorov 1984).

For Bhabha (1983a, 1983b: 19), conversely, representations of the Orient in

14

Western discourse evidence profound ambivalence towards "that otherness which is at once an object of dislike and derision." Colonial discourse, for him, is founded more on anxiety than arrogance, and colonial power has a conflictual economy – hence colonial stereotyping of subject peoples is complex, ambivalent, contradictory representational form, as anxious as it is assertive. So, for example, in an analysis of mimicry, Bhabha (1984) argues that when colonized people become "European" the resemblance is both familiar and menacing to the colonists, subverting their identities. The hybrid that articulates colonial and native knowledges may reverse the process of domination as repressed knowledges enter subliminally, enabling subversion, intervention, and resistance (Bhabha 1984). Similarly for Baudet (1965: vii): "the European's images of non-European man are not primarily, if at all, descriptions of real people, but rather projections of his own nostalgia and feelings of inadequacy."

One complex, controversial, and very much unresolved issue is whether discourse theory can recover the voices of colonized peoples. Something like this is the aim of the "subaltern studies group" (Guha and Spivak 1988). Guha's (1983: 2–3) original position was that colonial historiography denied the peasant recognition as a subject of history. Acknowledging peasants as makers of rebellion means attributing to them a consciousness (cf. Gramsci 1971: 53). Guha tries to identify the (recurring) elementary aspects in such a rebel consciousness, his main theme being that the peasant's subaltern identity includes an imposed negative consciousness from which, however, revolt often derives from inversion (as with the fight for prestige). Spivak (1987: 206–7) however sees the subaltern studies group's attempt to retrieve a subaltern or peasant consciousness as a strategic adherence to the essentialist and humanist notions of the Enlightenment. As long as such Western, modernist notions of subjectivity and consciousness are left unexamined, the subaltern will be narrativized in what appear to be theoretically alternative but politically similar ways (MacCabe 1987: xv). Spivak's alternative involves the structural notion of subject-positions, in which the "subject," for example of a statement, is not the immediate author but "a particular, vacant place that may in fact be filled by different individuals" (Foucault 1972: 95; see also Foucault 1980: 196–7). Here Spivak seeks to reinscribe the many, often contradictory, subject-positions assigned by multiple colonial relations of control and insurgency, so that a subaltern woman, for example, is subjected to three main domination systems, class, ethnicity, and gender. From this she reaches the extreme, and for us indefensible, position that subaltern women have no coherent subject-position from which to speak: "the subaltern cannot speak" (Spivak 1988: 308).

Regional discursive formations

These themes only indicate the potentials of discourse theory for understanding relations between geographical groups of people. We find these positions attractive in that here, at least, poststructural theory links with the causes of oppressed peoples, the geographical dimensions of power relations, and the relentless critique of everything, even notions usually considered to be emancipatory. We find particularly suggestive the hierarchical relations between centralized power articulated through hegemonic, rational, "truthful" discourses and the "mythological" discourses of peripheralized and dominated peoples. By criticizing the modern belief in rational humans speaking objective science, poststructural theory opens a space in which a wide range of beliefs, logics, and discourses can be newly valorized.

We theorize this in terms of *"regional discursive formations"* (cf. Lowe 1991; see also Peet 1996). Certain modes of thought, logics, themes, styles of expression, and typical metaphors run through the discursive history of a region, appearing in a variety of forms, disappearing occasionally, only to reappear with even greater intensity in new guises. A regional discursive formation also disallows certain themes, is marked by absences, silences, repressions, marginalized statements, allowing some things to be mentioned only in highly prescribed, "discrete," and disguised ways. Within a regional discursive formation even competing "opposite" notions often employ the same metaphors, perhaps even similar logics. Hence oppositional positions may be partly captured by hegemonic discourses which shift to incorporate particularly insightful and vivid images. We argue that regional discursive formations originate in, and display the effects of, certain physical, political-economic, and institutional settings. Hegemonic discursive formations grounded in material, political, or ideological power supremacies extend over spaces with greatly different physical characteristics and discursive traditions. As the previous discussion indicates, we find particularly relevant to the geographical imagination theoretical notions dealing with the power-saturated interactions and interchanges between people immersed in regional discursive formations, articulations which leave no discourse intact, which continually produce hybrids. We stress the theme of the discourse on nature as a powerful, almost primordial, element in discursive formation; here we see links with what we later call "environmental imaginaries."

The discourse of development

The world knows much better now what [development] policies work and what policies do not. . . . [Now] we almost [never] hear calls for alternative strategies based on harebrained schemes.

(World Bank Official, cited in Broad 1993: 154)

Such reconceptualizations of power–knowledge, discourse, and space see development as perhaps *the* main theme in the Western discursive formation; it is simply the case that, in the West, the passage of time is understood developmentally, that is, "Things are getting better all the time." By contrast, poststructuralism has increasingly come to see development efforts as "uniquely efficient colonizers on behalf of central strategies of power" – the apparent ability to "make things better" is *the* main way of achieving power (Dubois 1991: 19; Schuurman 1992; Slater 1992; Watts 1993). The pioneering work, by Escobar (1984–5, 1988, 1992a, 1995), thus finds modern development discourse to be the latest insidious chapter of the larger history of the expansion of Western reason; that is, he believes reasoned knowledge uses the developmental language of emancipation to create systems of power in a modernized world. Such hegemonic discourses appropriate societal practices, meanings, and cultural contents into the modern realm of explicit calculation, subjecting them to Western forms of power–knowledge. They ensure the conformity of the myriad peoples of the world to First World (especially American) types of economic and cultural behavior.

For Escobar, development has penetrated, integrated, managed, and controlled countries and populations in increasingly detailed ways. It has created a type of underdevelopment which is politically and economically manageable. Its power acts not by repression but normalization, the regulation of knowledges, the moralization of issues. The new space of the "Third World," carved out of the vast surface of global societies, is a new field of power dominated by development sciences accepted as positive and true. Yet, he says, political technologies which sought to erase underdevelopment from the face of the earth end up, instead, multiplying it to infinity.

Thus the Western, modernist discursive formation, formulated during momentous changes in global power relations, in control over nature, in science and technology, has as its dynamic theme the core concept of "development." This seizes control of the discursive terrain, subjugating alternative discourses which Third World people have articulated to express their desires for different societal objectives. People are controlled, or "discursively regulated" (Peet 1996), by replacing their aspirations with Western mimicry. Through critique, post-structural theory wants to liberate aspirations. In the following section we map in detail recent tendencies in the content and meaning of this contentious concept of "development."

Mapping development discourse: a cartography of power

The postindependence development efforts failed because the strategy was misconceived. Governments made a dash for "modernization", copying but not adapting Western models. . . . *This top down approach demotivated ordinary people, whose energies most needed to be mobilized in the development effort.* . . . *The strategy [after Independence] failed . . . because it was based on poorly adapted foreign models. The vision was couched in the idiom of modernization.* . . . In recent years, however, many elements of this vision have been challenged. Alternative paths have been proposed. They give primacy to agricultural development, and emphasize not only prices, markets and private sector activities but also *capacity building, grassroots participation, decentralization and sound environmental practices.* . . . The time has come to put them fully into practice.

(World Bank 1989: 3, 36, emphasis added)

Failed modernization, alternative visions, grassroots participation, people power, environmental sustainability: this is not a vocabulary typically associated with the most influential advocate of global capitalist development. Could the World Bank really have embraced the popular energies of "ordinary people" in the name of sustainable development alternatives? At the heart of its long-term strategy, says the Bank, is the desire to release energies that permit "*ordinary people* . . . to take charge of their lives" (World Bank 1989: 4, emphasis added). What is on offer is a recognition, indeed celebration, of democratization movements which have attended the frontal assault (led in large measure by global regulatory institutions like the IBRD [International Bank for Reconstruction and Redevelopment] and the IMF) on various state-centered development strategies (i.e. everything from government subsidies of food, to state provision of tertiary education, to import substitution industrialization strategies).

The "new" World Bank approach can be contested at many levels: its ability to rewrite history to suit the Bank's own ideological purposes, its unwillingness to assume accountability for its own failures (whether smallholder colonization schemes in Brazil or massive dam projects in India), its still flimsy commitment to the environment, its partial and limited interpretation of sustainability, and so on. But what is particularly striking is not the purported newness of what has been variously called the "Washington consensus" or the "new realism," but its historical antiquity; in other words the ease with which the Bank's new approach can be situated on a much larger map of development ideas, the links to what might be called a cartography of development discourses. Unlike the World Bank, which believes that the 1950s represents a historic watershed with the arrival of development thinking in Africa and elsewhere, development theorizing has a much deeper history and one characterized by a recycling of key development ideas which appear, disappear, and reappear in new guises under changed political-economic and ideological circumstances (what we referred to earlier as regional discursive formations). While these ideas may have real power and endurance as Hall (1989: 390) rightly notes in his discussion of the spread of Keynesian thinking, "they do not acquire political force independent of the constellation of institutions and interests already present there."

A genealogy of "development"

There is a genealogy of the Saint-Simonian doctrine [of development as a response to the faults of Progress] which runs from and through the nineteenth century to the present. One genealogical line from Comte to John Stuart Mill and then . . . to the Fabian socialists who domesticated [this] doctrine for Britain.

(Cowen and Shenton 1995: iii)

18

In his book *Keywords* Raymond Williams (1976: 104–6) notes that the complex genealogy of "development" in Western thinking can "limit and confuse virtually any generalizing description of the current world order." Rather it is in the analysis of the "real practices subsumed by development that more specific recognitions are necessary and possible." The history of these "real practices" is, however, long and complex. While "development" came into the English language in the eighteenth century with its root sense of unfolding, it was granted a new lease of life by the evolutionary ideas of the nineteenth century (Rist 1991; Williams 1976). As a consequence, development has rarely broken from organicist notions of growth or from a close affinity with teleological views of history, science, and progress in the West (Parajuli 1991). By the end of the nineteenth century, for example, it was possible to talk of societies in a state of "frozen development." Even radical alternative intellectual traditions, Marxisms among them, carried the baggage of historical stages, scientism, and modernization, forms of universalism which carried the appeal of secular utopias constructed with rationality and enlightenment. Development was modernity on a planetary scale in which the West was the "transcendental pivot of analytical reflection" (Slater 1992: 312).

There is another aspect to the genealogy, however, traced by Cowen and Shenton (1995) to eighteenth- and nineteenth-century notions of Progress, and specifically to development as a sort of theological discourse set against the disorder and disjunctures of capitalist growth. Classical political economy – including Smith, Ricardo, Malthus, and the like – is suffused with the tensions between the desire for unfettered accumulation on the one hand and unregulated desire as the origin of misery and vice (Herbert 1991). Development in Victorian England emerged in part, then, as a cultural and theological response to Progress. Christopher Lasch (1991), for example, has described a late nineteenth century obsessed by cultural instability and cataclysm. Saint-Simon devoted himself in his last years to a new creed of Christianity to accompany his industrial and scientific vision of capitalist progress. Trusteeship, mission, and faith were, according to Cowen and Shenton, the nineteenth-century touchstones of development.

Of course there is a more modern sense in which Third World development as state and multilateral policy harnessed to the tasks of championing economic growth, improving welfare, and producing governable subjects is of more recent provenance (Sachs 1992). These origins of development theory and practice as an academic and governmental enterprise – and of development economics as its hegemonic expression – are inseparable from the process by which the "colonial world" was reconfigured into a "developing world" in the aftermath of the Second World War. Africa, for example, became a serious object of planned development after the Great Depression of the 1930s. The British Colonial Development and Welfare Act (1940) and the French Investment Fund for Economic and Social Development (1946) both represented responses to the crises and challenges which imperial powers confronted in Africa, providing a means by which they could negotiate the perils of independence movements on the one hand and a perpetuation of the colonial mission on the other. The field of Development Economics which arose in the 1940s and 1950s – for example, the growth theories of Lewis, Hirschmann, and Rodenstein – sprouted in the soil of imperial planning initiatives, albeit propelled after 1945 by the establishment of a panoply of global development institutions (Bretton Woods, the United Nations) and President Truman's "program of development based on the concepts of democratic fair dealing" (20 January 1949, cited in Esteva 1992: 6).

A cartography of development

If development theory (and development economics in particular as its dominant expression) is a post-1945 construction rooted in growing U.S. hegemony on the one hand and the geopolitics of post-colonialism on the other, it nevertheless can be deposited on a much larger historical ground of ideas about comparative economic growth and sociopolitical transformation. One simple way to map development discourse historically in terms of its normative (i.e. goal-oriented) content, is to see development as a constant oscillation between the centrality of state, market, and civil society as means to secure key goals such as economic growth, social welfare, environmental sustainability, and national sovereignty (Table 1.1). This intellectual cartography is in no sense exhaustive – it refers largely to Eurocentric development theory associated with conventional development institutions and practices – and only refers to the *normative* (as opposed to the positive) aspects of development theory. As a heuristic device, however, it highlights a number of important points.

The first is to historicize one form of development itself, locating in the complex geopolitical environment of the inter- and post-war period, the construction, or more properly the invention, of development as planned social and economic improvement (Escobar 1992a; Watts 1993). A second is the recognition that development discourse is calibrated around the relative weight attributed in its normative vision to the role of the state, the market, and civil institutions. Typically, at any historical moment the prevailing or dominant development ideas – a particular center of intellectual gravity – are closely identified with one of these normative poles. For example, the 1980s counter-revolution, as Toye (1987) calls it, which shifted the market to center stage, a shift which stands in sharp contrast to the 1950s when there was widespread acceptance of some sort of state planning – a strange hybrid of a Gerschenkronian and Keynesian state – as a prerequisite for "catching up" and as a response to the maladies of relative backwardness. A third implication of Table 1.1 is that each *vertical* axis – state, market, civil society – is engaged in some sort of internal and external puzzle-solving. Internal because market-based theories, for example, are part of a tradition of market-based thinking which engages with itself as a prerequisite for developing new and different interpretations of the world. External in the double sense that particular ideas and theories within one of these vertical axes are always driven and shaped by their engagement with the other theoretical traditions external to it (certain forms of state-led theorizing are driven by their engagement with market-led explanations), but also because these same ideas are simultaneously part of a dialogue with the "external" world, that is to say by the problems, issues, and realities to which the theories must be made to speak. For example, the remarkable rise of the newly industrializing countries (NICs) – South Korea and Taiwan – in the post-1970 period acted as a major point of reference for debates over the relative significance of regulated or unregulated markets in their "catching up" and whether the East Asian NICs are free-market or "Leninist" success stories (Amsden 1989; Wade 1990).

No simple or direct relation exists between particular theoretical traditions – Marxism or modernization theory for example – and each axis. Marxism does not dismiss entirely the role of the market, for example, although the market nexus is defined in a particular way (Elson 1988); similarly, neo-liberalism rarely jettisons the state *in toto*, although it too is defined in a particular fashion. In this sense, theories tend to combine the normative content of development as particular configurations of state, market, and civil society, each constituted in ways peculiar to the core propositions of each theory. Different theoretical traditions tend naturally to weight these normative elements quite differently. In this sense, development theories may be distinguished in terms of the extent to which states, markets, and civil society fail. For example, whatever the purported virtues of markets, they may be monopolistic, imperfect, inflexible, or encourage externalities. Often seen as compensatory mechanisms for market failure, states may be rigid and inflexible mechanisms for allocating resources, they may be poorly co-ordinated, may create rents for particular classes, or may simply colonize civil society (Stern 1989). Civil society, often seen as a critical mediating space between state and market, a repository of rights, participation, and associational life, may equally be the crucible within which religious, ethnic, or other identifications impose strictures. It is important to emphasize, however, the *lateral* (i.e. diachronic) dimension to Table 1.1 in the sense that the intellectual and discursive traditions surrounding the market, state, and civil society engage each other, an engagement driven in some measure by the pressing development realities they seek to explain (for example Colclough and Manor's recent book [1991] is entitled *States or Markets?*).

These lateral and vertical dimensions vastly simplify the complexities of practical and theoretical differences in the field of development discourse. Individuals may shift locations on the map during the course of their careers – for example Albert Hirschmann moved from being a growth theorist in the 1950s to an institutionalist in the 1990s – and all theoretical traditions, almost by definition, contain particular definitions of states, markets, and civil society, which in some way reinforces the earlier point about the lack of correspondence between the vertical axis and theories of development *per se*. Last, it needs to be emphasized that development ideas are always *regionalized* into what we earlier called regional discursive formations: Latin American dependency theory is part of a particular

Table 1.1 A map of development discourse

| | Normative aspect of development theory | | |
	State[1]	Civil society[2]	Market[3]
Periodization	(Market failures, regulation for growth with equity, institutional and political capacity)	(Associational life, households, communities, lobbies, NGOs, non-state economic and cultural production)	(State failures, separability of equity and efficiency, harmony)
Phase 1 1760–1890 (first Industrial Revolutions)	*Relative backwardness and catching up* Protectionism, forced savings (Meiji reforms, Witte's Russia, List and Bismarck in Germany)	*Marx, proto-socialists and European populists* Artisanal production, small-scale co-operatives, collective control (Sismondi; Owen, Proudhon, Fourier, Herzen)	*Classical political economy* Laissez-faire, division of labor, comparative advantage (Smith, Ricardo, Malthus)
Phase 2 1890–1945 (classical imperialism)	**Keynesianism** State role in crisis regulation through fiscal and monetary policy (Keynes, Gerschenkron) **Soviet socialism** Nationalization, central planning, collectivization and primitive social accumulation (Preobrazhensky, Stalin)	**Gramsci, Arendt, neo-Marxist theory** Autonomy of civil society, solidarity, pluralistic rights **Neo-populism** Russian populists, Narodniks, East European Green Uprising, Gandhism (Chayanov, Gandhi, Chernyskevski)	**Neo-classical economics** Harmony and just returns, general equilibrium models (Marshall, Austrian School, Schumpeter, Pigou)

The *"Development" revolution*

	State[1]		Civil society[2]		Market[3]	
Phase 3 1945–1980 (Growth to crisis)	**Third World socialism and radical dependency** Maoism, Ho Chi Minh, Che, Debray, Indian Marxism/ Nehru, Kornai (delinking, basic needs, central planning)	**Development economics and growth with equity** ISI, protection, big push, linkages (Lewis, Myrdal, Mahalanobis, ECLA), redistributive strategies, basic needs	**Development populism** Agrarianism, small is beautiful, community development, informal sector, appropriate technology, African Socialism (Nyerere, Lipton, Mellor)	**Weberian modernization** Rationality, calculability, modernity, capacity (Hirschmann, Elias, Pye, Ong, Buttel)	**Neo-classical economic development** Pluralist state theory, agriculture and innovation, aid and trade (Bauer, Schultz, Ruttan, Myint, Jorgensen)	**Modernization theory** Human capital, capital formation, stages of growth, diffusion, savings, need achievement (Rostow, Chenery, Rodgers)
Phase 4 1980–1995 (Crisis, stabilization, adjustment)	**New political economy** **Developmental state** Neo-Weberian state capacity, relative autonomy, embeddedness (Wade, Chalmers, Johnson, Evans, Amsden)	**New growth theory** Endogenous government behavior, collective action theory, multiple equilibria (Becker, Bates, Krugman)	**New institutional economics** Transaction cost approaches, imperfect and asymmetrical information (Stiglitz, Bardham, Nugent, de Janvry, Williamson)	**The public sphere** Local knowledge/ peasant science, new social movements, NGO/ PVOs, feminisms, post-Marxism (Escobar, Shiva, Hettne, Offe, Laclau, Kothari, Habermas)	**Neo-liberal counter-revolution** Price distortions, rent-seeking, market strategies, trade theory (Timmer, Kreuger, Berg Report, Lal, Little, Balassa, Bauer)	

Source: This is a substantially revised and amended version of a figure originally prepared by Alain de Janvry, University of California, Berkeley, in connection with a course on development theory co-taught with Michael Watts.

[1] The state is understood as a set of institutions which act as a system of political domination/regulation, with specific effects on class and class struggle (see Jessop 1990: 28)

[2] Civil society is understood in the Gramscian sense as a non-state sphere of organizations – "the ensemble of organisms commonly called private" – where hegemony and consent are organized, and possessing the potential for rational self-regulation and freedom (Gramsci 1971).

[3] The market is understood as a nexus between buyers and sellers (but an institutional nexus that has to be made; cf. Elson 1988) – that is to say, an auction in which buyers and sellers bid against one another or as a broker-organized market.

regional discursive formation whose genesis and character was very much wrapped up with intellectual figures and activists associated with the Economic Commission for Latin America and subsequently with a number of Chilean and Brazilian universities. Likewise, some traditions of Marxist theorizing and forms of state planning have a distinctive Indian or South Asian character.

Development theory in the 1980s and 1990s

The Cold War is over and Communism and the socialist bloc have collapsed. The United States and Capitalism have won, and in few areas of the globe is that victory so clear cut . . . as Latin America. Democracy, free-market economics and pro-American outpourings of sentiment and policy dot the landscape of a region where until recently left–right confrontation and the potential for social revolution and progressive reform were widespread.

(Castañéda 1993: 3)

Poverty is the leading cause of premature death and ill health across the planet and the gaps between rich and poor are widening not closing the World Health Organization warned this week. . . . In a Foreword, Hirosho Nakajima, WHO's director-general, describes the report as "a devastating portrait of our times. . . . Poverty . . . wields its destructive influence at every stage of human life and for most of its victims the only escape is an early grave. Poverty provides that too."

(*Guardian* 7 May 1995: 1)

In relation to this simple map, the 1980s represented a period of retrenchment and restructuring in which recession and the debt crisis focused attention on short-term management ("disequilibria"). The literature was dominated by questions of stabilization and adjustment, driven increasingly by a neo-liberal orthodoxy which sought to reaffirm the necessity of reintegration into a global market and emphasized a "back to the future" strategy (i.e. a return to the colonial model of comparative advantage and export-oriented commodity production). The East Asian NICs were studied as success stories in the context of widespread failure (stagnation, corruption, de-industrialization) of debt- or state-led development models. State-centered analysis focused both on the so-called relative autonomy (or "embeddedness") of the developmental state in Taiwan and South Korea and the problems of state accountability, credibility, and rent-seeking in Latin America and Africa, not least in relation to the 1980s reform packages for stabilization. Ironically, state- and market-centered theories converged at the level of analytics in development economics, largely through transaction cost and collective action theory and the so-called "new institutional economics" (Bardhan 1989). By the 1990s in a rather different geopolitical and economic environment – the end of the Cold War, a declining debt burden, new social actors – development seemed to gravitate around the "balance" between state, market, and civil organizations, each with different incentive schemes and compliance–co-operation mechanisms (de Janvry *et al.* 1991).

For both theoretical and empirical reasons, then, the 1980s saw a growing concern with *institutions*, whether expressed in terms of agrarian social relations (Bardhan 1989), state–society relations (Migdal 1989), or new social movements (Melucci 1988). Moreover, criticisms levelled at the failings of both neo-liberal and authoritarian–bureaucratic development provided considerable momentum for a focus on institutions within civil society, especially agreements based on bargaining, co-operation, and persuasion. As de Janvry *et al.* (1991: 4) note:

> When the state fails to deliver public goods, insurance, management of externalities, minimum basic needs and democratic rights, civil organizations may fill the vacuum. The same holds for the market where market failures lead to the emergence of [civil] institutions, many of which take the form of organizations.

Of particular interest are development strategies that build relations of complementarity between civil organizations and the market and the state.

This resurgence of civil society in development thinking has been driven by a complex set of political forces and intellectual confluences. We have already referred to the impact of "people's power" in the overthrow of various Stalinisms in Eastern Europe but one should take note also of the proliferation of new social actors and civic movements, in part as a response to the austerity of the 1980s, in Latin America, South Africa, the Philippines, India, and more recently in parts of sub-Saharan Africa (we discuss this in more detail later). But there has also been a rethinking of the relations between culture and development by returning to the modernization theory of Shils, Geertz, and Weber (Hoben and Hefner 1991), in the role of grassroots organizations in the context of diminishing states and expanding markets (Uphoff 1991), in the social embeddedness of states and markets (Evans 1991; Friedland and Robertson 1991), in the endogeneity of development institutions and social norms (de Janvry *et al.* 1991), and in the promotion of local knowledge systems and resource management (Richards 1985; Warren 1991). All of these quite different tendencies nonetheless reaffirm the confluence of analytics noted by Bardhan (1989) in his observation that the analysis of institutions has emerged as a central problematic, whether expressed in terms of analytical Marxism, the contract theory of the neo-institutionalists, or the anthropological study of common property regulation.

Environment, development, and civil society: "populism" and sustainability

> To throw some light on discussions about "the people" and "the popular",
> one need only to bear in mind that the "people" or "the popular" . . . is first
> of all one of those things at stake in the struggle between intellectuals.
>
> (Bourdieu 1990: 150)

According to the United Nations Development Program (UNDP 1992) the polarization of global wealth doubled between 1960 and 1989. In the *fin de siècle* world economy, 82.7 per cent of global income is accounted for by the wealthiest 20 per cent, while the poorest 20 per cent account for 1.4 per cent of world income. In 1960, the top fifth of the world's population made thirty times more than the bottom fifth; by 1989 the disparity had grown to sixty times. The growing bi-modal character of relations between the North and South (indeed within Third World states, as Brazil, the Philippines, and India testify) is unquestionably rooted in the period of adjustment and stabilization since the oil crisis of the 1970s. For good reason, then, have many intellectuals and activists from the South come to see development discourse as a cruel hoax, a "blunder of planetary proportions" (Sachs 1992: 3). "You must be either very dumb or very rich if you fail to notice," notes Mexican activist Esteva (1992: 7) "that 'development' stinks." It is precisely the groundswell of *anti-development thinking*, oppositional discourses that have as their starting point the rejection of development, of rationality, and the Western modernist project (see Escobar, Chapter 2 in this volume), at the moment of a purported Washington consensus and free-market triumphalism, that represents one of the striking paradoxes of the 1990s. Ironically, however, both of these discourses – whether the World Bank line or its radical alternative – look to *civil society, participation, and ordinary people* for their development vision for the next millennium.

It is perhaps unsurprising that the enhanced emphasis within current development discourse on consolidating and promoting civil society has often drawn from the various strains of populism, in other words ideas about the power of what the World Bank called "ordinary people." Populism here implies not only a broadly specified development strategy – that is to say, the promotion of small-scale, owner-operated, anti-urban programs which stand against the ravages of industrial capitalism (Kitching 1980) – but also a particular sort of politics, authority structure, and ideology in which an effort is made to manufacture a collective popular will and an "ordinary" subject (Laclau 1977). In general populism: "is . . . based on the following major premiss: *virtue resides in the simple people, who are in the overwhelming majority, and in their collective traditions*" (Wiles 1969: 166, original emphasis). The recycling of populisms in development discourse, therefore, contains both an historical *continuity* – the recurrent motif of "the people" and "the ordinary" in development policy and practice – and an historical *difference* insofar as populist claims are always rooted in specific and local configurations of political and ideological discourses and practices.

Populism in no sense exhausts discussions of the role of civil society and civic traditions in development (see Gramsci 1971; Keane 1988; McGuigan 1992; Watts 1995) but it represents an important line of thinking and theorizing from the early nineteenth century to the present. Indeed a distinctive feature of populism – which perhaps explains its current appeal – is its flexible ability to draw on liberalism, nationalism, and socialism in fashioning its pragmatic, rather than political, agenda:

26

[Populism] . . . is profoundly a-political. . . . It goes beyond democracy
to consensus. . . . It calls on the state to inaugurate restoration, but it
distrusts the state and its bureaucracy and would minimize them before
the rights and virtues of local communities and the populist individual.

(Macrae 1969: 162)

But when "people" are invoked in developmental discourses about civil society
– whether the World Bank singing the praises of the ordinary African worker or
the geographer lauding peasant science – who the people are, and how they are
constructed, are precisely political questions (Bennett 1986).

Populist strategies, and the language of populism more generally, rest on what
Laclau (1977: 193) calls the "double articulation of discourse": that is, on the
one hand the tensions between "the people" and those who rule (the power bloc),
and on the other the various ways in which "the people" and their interests are
articulated or aligned with specific classes. How, in other words, does particular
populist language articulate with a particular power bloc and how is a particular
populist subject "interpolated" – for example, the ordinary peasant possessed of
local knowledge and resource management capability, or the informal sector
worker equipped with the entrepreneurial skills for appropriate technology
or flexible specialization? There is little doubt that the confluence of social
movements in the former socialist bloc (the 1989 "revolutions") with a neo-
liberal conservatism which advertises individual agency in the marketplace
(for example the authoritarian populism of Mrs Thatcher) has helped sustain a
developmental populism for the 1990s reflected in the uncritical promotion of
NGOs, civil institutions, and the power of ordinary people.

Current populist development thinking, therefore, can and should be located
on a larger historical canvas, but its particular character and specificity must be
rooted in the *realpolitik* of the end of the Cold War, a widespread disenchantment with state-administered politics, and in the self-interested, freedom-loving
individual of the neo-liberal counter-revolution (Biersteeker 1990; Fukuyama
1990). But as we shall see, populism is also an important ingredient in the
development–environment debate, whether expressed in terms of grassroots
green movements, indigenous technical knowledge for sustainable development,
or the calls for administrative decentralization in local resource management.
Civil society and populist thinking, in other words, cut across many of the issues
which are treated in this volume (see Chapters 4, 5, 9, and 10 by Bebbington,
Zimmerer, Schroeder and Suryanata, and Rangan in particular).

POLITICS, MOVEMENTS, CIVIL SOCIETY:
A LIBERATION ECOLOGY?

Development can only occur when the people it affects participate in the
design of the proposed policies, and the model which is implemented
thereby corresponds to the local people's aspirations. . . . The indigenous

people of the Amazon have always lived there; the Amazon is our home. We know its secrets, both what it can offer us, and what its limits are . . . (Statement by the Co-ordinating Body for the Indigenous Organizations of the Amazon Basin, 1989)

Political economy and political ecology have long been interested in social movements of many kinds. Much of the work theorizing social movements begins with Marxism, historical materialism, and a dialectical theory of social and environmental change. In the materialist view the productive transformation of nature is the primary activity making possible the whole structure of human existence. The productive forces (labor and means of production such as tools, machines, infrastructure) are organized by social relations (kinship, lineage, class) fundamentally characterized by inequalities of power – for example a minority class owning nature and the means by which continued life is made possible. The idea more generally is that modes of production create appropriate forms of consciousness, ideologies, and politics and have a certain level and type of effect on natural environment. From a dialectical view, societal dynamics emerge from contradictory oppositions in the material reproduction of existence, conflicts between the forces of production and a limited natural environment for example, which result in crises. These moments of contradictory crisis are, for classical Marxists, the contexts in which class existing "in-itself" engages in intensified political struggle and becomes class "for-itself," that is a group with collective identity, a collective agent which forces necessary social and environmental transformations. In Marx's own works, class is the main form of social engagement, and control of the means of production its primary terrain of struggle (Marx 1970).

Critique of classical Marxism

This economistic theory of society is open to severely restrictive "readings" or interpretations: notions of the "iron laws of history" in which technology creates change; reducing people to being passive "bearers" of social relations; and focusing on class to the exclusion of other social relations, are prominent examples. The Italian Marxist Antonio Gramsci (1971) argued for two modifications to classical Marxism. First, social strife endemic to capitalist and other modes of production may be countered by state force (army, police) but also by an ideological and cultural hegemony operating through traditions, myths, conventional morality, and "common sense," all of which are significant terrains of struggle. Second, transformative human actions do not result automatically from material contradictions; they are mediated by subjective meanings and conscious intentions. Material changes, such as resource deprivation or environmental crisis, may create higher propensities for transformative action and limit the range of its possible outcomes, but ideological and political practices are relatively autonomous and are literally the decisive moments in the transformation of material conditions

28

into political practices. Gramsci believed that capitalism could be transformed only through a range of cóunter-hegemonic movements using new strategies in several realms of social, political, and cultural life.

Similarly, various "neo-" and "post-" Marxist critics often accept Marxian principles of class stratification and social antagonism, but also challenge parts of the classical Marxist account (Cohen 1982). Two main aspects of Marxism, an evolutionary unfolding of the objective contradictions between the forces and relations of production (Marx 1970), and "the history of class struggles" (Marx and Engels 1974: 84), are found not easily to cohere in a single theory (Habermas 1971). Similarly Marxist theory is said to find all spheres of social life penetrated by a single, productivist logic which privileges economy and identifies class relations as key to the structure of domination and the forms of resistance; for Cohen (1982: xiii) this occludes other aspects of society and precludes an understanding of the novelty of recent social movements.

Neo-Marxist theorists modify the classical formulation. Theoreticians like Marcuse (1964) search for a substitute revolutionary subject to play the leading role previously assigned to the proletariat. "New working class" theorists (Aronowitz 1973; Gorz 1967; Mallet 1969) see welfare state capitalism providing a new strategy for labor. Structural Marxist class analysis (Poulantzas 1973; Wright 1979) rejects many of the features stressed by humanist Marxism to concentrate on classes defined as effects of structures. Theorists of the "new intellectual class" (Gouldner 1979; Szelenyi and Konrad 1979) transfer attention from workers to critical intellectuals. Despite such modifications, for critics like Cohen (1982: 3), the presupposition of Marxism remains that production relations are key to the logic of society and radical social movements. Post-Marxists, by comparison, argue that production is only one arena for collective resistance, that groups other than the working class are now significant sources of social movements, that greater attention has to be given to active processes of human agency.

Cohen (1985) also criticizes the (non-Marxist) "resource-mobilization paradigm" based in conflict models of collective action (Gamson 1975; Oberschall 1973; Tilly et al. 1975). Here the assumption is that conflicts of interest are built into institutionalized power relations. Collective actions involve the rational pursuit of interests by conflicting groups (Olson 1965). The mobilization of groups depends on their resources, especially the social networks in which they are embedded and their organizational structures. This approach assumes that individuals join groups when the benefits from so doing exceed the costs. Yet it remains unclear from this point of view why individuals acting rationally in pursuit of their interests get involved in groups (the "free rider problem" [Miller 1992]) and what gives groups their solidarity in the first place. Many theorists therefore maintain that such neo-utilitarian, rational-actor models are inapplicable, for collective action involves something other than strategic or instrumental kinds of rationality. Thus Habermas (1984) differentiates *system*, in which people operate under strategic rationalities following technical rules, and *lifeworld*, with its communicative rationality oriented towards consensus, under-standing, and collective action. For Habermas social movements of resistance emerge when commodifying systems colonize lifeworlds; resistance struggles are as much against dominant rationalities as they are against institutional control.

The result of such criticisms is a position which analyzes the conditions and processes by which structural change is transformed into collective action (Klandermans and Tarrow 1988). Here geography is both part of the structure (as with control of space, environments, resources, etc.) and part of the process by which structures are transformed into collective actions (the influence of terrains of struggle on the forms and intensities of struggles [Ackelsberg and Breitbart 1987–8]). Group consciousness and collective identities are made through the sharing of confined spaces in places with definite environmental conditions, and have tendencies towards common environmental imaginaries, an idea we develop in our conclusion (Chapter 12). Even the clarification of terms like "terrains," "fields of action," "arenas," etc. (Rucht 1988) is only just beginning. Clearly, however, this is rich in potential, especially in the area of social struggles over natural environments – what we refer to as liberation ecologies.

Urban social movements

Drawing on the Marxist tradition, but again differing significantly from it, a series of works explores the connections between contradictions, crises, and urban social conflicts. This work was precipitated by the rise of protest movements (civil rights, student, feminist, environmental, etc.) often centered on identity politics which came to be referred to as the "new social movements" – as compared with "old" movements, such as working-class organizations. In Castells's (1977) early work, urban social movements respond to the structural contradictions of the capitalist system; but these contradictions are of a plural-class and secondary nature, involving various deprivations, rather than the working class struggling to control the productive apparatus. Thus protest movements organize around common interests on a variety of terrains of struggle, often in opposition to the state and other political and sociocultural institutions, rather than the econ-omically ruling class directly. Indeed, Castells (1983: 299) came to believe (wrongly in our opinion) that "the concept of social movement as an agent of social transformation is strictly unthinkable in the Marxist theory." He argues instead that social change happens when a new urban meaning is produced through conflict, domination, and resistance to domination. For Castells (1983: 311) "the new emerging social movements call for the pre-eminence of human experience over state power and capitalist profit."

Another sequence of works in the post-Marxist vein stems from collaboration between Laclau and Mouffe. Mouffe (1984) argues that the commodification of social life, bureaucratization, and "cultural massification" create new forms of subordination to which new social movements respond. Laclau and Mouffe (1985) find the common denominator of all the new social movements (urban, ecological, feminist, anti-racist, regional, or sexual minorities) to be their differentiation from workers' struggles considered as class struggles. Indeed Laclau (1985: 29) argues that:

Categories such as "working class", "petit bourgeois", etc. [have become] less and less meaningful as ways of understanding the overall identity of social agents. The concept of "class struggle" for example, is neither correct nor incorrect – it is, simply, totally insufficient as a way of accounting for contemporary social conflicts.

For Laclau (1985: 27) the rise of the new social movements precipitated a crisis in the way social agents and conflicts are theorized. It became increasingly difficult to identify social groups with a coherent system of "subject positions." The social transformations of the twentieth century weakened the ties between the subject's various identities so that, for example, the worker's position in the relations of production and his/her position as consumer, resident, or political participant, are increasingly autonomous: this autonomy specifies the new social movements. For Laclau, also, subject-positions always display openness and ambiguity and there is no fully acquired social identity. Further, the social contradictions to which social agents respond cannot be reduced to moments of an underlying societal logic – "the social is in the last instance groundless" (Laclau 1985: 34). This leads to a differing conception of radical politics. In the nineteenth century, Laclau says, crises involved a total model of society and social struggles developed a unified political imaginary. In the twentieth century, a multiplication of points of rupture in society leads to a proliferation of antagonisms, each tending to create its own space and politicize a specific area of social relations. What Laclau (1985: 39) calls the "moment of totalization" in the political imaginary is now restricted to specific demands in particular circumstances. Rather than finding this a political retreat, Laclau finds the democratic potential of the new social movements lying precisely in their implicit demands for a radically open and indeterminate view of society.

The self-production of society

Given such (partly valid) criticisms some recent theorizing has drawn, instead, on a tradition in French social theory initiated by Castoriadis, drawing on the French and German phenomenological traditions and continued, in modified form, by Touraine. Like Marx, Castoriadis begins with the physical environment, the biological properties of human beings, and the necessity of material and sexual reproduction, for which fragments of logic and applied knowledge must be created. But he claims this is as true for apes as it is for humans. Instead for Castoriadis (1991: 41):

The construction of its own world by each and every society is, in essence, the creation of a world of meanings, its social imaginary significations,

31

which organize the (pre-social, "biologically given") natural world, institute a social world proper to each society (with its articulations, rules, purposes, etc.), establish the ways in which socialized and humanized individuals are to be fabricated, and insaturate the motives, values, and hierarchies of social (human) life. Society *leans upon* the first natural stratum, but only to erect a fantastically complex (and amazingly coherent) edifice of significations which vest any and every thing with *meaning*.

Knowing a society therefore entails reconstituting the world of its social imaginary significations. Furthermore, for Castoriadis (1991: 34): "History does not happen to society: history is the self-deployment of society." His notion is that the elements of social-historical life are created each time (in terms of relevance, meaning, connections, etc.) in and through the particular institution of society to which they "belong." Thus each social-historical instance has an essential singularity: phenomenologically specific in the social forms and individuals it creates, ontologically specific in that it can put itself into question, explicitly alter itself through self-reflective activity.

Similarly Touraine (1988) replaces the construct of society as a system driven by an inner logic with society as a "field of action." His stress lies more on the social praxis involved in the genesis of norms and conflicts over their interpretations. Whereas in Marxism classes are defined structurally by positions in the production process, for Touraine they are defined more directly in terms of social action. Touraine distinguishes himself from the main message of structural/poststructural social theory. From Marcuse to Althusser to Foucault and Bourdieu, the claim is that social life is nothing more than "the system of signs of an unrelenting domination" (Touraine 1988: 71) – in such systems radical social movements would be quickly shunted to the margins. For Touraine, by comparison, the necessary decomposition of society, the passage from one cultural and societal field to another, makes possible the entry of social movements with transformative capabilities.

At the core of his analysis lie conflicts over "cultural orientations," between an innovative ruling class which manages culture and people subordinated to its domination (Touraine 1988: 155). For Touraine (1985: 750–4) social conflicts involve the competitive pursuit of collective interests but also the reconstitution of social, cultural, or political identities; above all, conflict occurs over control of the main cultural patterns through which relationships with the environment are normatively organized. Most significantly, for Touraine, class struggles and social movements express conscious contestation over the "self-production of society," by which he means the work society performs on itself in terms of reinventing its norms, institutions, and practices. Struggles over historicity lie at the center of the functioning of society and the processes by which society is created.

New social movements in the Third World

Recent social movements theory has therefore moved away from what are frequently found to be the restrictions of classical (Marxist) theories. But also the geographic focus of research has tended to shift towards new social movements in the Third World, particularly Latin America. A multiplication of groups independent of traditional trade unions and political parties – squatter movements and neighborhood councils, base-level communities within the Catholic Church, indigenist associations, women's associations, human rights committees, youth meetings, educational and artistic activities, coalitions for the defense of regional traditions and interests, self-help groupings among unemployed and poor people – created a new social reality which, in Evers's (1985: 44) terms, "lies beyond the realm of traditional modes of perception and instruments of interpretation." Radical theorists found in these movements potential for a new political hegemony constructed through the direct action of the masses. This radical optimism has more recently been tempered as some movements declined and their limited potential was realized. Nevertheless it remains the case that Third World people's movements rather than First World workers' movements are seen as potentially transformative of the existing social structures.

Although not strictly in the social movements tradition, some of the more interesting ideas in this vein derive from the work of Scott on everyday resistance (1985, 1990). Scott, too, criticizes structuralist variants of Marxism for assuming that class relations can be inferred from a few diagnostic features like the dominant mode of production. While economic factors structure the situations faced by human actors, people fashion their own responses within these, based on their experiences and histories. Also, class does not exhaust the total explanatory space of social actions, especially in peasant villages, where kinship, neighborhood, faction, and ritual links are competing foci of human identity and solidarity: "the messy reality of multiple identities [is] the experience out of which social relations are conducted" (Scott 1985: 43).

Drawing on phenomenology and ethnomethodology, Scott (1985: 80) argues that subordinate classes "have rarely been afforded the luxury of open, organized, political activity" which is the preserve of the middle classes and intelligentsia. Instead he focuses on:

> *everyday* forms of peasant resistance – the prosaic but constant struggle between the peasantry and those who seek to extract labor, food, taxes, rents and interest from them. Most of the forms this struggle takes stop well short of outright collective defiance. Here I have in mind the ordinary weapons of relatively powerless groups: foot dragging, dissimulation, false compliance, pilfering, feigned ignorance, slander, arson, sabotage, and so forth.
>
> (Scott 1985: 29)

In struggles over land, everyday resistance might entail piecemeal peasant squatting on plantation or state forest land; open defiance, by contrast, would

33

be a mass invasion that challenges property rights. Everyday forms of resistance are often the most significant and effective over the long run.

Drawing more directly on poststructural themes, Escobar (1992b) sees social movements equally as cultural struggles over meaning as over material conditions and needs. Escobar draws a number of themes which might make this cultural dimension more visible. First is a more accurate theorization of the practice of everyday life through which culture is created and reproduced – the idea is to locate daily life at the intersection of the articulation of meaning through practice on the one hand and macro-processes of domination on the other. Second he finds it necessary to rethink the relations between everyday life, culture, and politics: in terms, for example, of Touraine's (1981, 1988) notion of historicity; or Melucci's (1988) proposition that networks of relationships submerged in everyday life lie behind the creation of cultural models and symbolic challenges by movements; or Laclau and Mouffe's (1985) argument that politics is a discursive articulatory process. Third, there is developing a micro-sociology and ethnography of popular resistance: de Certeau's (1984) notion that the "marginal majority" effect multiple, infinitesimal changes in the dominant forms under which they live; Fiske's (1989: 10) claim for "semiotic resistance" which originates in the "desire of the subordinate to exert control over the meaning of their lives"; and Williams' (1980) insistence on the continuation of residual practices that have a collective character and which can provide a basis for resistance and action. Generally the idea for Escobar (1992b) is to relate structural theories of global transformation to the "subjective mapping of experience."

The notion of everyday resistance may be combined with a poststructural interest in discourses of protest. A wide array of popular statements which often appear only at the local level can be read, for example, as evidences of environmental resistance. Academic work, displayed in several of the chapters of this book, can usefully compare "documentary" evidence of resistance with a critique of hegemonic discourses on development and environment. Here the mission is to pose alternatives in stronger terms. Rather than "speaking for" subaltern peoples, the idea is to help uncover discourses of resistance, put them into wider circulation, create networks of ideas. Rather than saying what peasant consciousness should be, were it to be "correct," the idea is to allow discourses to speak for themselves.

Environmental movements, environmental security

This literature suggests several themes for a liberation ecology. As we have seen, critiques of classical Marxism widen the spectrum of social movements opposing a hegemony extending far beyond control of productive resources to include culture, ideology, way of life; there is a particular interest, stemming from poststructural work on discourse, in examining the thoughts, imaginaries, statements, and institutions of both dominant and subaltern groups. The mediations between structural contradictions, deprivations, and various forms of socio-political actions are now seen as highly significant, rather than contradiction automatically producing organized opposition. These mediations seem to consist of at least five moments or types: (1) perceptions and interpretations which place the adverse situations faced by people into their meaning systems; (2) the sense of collective identity, commonality with others, often place-based or environmentally structured; (3) conditions which spur deprived, injured, or aggrieved people to different levels or types of actions ranging from sullen individual resistance to organized social movements; (4) linkages between social movements which create broad-based political forces; (5) the possibility of joining "old" social movements, such as unions and leftist parties, to new movements, such as organizations advocating environmental justice.

Many of these ideas have been deployed in the analysis of burgeoning Third World "environmental movements" (Ghai and Vivian 1992). While some work, often by political scientists working in the "new" domain of environmental security, posits simple and unmediated relationships between environmental change and social instability/civil strife (see Homer-Dixon *et al.* 1993), much of the environmental movements' literature tends to the local in its purview and often focuses on efforts to take resources out of the marketplace, to construct a sort of moral economy of the environment (see Martinez-Alier 1990). Broad (1993), for example, documents what she calls a new citizens' movement in the Philippines (5–6 million strong) consisting of "mass-based organizations" which arise from the intersection of political-economic plunder and local demands for participation and justice. In much of this literature the label "environmental" is hardly appropriate, since the proliferation of grassroots and NGO movements often focus more broadly on livelihoods and justice. Indeed, it is striking how indigenous rights movements, conservation politics, food security, the emphasis on local knowledges and calls for access to, and control over, local resources (democratization, broadly put) crosscut the environment–poverty axis. This multi-dimensionality is, according to some (Escobar 1992a), indicative of a new mode of doing politics, so-called autopoietic (that is to say, self-producing and self-organizing) movements which exercise power outside the state arena and which seek to create "decentered autonomous spaces."

These are ambitious claims which require careful scrutiny, though we are in general in agreement with Antje Linkenbach (1994: 81–2) when she says that:

> Ecological movements are not creating a new economics for a new civilization, they are not presenting a solution for the crisis of the modern world, and they do not have the capacity . . . for ending development. But they can show the difficulties, shortcomings and limited scopes of the dominant as well as the alternative models for development at the level of action.

However, if reasonable people can disagree over the potential and scope of these movements, it is striking how little is said in the "environment as social movement" literature about the conditions under which local movements transcend their locality, and hence contribute to the building of a robust civil society, and about the continuing problems of productivity and growth in the face of mass poverty. Whatever their shortcomings analytically, however, the existence of such grassroots livelihood movements – rubber tappers in eastern Amazonia, tree huggers in North India, or Indian communities fighting transnational oil companies in Ecuador – represents for the new social movements community the building blocks for an "alternative to development" (Sachs 1992).

CONCLUSION: POLITICAL ECOLOGY REDUX

Like other ideologies environmentalism is socially constructed but it is an especially indeterminate, malleable ideological form.

(Buttel 1992: 15)

[A]mbiguity runs through all of the most important discourses on economy and the environment today. . . . Precisely this obscurity leads so many people so much of the time to talk and write about "sustainability": the word can be used to mean almost anything . . . which is part of its appeal.

(O'Connor 1994: 152)

Political ecology is changing as the underlying social theory moves in post-structural directions and as new developments and tendencies occur in the politics of environment. In terms of the former, social theory is itself an arena of struggle, changing in the sources it draws upon, the themes it stresses, the shape of its contemplative imagination, and the directions and significance of its political outcomes. In terms of the latter, social struggles over land and resources, the environmental conditions of human existence, erupt in a profusion of styles and intensities, sometimes becoming full-fledged social movements, sometimes remaining as more prosaic and circumscribed individual resistance. As these two tendencies interact, political ecology as a specialized branch of critical social theory undergoes its own partly autonomous shift in thematic structure and theoretical style, very much as earlier versions also are critiqued and revised. Earlier in this chapter we surveyed the origins of political ecology and its concern to link the political economy of development – typically expressed through the ability of the land manager to manage his/her resources in the context of particular relations of production and circulation – with the traditional environmental concerns of ecological anthropologists and geographers. We then placed political ecology in a broader context which encompasses debates over the nature of modernity itself. The renewal of long-standing skepticism about modernity and its rationality, this time as poststructural and postmodern philosophies, leads social theory in fascinating directions, many of which are worth pursuing for the challenges they pose and the insights they stimulate. The simple notion that truth is socially and culturally constructed, rather than discovered already existing as a quality inherent in things, reverberates through social theory, revealing itself in remarkably diverse places. In particular the critique of reason as the discovery of eternal verities, re-emphasizes the imaginative and discursive aspects of reason*ing* as a creative, constituting act which transforms realizations about what already exists into projects of how to make new things exist. Thus environmental crises do not project truth into consciousness, on the basis of which people act in appropriate ways. Rather, multiple realizations about all levels of environmental problems are one main source stimulating a series of creative reactions which may (or may not) emerge as fully formed social movements. Furthermore, these movements are collectivities organized around common concerns and oppressions. But as well as being practical struggles over livelihood and survival, they contest the "truths," imaginations, and discourses through which people think, speak about, and experience systems of livelihood.

Hence in terms of social theory and political ecology we find both *a broader conception of the forms of contention* (from class struggle to social movements to everyday resistance) and *a deeper conception of what is contended* (from ownership of productive resources to control over the human imagination). We suggest that the social imaginaries and discourses which environmental and other social movements contend, do not arise on the head of a pin or in a de-natured ivory tower. Rather, the environment itself is an active constituent of imagination, and the discourses themselves assume regional forms that are, as it were, thematically organized by natural contexts. In other words, there is not an imaginary made in some separate "social" realm, but an environmental imaginary, or rather whole complexes of imaginaries, with which people think, discuss, and contend threats to their livelihoods – a claim we discuss and elaborate in more detail in our concluding chapter. Notions like "environmental imaginary," which draw on the Marxist conception of consciousness, poststructural ideas about imagination and discourse, and, dare we add, environmental determinism from early-modern geography, open political ecology to considerations so different that we propose a new term to describe them – liberation ecology. The intention is not simply to *add* politics to political ecology, but to raise the emancipatory potential of environmental ideas and to engage directly with the larger landscape of debates over modernity, its institutions, and its knowledges.

Liberation ecology is not set in concrete as an already formed structure of ideas. It is a discourse about nature, Marxist in origin, poststructural in recent influence, politically transformative in intent, but subject still to the fiercest of debates. These concern vital, fundamental issues, such as attitude towards modernity, rationality, and emancipation. Compare Chapters 2 and 4 by Escobar and Yapa, so critical of such basic tenets of modernism as developmentalism or environmentalism that they advocate their abandonment, with Chapter 10 by Rangan, critical of the Chipko movement, that darling of the anti-modernists, or Chapter 4 by Bebbington, which shows that Ecuadorean movements defend their indigenous cultures by self-consciously embracing modern techniques. Our own position – certainly not endorsed by everyone in this volume – tends towards a critical modernism in which rationality is contended rather than abandoned. But the main point is that liberation ecology is a discursive arena rather than a doctrine, a site where the broad issues of politics and thought that shape and mark our time are freely and audaciously discussed in terms of their environmental applications.

The poststructural and postmodern critiques of Western science as rationality in its pure and universal form thus open the way for a fuller understanding of the multiplicity of ways of comprehending the extraordinarily complex nexus of development–environment relations. A poststructural ecology may begin with the devastating environmental consequences of modernity but it deepens this practical critique by arguing for its path-dependency, substantially different local discourses about environment, each marked by its own contradictions, each with lessons to teach and problems to avoid. In this sense a retrieval of peasant and indigenous discourses on nature, land use, and ecological regulation and management need not romanticize pre-capitalist or non-Western relations between society and nature as Shiva (1991) does. Furthermore, one of the great merits of the turn to discourse, broadly understood, within political ecology, is the demands it makes for nuanced, richly textured empirical work (a sort of political-ecological thick description) which matches the nuanced beliefs and practices of the world. Some of the contributions to this book precisely capture this fine-grained and culturally sensitive analysis.

In our view, accounts of environment and development should begin still with the overall contradictory character of relations between societies and natural environments and recognize that dialectics remains a compelling theory of contradiction, crisis, and change (see Harvey 1993). But we would argue that poststructural theory, which owes much of its appeal to the deconstruction of the Western myths of science, truth, and rationality, itself has fabricated a mythology about the dialectic or, rather, has taken the "dialectic" of Stalin's iron laws of history as its prevailing model. The dialectic, then, is portrayed as an idealist device in which thesis incorporates antithesis during teleological passage to an already given synthesis, allowing no room for contingency, difference, or, for that matter, the new. In our view, dialectical analysis instead imagines a system of relations which does not consume the autonomy of the particular; it is one in which a number of dynamic tendencies in shifting hierarchical arrangements are constantly disturbed and dislocated by new sequences of different events, a dynamic which has pattern, order, and determination without being teleological. It is a theory of totalities which, because it values their unique aspects, is not *totalizing*.

This body of work at its best locates specific sorts of movements emerging from the tensions and contradictions of under-production crises, understands the imaginary basis of their oppositions and visions of a better life and the discursive characters of their politics, and sees the possibilities for broadening environmental issues into a movement for livelihood, entitlements, and social justice. This is a tall order, and in a sense the theoretical work has only just begun. We believe that this book represents, from a multiplicity of vantage points, a common effort to refine and deepen the political, and in so doing pushes toward what we have termed a liberation ecology. Furthermore, we also believe that the contributions to this volume – to return to the quotations by Buttel and O'Connor cited at the beginning of the conclusion – provide building blocks on which environmentalism and sustainability can be critically assessed and hopefully reconstructed.

REFERENCES

Ackelsberg, M. A. and M. M. Breitbart. 1987–8. "Terrains of protest: striking city women," *Our Generation* 19, 1: 151–75.

Adams, William. 1991. "The greening of development?" paper delivered to the Canadian Association of Geographers, Annual Meeting, Toronto.

Agarwal, Bina. 1992. "The gender and environment debate," *Feminist Studies* 18: 119–59.

Altieri, Miguel and Susanna Hecht (eds.). 1990. *Agroecology and Small Farm Development.* Boca Raton, FL: CRC Press.

Amsden, Alice. 1989. *Asia's Next Giant.* Oxford: Oxford University Press.

Aronowitz, Stanley. 1973. *False Promises.* New York: McGraw-Hill.

Bardhan, Pranab (ed.). 1989. *The Economic Theory of Agrarian Institutions.* Oxford: Oxford University Press.

Barnes, Trevor and James Duncan (eds.). 1992. *Writing Worlds: Discourse, Text and Metaphor in the Representation of Landscape.* London: Routledge.

Baudet, Henry. 1965. *Paradise on Earth: Some Thoughts on European Images of Non-European Man.* New Haven, CT: Yale University Press.

Beck, Ulrich. 1994. *Risk and Modernization.* Boulder, CO: Westview Press.

Beinart, William and Peter Coates. 1995. *Environment and History.* London: Routledge.

Bennett, T. 1986. "The politics of the popular and popular culture," in T. Bennett, G. Marton, C. Mercer and J. Woollacott (eds.) *Popular Culture and Social Relations.* Milton Keynes: Open University Press, pp. 6–21.

Benton, Ted. 1989. "Marxism and natural limits," *New Left Review* 178: 51–86.

Bhabha, Homi K. 1983a. "Difference, discrimination and the discourse of colonialism," in Francis Barker *et al.* (eds.) *The Politics of Theory.* Colchester: University of Essex Press.

—— 1983b. "The Other question," *Screen* 24, 6: 18–35.

—— 1984. "Of mimicry and man: the ambivalence of colonial discourse," *October* 28: 125–33.

Biersteker, Thomas. 1990. "The triumph of neoclassical economics in the developing world," in James Roseneau and Ernst-Otto Czempiel (eds.) *Governance without Government.* Cambridge: Cambridge University Press, pp. 102–32.

Blaikie, Piers. 1985. *The Political Economy of Soil Erosion.* London: Methuen.

Blaikie, Piers and Harold Brookfield. 1987. *Land Degradation and Society.* London: Methuen.

Booth, David (ed.). 1995. *Rethinking Social Development.* London: Methuen.

Botkin, D. 1990. *Discordant Harmonies.* New York: Oxford University Press.

Bourdieu, Pierre. 1990. *In Other Words.* London: Polity.

Bramwell, Anna. 1989. *Ecology in the Twentieth Century.* New Haven, CT: Yale University Press.

Broad, Robin. 1993. *Plundering Paradise.* Berkeley: University of California Press.

Bryant, Raymond. 1992. "Political ecology," *Political Geography* 11: 12–36.

Bunker, Stephen. 1985. *Underdeveloping the Amazon.* Chicago: University of Illinois Press.

Burawoy, Michael. 1985. *The Politics of Production.* London: Verso.

Buttel, Fred. 1992. "Environmentalization," *Rural Sociology* 57, 1: 1–27.

Castaneda, Jorge. 1993. *Utopia Unarmed.* New York: Vintage.

Castells, Manuel. 1977. *The Urban Question: A Marxist Approach.* Cambridge, MA: MIT Press.

—— 1983. *The City and the Grass Roots: A Cross Cultural Theory of Urban Social Movements.* Berkeley: University of California Press.

Castoriadis, Cornelius. 1991. "The social historical: mode of being, problems of knowledge," in C. Castoriadis (ed.) *Philosophy, Politics, Autonomy.* New York: Oxford University Press.

Cockburn, Alex and James Ridgeway (eds.). 1979. *Political Ecology.* New York: Times Books.

Cohen, Jean L. 1982. *Class and Civil Society: The Limits of Marxian Critical Theory.* Amherst: University of Massachusetts Press.

—— 1985. "Strategy or identity: new theoretical paradigms and contemporary social movements," *Social Research* 52, 4 (Winter): 663–716.

Colclough, C. and J. Manors (eds.). 1991. *States or Markets?* Oxford: Clarendon Press.

Collins, Jane. 1987. *Unseasonal Migrations.* Princeton, NJ: Princeton University Press.

Cowen, Michael and Robert Shenton. 1995. *Doctrines of Development.* London: Routledge.

Cronon, William. 1992. *Nature's Metropolis.* New Haven, CT: Yale University Press.

de Certeau, Michel. 1984. *The Practice of Everyday Life.* Berkeley: University of California Press.

de Janvry, Alain and Raoul Garcia. 1988. "Rural poverty and environmental degradation in Latin America," paper presented to the IFAD Conference on Smallholders and Sustainable Development, Rome.

de Janvry Alain, Elizabeth Sadoulet, and E. Thornebecke. 1991. "States, markets and civil institutions," paper prepared for the ILO–Cornell–Berkeley Conference on States, Markets, and Civil Society, Ithaca.

Derrida, Jacques. 1971. "White mythology," in *Margins of Philosophy.* Chicago: University of Chicago Press.

Dreyfus, Hubert L. and Paul Rabinow. 1982. *Michel Foucault: Beyond Structuralism and Hermeneutics.* Chicago: University of Chicago Press.

Dubois, M. 1991. "The governance of the Third World: a Foucauldian perspective on power relations in development," *Alternatives* 16: 1–30.

Elson, Diane. 1988. "Market socialism or socialization of the market?" *New Left Review* 172: 3–42.

Escobar, Arturo. 1992a. "Imagining a post-development era? Critical thought, development and social movements," *Social Text* 31/32: 20–56.

—— 1992b. "Culture, economics, and politics in Latin American social movements theory and research," in A. Escobar and S. E. Alvarez (eds.) *The Making of Social Movements in Latin America.* Boulder, CO: Westview Press, pp. 62–85.

—— 1995. *Encountering Development.* Princeton, NJ: Princeton University Press.

Esteva, Gustavo. 1992. "Development," in Wolfgang Sachs (ed.) *The Development Dictionary: A Guide to Knowledge as Power.* London: Zed Books, pp. 6–25.

Esty, Daniel. 1994. *Greening the GATT.* Washington, DC: Institute of International Economics.

Evans, Peter. 1995. *Embedded Autonomy.* Princeton, NJ: Princeton University Press.

Evers, Tilman. 1985. "Identity: the hidden side of new social movements in Latin America," in David Slater (ed.) *New Social Movements and the State in Latin America*. Amsterdam: CEDLA, pp. 43–71.

Faber, Daniel. 1992. *Environment Under Fire*. New York: Monthly Review Press.

Fairhead, James and Melissa Leach. 1994. "Contested forests," *African Affairs* 93: 481–512.

Feshbach, Murray and Alfred Friendly. 1992. *Ecocide in the USSR*. New York: Basic Books.

Fiske, John. 1989. *Reading the Popular*. Boston, MA: Unwin Hyman.

Foucault, Michel. 1972. *The Archaeology of Knowledge*. New York: Harper & Row.

—— 1973. *The Order of Things*. New York: Vintage Press.

—— 1980. *Power/Knowledge: Selected Interviews and Other Writings 1972–1977*, edited by Colin Gordon. New York: Pantheon Books.

Friedland, Roger and Andrew Robertson. 1991. *Beyond the Marketplace*. New York: de Gruyter.

Fukuyama, Francis. 1990. *The End of History and the Last Man*. New York: Free Press.

Gadgil, Madhav and Ramachandra Guha. 1992. *This Fissured Land*. New Delhi: Oxford University Press.

Gamson, W. 1975. *The Strategy of Social Protest*. Homewood, IL: Dorsey.

Garcia Barrios, R. and L. Garcia Barrios. 1990. "Environmental and technological degradation in peasant agriculture," *World Development* 18: 1569–85.

Ghai, Dharam. 1992. "Conservation, livelihood and democracy," Discussion Paper #33. Geneva: UNRISD.

Ghai, Dharam and Jessica Vivian (eds.). 1992. *Grassroots Environmental Action*. London: Routledge.

Gleissman, Stephen (ed.). 1990. *Agroecology*. Berlin: Springer Verlag.

Goodman, David, Bernardo Sorj, and John Wilkinson. 1990. *From Farming to Biotechnology*. Oxford: Blackwell.

Gorz, André. 1969. *Strategy for Labor*. Boston, MA: Beacon Press.

Gouldner, Alvin. 1979. *The Future of Intellectuals and the Rise of the New Class*. New York: Seabury.

Gramsci, Antonio. 1971. *Selections from the Prison Notebooks*. New York: International Publishers.

Grossman, Larry. 1984. *Peasants, Subsistence Ecology and Development in the Highlands of Papua New Guinea*. Princeton, NJ: Princeton University Press.

Grove, Richard. 1993. *Green Imperialism*. Cambridge: Cambridge University Press.

Grundemann, Reiner. 1991. *Marxism and Ecology*. Oxford: Oxford University Press.

Guha, Ramachandra. 1990. *Unquiet Woods*. Berkeley: University of California Press.

Guha, Ranajit. 1983. *Elementary Aspects of Peasant Insurgency in Colonial India*. Delhi: Oxford University Press.

Guha, R. and G. Spivak (eds.). 1988. *Selected Subaltern Studies*. Delhi: Oxford University Press.

Habermas, Jürgen. 1971. *Knowledge and Human Interests*. Boston, MA: Beacon Press.

—— 1984. *The Theory of Communicative Action: Vol. 1*. Boston, MA: Beacon Press.

Hall, Peter. 1989. *The Political Power of Economic Ideas*. Princeton, NJ: Princeton University Press.

Harvey, David. 1993. "The nature of environment," *Socialist Register*, London: Merlin, pp. 1–51.

Hass, Peter. 1993. "Epistemic communities and the dynamics of international environmental co-operation," in Volker Rittberger (ed.) *Regime Theory and International Relations*. London: Oxford University Press.

Hecht, Susanna and Alex Cockburn. 1989. *The Fate of the Forest*. London: Verso.

Herbert, Christopher. 1991. *Culture and Anomie.* Chicago: University of Chicago Press.

Herskovitz, Lawrence. 1993. "Political ecology and environmental management in loess plateau, China," *Human Ecology* 21, 4: 327–53.

Hoben, Allen and Robert Hefner. 1991. "The Integrative Revolution Revisited," *World Development* 19: 17–30.

Homer-Dixon, Thomas, Jeffrey Boutwell, and George Rathjens. 1993. "Environmental change and violent conflict," *Scientific American* February: 38–45.

Horkheimer, Max and Theodore W. Adorno. 1991. *Dialectics of Enlightenment.* New York: Continuum.

Hulme, Peter. 1986. *Colonial Encounters: Europe and the Native Caribbean, 1692–1797.* London: Methuen.

IICQ. 1992. "Indigenous Vision," *India International Center Quarterly* 1–2.

Jessop, Bob. 1990. *State Theory.* Cambridge: Polity.

Kautsky, Karl. 1906. *La Question agraire.* Paris: Maspero.

Keane, John. 1988. *Democracy and Civil Society.* London: Verso.

Kirby, Andrew (ed.). 1990. *Nothing to Fear.* Tucson: University of Arizona Press.

Kitching, Gavin. 1980. *Development and Underdevelopment. in Historical Perspective.* London: Methuen.

Klandermans, B. and S. Tarrow. 1988. "Mobilization into social movements: synthesizing European and American approaches," *International Social Movements Research* 1: 1–38.

Kloppenberg, Jack. 1989. *First the Seed.* New York: Cambridge University Press.

Kornai, Janos. 1980. *The Economics of Shortage.* Amsterdam: North Holland.

Laclau, Ernesto. 1977. *Politics and Ideology in Marxist Theory.* London: Verso.

—— 1985. "New social movements and the plurality of the social," in David Slater (ed.) *New Social Movements and the State in Latin America.* Amsterdam: CEDLA, pp. 27–42.

Laclau, Ernesto and Chantal Mouffe. 1985. *Hegemony and Socialist Strategy: Towards a Radical Democratic Politics.* London: Verso.

Lasch, Christopher. 1991. *The True and Only Heaven.* New York: Norton.

Leff, Enrique. 1995. *Green Production.* New York: Guilford.

Levins, Richard and Richard Lewontin. 1985. *The Dialectical Biologist.* Cambridge, MA: Harvard University Press.

Linkenbach, Antje. 1994. "Ecological movements and the critique of development," *Thesis Eleven* 39: 63–85.

Lipietz, Alain. 1988. *Mirages and Miracles.* London: Verso.

Little, Peter and Michael Horowitz (eds.). 1987. *Lands at Risk in the Third World.* Boulder, CO: Westview Press.

Lowe, Linda. 1991. *Contested Terrains. British and French Orientalisms.* Berkeley: University of California Press.

MacCabe, Colin. 1987. "Foreword," in G. Spivak, *In Other Worlds: Essays in Cultural Politics.* New York: Methuen, pp. ix–xix.

Macdonell, Diane. 1986. *Theories of Discourse: An Introduction.* Oxford: Basil Blackwell.

McGrane, Bernard. 1989. *Beyond Anthropology: Society and the Other.* New York: Columbia University Press.

McGuigan, Jim. 1992. *Cultural Populism.* London: Routledge.

MacKenzie, Fiona. 1991. "Agrarian political economy, land use and environmental change," *Journal of Peasant Studies* 15: 236–59.

Macrae, D. 1969. "Populism as an ideology," in Ghita Ionescu and Ernest Gellner (eds.) *Populism.* London: Macmillan, pp. 153–66.

Mallet, Serge. 1969. *La Nouvelle Classe ouvrière.* Paris: Editions du Seuil.

Marcuse, Herbert. 1964. *One-dimensional Man.* Boston, MA: Beacon Press.

Martinez-Alier, Juan. 1990. "Poverty as a cause of environmental degradation," report prepared for the World Bank, Washington, DC.

Martinez-Alier, Juan and K. Schluepmann. 1987. *Ecological Economics.* Oxford: Blackwell.

Marx, Karl. 1970. Preface to *A Contribution to the Critique of Political Economy.* Moscow: Progress Publishers.

Marx, Karl and Friedrich Engels. 1974. "Manifesto of the Communist Party," in K. Marx, F. Engels, and V. Lenin, *On Historical Materialism.* New York: International Publishers.

Mellor, John. 1988. "Environmental problems and poverty," *Environment* 30, 9: 8–13.

Melucci, Alberto. 1988. "Getting involved: identity and mobilization in social movements," in H. Kriesi, S. Tarrow, and B. Vui (eds.) *International Social Movements Research Vol. 1.* London: JAI Press.

Merchant, Carolyn. 1993. *Major Problems in American Environmental History.* Boston: Heath.

Meyer, William and Billie Turner. 1992. "Human population growth and global landuse/cover change," *Annual Review of Ecology and Systematics* 23: 39–61.

Migdal, Joel. 1989. *Strong States and Weak Societies.* Princeton, NJ: Princeton University Press.

Miller, Byron. 1992. "Collective action and national choice: place, community, and the limits to individual self-interest," *Economic Geography* 68, 1 (January): 22-47.

Mouffe, Chantal. 1984. "Towards a theoretical interpretation of 'New Social Movements'," in S. Hannienen and L. Paldan (eds.) *Rethinking Marx.* Berlin: Argument-Sonderbond, pp. 139–43.

Neumann, Rod. 1992. "Resource use and conflict in Arusha National Park, Tanzania." Ph.D. dissertation, University of California, Berkeley.

Nietschmann, Bernard. 1973. *Between Land and Water: The Subsistence Ecology of the Miskito Indians.* New York: Academic Press.

Oberschall, A. 1973. *Social Conflicts and Social Movements.* Englewood Cliffs, NJ: Prentice-Hall.

O'Connor, James. 1988. "Capitalism, nature, socialism: a theoretical introduction," *Capitalism, Nature, Socialism* 1: 11–38.

O'Connor, Martin (ed.). 1994. *Is Capitalism Sustainable?* New York: Guilford Press.

Olson, Mancur. 1965. *The Logic of Collective Action.* Cambridge, MA: Harvard University Press.

Ostrom, Elinor. 1990. *Governing the Commons.* Cambridge: Cambridge University Press.

Parajuli, P. 1991. "Power and knowledge in development discourse," *International Social Science Journal* 127: 173–90.

Peet, Richard. 1996. "A sign taken for history: Daniel Shays' Memorial in Petersham, Massachusetts," *Annals of the Association of American Geographers* 86, 1: 21–43.

Peluso, Nancy. 1993a. *Rich Forests, Poor People.* Berkeley: University of California Press.

—— 1993b. "Coercing conservation?" *Global Environmental Change* June: 199–217.

Perrings, Charles. 1987. *Economy and the Environment.* Cambridge: Cambridge University Press.

Poulantzas, Nicos. 1973. *Political Power and Social Classes.* London: New Left Books.

Rabinow, Paul. 1986. "Representations are social facts: modernity and post-modernity in anthropology," in James Clifford and George E. Marcus (eds.) *Writing Culture: The Poetics and Politics of Ethnography.* Berkeley: University of California Press, pp. 234–61.

Rappaport, Roy. 1967. *Pigs for the Ancestors.* New Haven, CT: Yale University Press.

Redclift, Michael. 1987. *Sustainable Development.* London: Methuen.

Rich, Bruce. 1994. *Mortgaging the Earth.* Boston: Beacon Press.

Richards, Paul. 1985. *Indigenous Agricultural Revolution.* London: Hutchinson.

Rist, Gerhard. 1991. "'Development' as part of the modern myth," *European Journal of Development Research* 2: 10–21.

Rorty, Richard. 1979. *Philosophy and the Mirror of Nature.* Princeton: Princeton University Press.

Rucht, D. 1988. "Themes, logics and arenas of social movements: a structural approach," *International Social Movements Research* 1: 305–28.

Sachs, Wolfgang (ed.). 1992. *The Development Dictionary.* London: Zed Books.

Said, Edward W. 1979. *Orientalism.* New York: Vintage Books.

Sand, Peter. 1995. "Trusts for the earth," in W. Lang (ed.) *Sustainable Development and International Law.* London: Graham & Trotman, pp. 167–84.

Schuurman, F. J. (ed.). 1992. *Beyond the Impasse: New Directions in Development Theory.* London: Zed Books.

Scott, J. C. 1985. *Weapons of the Weak: Everyday Forms of Peasant Resistance.* New Haven, CT: Yale University Press.

—— 1990. *Domination and the Arts of Resistance: Hidden Transcripts.* New Haven, CT: Yale University Press.

Shiva, Vandana. 1989. *Staying Alive.* London: Zed Books.

Shrader-Frechette, N. 1990. *Risk and Rationality.* Berkeley: University of California Press.

Sivaramakrishnan, K. 1995. "Colonialism and forestry in India," *Comparative Studies in Society and History* 37, 1: 3–40.

Slater, David. 1992. "On the borders of social theory," *Society and Space* 10: 307–27.

Socialist Review. 1992. "Environment as politics," *Socialist Review,* Special Issue.

Spivak, Gayatri Chakravorty. 1987. *In Other Worlds: Essays in Cultural Politics.* New York: Routledge.

—— 1988. "Can the subaltern speak?" in Cary Nelson and Lawrence Grossberg (eds.) *Marxism and the Interpretation of Culture.* Urbana: University of Illinois Press, pp. 271–313.

Stern, Nicholas. 1989. "The economics of development: a survey," *The Economic Journal* 397: 597–686.

Stonich, Susan. 1989. "The dynamics of social processes and environmental destruction," *Population and Development Review* 15: 269–96.

Szelenyi, Ivan and Georg Konrad. 1979. *The Intellectuals on the Road to Class Power.* New York: Harcourt, Brace, Jovanovich.

Taylor, Peter and Fred Buttel. 1992. "How do we know we have environmental problems?" *Geoforum* 23: 405–16.

Tiffen, Mary and Michael Mortimore. 1994. *More People, Less Erosion.* New York: Praeger.

Tilly, C., L. Tilly and R. Tilly. 1975. *The Rebellious Century: 1830–1930.* Cambridge, MA: Harvard University Press.

Todorov, Tzvetan. 1984. *The Conquest of America: The Question of the Other.* New York: Harper & Row.

Toulmin, Camilla. 1992. *Water, Women and Wells.* Oxford: Oxford University Press.

Touraine, Alain. 1981. *The Voice and the Eye.* New York: Cambridge University Press.

—— 1985. "An introduction to the study of social movements," *Social Research* 52, 4 (Winter): 749–87.

—— 1988. *The Return of the Actor.* Minneapolis: University of Minnesota Press.

Toye, John. 1987. *Dilemmas of Development.* Oxford: Blackwell.

Turner, Billie Lee, W. C. Clark, R. W. Kates, J. F. Richards, J. T. Mathews and W. Meyer (eds.) 1990. *The Earth as Transformed by Human Action.* Cambridge: Cambridge University Press.

UNDP. 1992. *The Human Development Report.* Geneva: UNDP.

Uphoff, Norman. 1991. "Grassroots organizations and NGOs in rural development," unpublished manuscript, Cornell University, Ithaca.

Utting, Peter. 1994. "Social and political dimensions of environmental protection in central America," *Development and Change* 25: 231–59.

Wade, Robert. 1990. *Governing the Market.* Princeton, NJ: Princeton University Press.

Warren, Michael (ed.). 1991. "Indigenous knowledge systems and development," Special Issue on Indigenous Knowledge Systems and Development, *Agriculture and Human Values* 7.

Watts, Michael. 1983. *Silent Violence: Food, Famine and Peasantry in Northern Nigeria.* Berkeley: University of California Press.

—— 1987. "Drought, environment and food security," in Michael Glantz (ed.) *Drought and Hunger in Africa.* Cambridge: Cambridge University Press, pp. 171–212.

—— 1993. "Development I: power, knowledge, discursive practice," *Progress in Human Geography* 17, 2: 257–72.

—— 1995. "A new deal of the emotions," in Jonathan Crush (ed.) *Power and Development.* London: Routledge.

Wiles, Peter. 1969. "A syndrome not a doctrine," in Ghita Ionescu and Ernest Gellner (eds.) *Populism.* London: Macmillan, pp. 166–80.

Williams, Raymond. 1976. *Keywords: A Dictionary of Culture and Society.* Oxford: Oxford University Press.

—— 1980. *Problems in Materialism and Culture.* London: New Left Books.

Wolf, Eric. 1972. "Ownership and political ecology," *Anthropological Quarterly* 45: 201–5.

World Bank. 1981. *Toward Accelerated Development in Sub-Saharan Africa.* Washington, DC: World Bank.

—— 1989. *Sustainable Growth with Equity.* Washington, DC: World Bank.

—— 1992. *The Development Report.* Washington, DC: World Bank.

World Health Organization. 1995. *State of the World Report.* Geneva: WHO.

Worster, Donald. 1977. *Nature's Economy.* Cambridge: Cambridge University Press.

Wright, Eric Olin. 1979. *Class, Crisis and the State.* London: Verso.

Young, Robert. 1990. *White Mythologies: Writing History and the West.* London: Routledge.

Zimmerer, Karl. 1994. "Human geography and the new ecology," *Annals of the Association of American Geographers* 84, 1: 108–25.

2

CONSTRUCTING NATURE

Elements for a poststructural political ecology

Arturo Escobar

INTRODUCTION: THE DISCOURSE OF NATURE AND THE NATURE OF DISCOURSE

This chapter argues for a poststructural political ecology. The project reflects a growing belief that nature is socially constructed, something entirely different from saying "There is no real nature out there"; but it takes the further step of insisting that the constructs of political economy and ecology as specifically modern forms of knowledge, as well as their objects of study, must be analyzed discursively. It is necessary to reiterate the connections between the making and evolution of nature and the making and evolution of the discourses and practices through which nature is historically produced and known. The relationship between nature and capital has been articulated historically by different discursive regimes, including in recent times – as we shall see – the discourses of sustainable development and biodiversity conservation. The argument developed here is thus a reflection on the discourses of nature from the vantage point of recent theories of the nature of discourse.

From a certain poststructural perspective (Foucaultian and Deleuzian in particular) there cannot be a materialist analysis which is not, at the same time, a discursive analysis. The poststructural analysis of discourse is not only a linguistic theory; it is a social theory, a theory of the production of social reality which includes the analysis of representations as social facts inseparable from what is commonly thought of as "material reality." Poststructuralism focuses on the role of language in the construction of social reality; it treats language not as a reflection of "reality" but as constitutive of it. That was the whole point, for instance, of Said's (1979) *Orientalism*. For some, there is no materiality unmediated by discourse, as there is no discourse unrelated to materialities (Laclau and Mouffe 1985). Discourse, as used in this chapter, is the articulation of knowledge and power, of statements and visibilities, of the visible and the expressible. Discourse is the process through which social reality inevitably comes into being.

Anthropologists have recently incorporated these insights into analyses of systems of production and systems of signification, systems of meanings of nature, and systems of use of resources, as inextricably bound together (Comaroff and Comaroff 1991; Gudeman and Rivera 1990; Hvalkof 1989). Political ecologists are beginning to emulate this fruitful trend. Space, poverty, and nature are seen through the lens of a discursive materialism "where ideas, matter, discourse, and power are intertwined in ways that virtually defy dissection" (Yapa 1995: 1). Insisting that we look at the way local cultures process the conditions of global capital and modernity (Pred and Watts 1992) is another important step in this direction.

In this chapter, I also take as point of departure a recent claim in political economy: the suggestion that capital, undergoing a significant change in form, enters an "ecological phase." No longer is nature defined and treated as an external, exploitable domain. Through a new process of capitalization, effected primarily by a shift in representation, previously "uncapitalized" aspects of nature and society become internal to capital. "Correspondingly, the primary dynamic of capitalism changes form, from accumulation and growth feeding on an external domain, to ostensible self-management and conservation of the system of capitalized nature closed back upon itself" (M. O'Connor 1993: 8). This transformation is perhaps most visible in discussions of rainforest biodiversity: the key to the survival of the rainforest is seen as lying in the genes of the species, the usefulness of which could be released for profit through genetic engineering and biotechnology in the production of commercially valuable products, such as pharmaceuticals. Capital thus develops a conservationist tendency, significantly different from its usual reckless, destructive form.

This proposal significantly qualifies views of the dialectic of nature and capital. The argument has been that capitalist restructuring takes place at the expense of production conditions: nature, the body, space. Driven by competition and cost-shifting among individual capitals/capitalists, this restructuring signifies a deepening of the encroachment of capital on nature and labor, an aggravation of the ecological crisis, and an impairment of capital's conditions of production – what James O'Connor (1988, 1989) calls "the second contradiction" of capitalism. For M. O'Connor, the expansionist drive of capital onto external nature implied by the second contradiction is only one tendency. Another entails a more pervasive discursive incorporation of nature as capital. This calls not for exploitative accumulation – with the concomitant impairment of production conditions – but, on the contrary, the sustainable management of the system of capitalized nature. Although the two forms may coexist, the first is prelude to the second, which appears when brute appropriation is contested by social movements. To the extent that the second entails deeper cultural domination – even the genes of living species are seen in terms of production and profitability – we are led to conclude that it will continue to achieve dominance in the strategies of both capital and social movements.

The present chapter is a contribution to the understanding of the articulations established by capital between natural and social systems. It argues that both forms of capital – exploitative and conservationist, modern and postmodern, let us say – are necessary given current conditions in the Third and First Worlds; both – not only the second form – require complex cultural and discursive articulations; both take on different but increasingly overlapping characteristics in the Third and First Worlds, and must be studied simultaneously; both can be studied by appealing to a poststructural political ecology; and that social movements and communities increasingly face the double task of building alternative productive rationalities and strategies, and culturally resisting the inroads of new forms of capital and technology into the fabric of nature and culture.

The first part of the chapter develops a nuanced reading of the discourse of sustainable development to bring out its mediation between nature and capital, particularly in the Third World. The second part elaborates the two forms of ecological capital; a brief example from the Pacific Coast region of Colombia is presented to show their respective rationalities and modes of operation. The third part analyzes the discourses of technoscience and biotechnology through which a veritable reinvention of nature is being effected, most clearly in the most industrialized countries, but increasingly in the Third World as well. The fourth part discusses the implications of the analysis for social practice; it focuses on the possibility of building alternative production rationalities by social movements faced with the two logics of ecological capital. The conclusion restates the case for the development of a poststructuralist political ecology as a means of ascertaining the types of knowledge that might be conducive to eco-socialist strategies.

"SUSTAINABLE DEVELOPMENT": DEATH OF NATURE, RISE OF ENVIRONMENT

By starting with the contemporary discourse that most forcefully seeks to articulate our relation to nature, we can "unpack" dominant assumptions about society and nature, and the political economy that makes such assumptions possible: the discourse of "sustainable development," launched globally in 1987 with the report of the World Commission on Environment and Development convened by the United Nations under the chair(wo)manship of Norway's former prime minister, Gro Harlem Brundtland. That report, published under the title *Our Common Future*, begins as follows:

> In the middle of the twentieth century, we saw our planet from space for the first time. Historians may eventually find that this vision had a greater impact on thought than did the Copernican revolution of the sixteenth century, which upset the human self-image by revealing that the earth is not the center of the universe. From space, we saw a small and fragile ball dominated not by human activity and edifice, but by a pattern of clouds, oceans, greenery, and soils. Humanity's inability to fit its doings into that pattern is changing planetary systems, fundamentally. Many such

changes are accompanied by life-threatening hazards. This new reality, from which there is no escape, must be recognized – and managed.

(World Commission on Environment and Development 1987: 1)

The category "global problems," to which *Our Common Future* belongs, is of recent invention. Its main impetus comes from the ecological fervor fostered by the Club of Rome reports of the 1970s, which provided a distinctive vision of the world as a global system where all parts are interrelated, thus demanding management on a planetary scale (Sachs 1988). The notion that nature and the Earth can be "managed" is an historically novel one. Like the earlier scientific management of labor, the management of nature entails its capitalization, its treatment as commodity. Moreover, the sustainable development discourse purports to reconcile two old enemies – economic growth and the preservation of the environment – without significant adjustments to the market system. This reconciliation is the result of complex discursive operations involving capital, representations of nature, management, and science. In the sustainable development discourse, nature is reinvented as environment so that capital, not nature and culture, may be sustained.

Seeing the Earth from space was not so great a revolution. This vision only re-enacted the scientific gaze established in clinical medicine at the end of the eighteenth century. The representation of the globe from space is but another chapter of the alliance which, two centuries ago, "was forged between words and things, enabling one *to see* and *to say*" (Foucault 1975: xii). Twentieth-century space exploration belongs to the paradigm defined by the spatialization and verbalization of the pathological, effected by the scientific gaze of the nineteenth-century clinician. As with the gaze of the clinician at an earlier time, environmental sciences today challenge the Earth to reveal its secrets to the positive gaze of scientists. This operation only ensures, however, that the degradation of the Earth be redistributed and dispersed, through the professional discourses of environmentalists, economists, geographers, and politicians. The globe and its "problems" have finally entered rational discourse. Disease is housed in nature in a new manner. As the medicine of the pathological led to a medicine of social space (the healthy biological space was also the social space dreamt of by the French Revolution), so will the "medicine of the Earth" result in new constructions of the social that allow some version of nature's health to be preserved.

In the Brundtland Report, we find a reinforcing effect between epistemology and the technologies of vision.

The instruments of visualization in multinationalist, postmodernist culture have compounded [the] meanings of disembodiment. The visualizing technologies are without apparent limit. . . . Vision in this technological feast becomes unregulated gluttony; all seems not just mythical about the god trick of seeing everything from nowhere, but to have put the myth into ordinary practice.

(Haraway 1988: 581)

49

The Report thus inaugurated a period of unprecedented gluttony in the history of vision and knowledge with the concomitant rise of a global ecocracy.

This might sound too harsh a judgment. We should construct the argument step by step. To begin with, management is the sibling of gluttonous vision, particularly now when the world is theorized in terms of global systems. The narrative of management is linked to the visualization of the Earth as a "fragile ball." Carrying the baton from Brundtland, *Scientific American*'s September 1989 Special Issue on "Managing Planet Earth" reveals the essence of the managerial attitude. At stake for this group of scientists (all either male academics or businessmen) is the continuation of the existing models of growth and development through appropriate management strategies. "What kind of planet do we want? What kind of planet can we get?" asks the opening article (Clark 1989: 48). "We" have responsibility for managing the human use of planet earth. "We" "need to move peoples and nations towards sustainability" by effecting a change in values and institutions that parallel the agricultural or industrial revolutions of the past.

The question in this discourse is what new manipulations can we invent to make the most out of nature and "resources." But who is this "we" who knows what is best for the world as a whole? Once again, we find the familiar figure of the (white male) Western scientist-turned-manager. A full-page picture of a young Nepalese woman "planting a tree as part of a reforestation project" is exemplary of the mind-set of this "we." Portrayed are not the women of the Chipko movement in India, with their militancy, their radically different forms of knowledge and practice of forestry, defending their trees politically and not through carefully managed "reforestation" projects. Instead there is a picture of an ahistorical young dark woman, whose control by masculinist and colonialist sciences, as Shiva (1989) has shown, is assured in the very act of representation. This regime of representation assumes that the benevolent hand of the West should save the earth; the Fathers of the World Bank, mediated by Gro Harlem Brundtland, the matriarch-scientist, and the few cosmopolitan Third Worlders who made it to the World Commission, will reconcile "humankind" with "nature." It is still the Western scientist that speaks for the Earth.

But can reality be "managed"? The concepts of planning and management embody the belief that social change can be engineered and directed, produced at will. The idea that poor countries could more or less smoothly move along the path of progress through planning has always been held to be an indubitable truth by development experts. Perhaps no other concept has been so insidious, no other idea gone so unchallenged, as modern planning. The narratives of planning and management, always presented as "rational" and "objective," are essential to developers (Escobar 1992). A blindness to the role of planning in the normalization and control of the social world is present also in environmental managerialism. As they are incorporated into the world capitalist economy, even the most remote communities of the Third World are torn from their local context, redefined as "resources" to be planned for, managed.

The rise of sustainable development is related to complex historical processes, including modifications in various practices (of assessing the viability and impact of development projects, obtaining knowledge at the local level, development assistance by NGOs); new social situations (the failure of top–down development projects, new social and ecological problems associated with that failure, new forms of protest, deficiencies that have become accentuated); and international economic and technological factors (new international divisions of labor with the concomitant globalization of ecological degradation, coupled with novel technologies that measure such degradation). What needs to be explained, however, is precisely why the response to this set of conditions has taken the form of "sustainable development," and what important problems might be associated with it. Four aspects are involved in answering this question.

First, the emergence of the concept of "sustainable development" is part of a broader process of the problematization of global survival, a process which induces a re-working of the relationship between nature and society. This problematization appeared as a response to the destructive character of development, on the one hand, and the rise of environmental movements in both the North and the South, on the other, resulting in a complex process of internationalization of the environment (Buttel *et al.* 1990). What is problematized is not the sustainability of local cultures and realities, but rather that of the global ecosystem, "global" being defined according to a perception of the world shared by those who rule it. Ecosystems professionals tend to see ecological problems as the result of complex processes that transcend cultural and local contexts. The slogan "Think globally, act locally" assumes not only that problems can be defined at a global level, but also that they are equally compelling for all communities. The professionals believe that since all people are passengers of spaceship earth, all are responsible for environmental degradation. They do not always see, in short, that there are great differences and inequities in resource problems between countries, regions, communities, and classes.

Second, the sustainable development discourse is regulated by a peculiar economy of visibilities. Ecosystems analysts have discovered the "degrading" activities of the poor, but have seldom recognized that such problems were rooted in development processes that displaced indigenous communities, disrupted people's habitats and occupations, and forced many rural societies to increase their pressures on the environment. Now the poor are admonished not for their lack of industriousness but for their "irrationality" and lack of environmental consciousness. Popular and scholarly texts alike are filled with representations of dark and poor peasant masses destroying forests and mountainsides with axes and machetes, thus shifting visibility and blame away from the large industrial polluters in North and South, and the predatory way of life fostered by capitalism and development, to poor peasants and "backward" practices such as slash-and-burn agriculture.

Third, the ecodevelopmentalist vision expressed in mainstream versions of sustainable development reproduces central aspects of economism and

51

developmentalism. The sustainable development discourse redistributes in new fields many of the concerns of classical development: basic needs, population, resources, technology, institutional co-operation, food security and industrialism are found reconfigured and reshuffled in the sustainable development discourse. That discourse upholds ecological concerns, although with a slightly altered logic. By adopting the concept of "sustainable development," two old enemies, growth and the environment, are reconciled (Redclift 1987), unfolding a new field of social intervention and control. Given the present visibility of ecological degradation, this process necessitates an epistemological and political reconciliation of ecology and economy, intended to create the impression that only minor corrections to the market system are needed to launch an era of environmentally sound development, and hiding the fact that the economic framework itself cannot hope to accommodate environmental concerns without substantial reform (Marglin 1992; Norgaard 1991a, 1991b). The sustainable development strategy, after all, focuses not so much on the negative consequences of economic growth on the environment, as on the effects of environmental degradation on growth and the potential for growth. Growth (i.e. capitalist market expansion) and not the environment has to be sustained. Poverty is believed to be a cause, as well as an effect, of environmental problems; growth is needed to eliminate poverty and, in turn, protect the environment. Unlike the discourse of the 1970s, which focused on "the limits to growth," the discourse of the 1980s was fixated on "growth of the limits" (Sachs 1988).

Fourth, the reconciliation of growth and environment is facilitated exactly by the new concept of the "environment," the importance of which has grown steadily in the post-Second World War ecological discourse. The development of ecological consciousness accompanying the rapid growth of industrial civilization also effected the transformation of "nature" into "environment" (Sachs 1992). No longer does nature denote an entity with its own agency, a source of life and discourse, as was the case in many traditional societies, and with European Romantic literature and art of the nineteenth century. For those committed to the world as resource, the "environment" becomes an indispensable construct. As the term is used today, "environment" includes a view of nature from the perspective of the urban-industrial system. Everything that is relevant to the functioning of this system becomes part of the environment. The active principle of this conceptualization is the human agent and his/her creations, while nature is confined to an ever more passive role. What circulates are raw materials, industrial products, toxic wastes, "resources"; nature is reduced to stasis, a mere appendage to the environment. Along with the physical deterioration of nature, we are witnessing its symbolic death. That which moves, creates, inspires – that is, the organizing principle of life – now resides in the environment.

The danger of accepting uncritically the sustainable development discourse is highlighted by a group of environmental activists from Canada:

A genuine belief that the Brundtland Report is a big step forward for the environmental/green movement . . . amounts to a selective reading, where the data on environmental degradation and poverty are emphasized, and the growth economics and "resource" orientation of the Report are ignored or downplayed. This point of view says that given the Brundtland Report's endorsement of sustainable development, activists can now point out some particular environmental atrocity and say, "This is not sustainable development". However, environmentalists are thereby accepting a "development" framework for discussion.

(Green Web 1989: 6)

Becoming a new client of the development apparatus by adopting the sustainable development discourse means accepting the scarcity of natural resources as a given fact; this leads environmental managers into stressing the need to find the most efficient forms of using resources without threatening the survival of nature and people. As the Brundtland Report put it, the goal should be to "produce more with less" (World Commission on Environment and Development 1987: 15). The World Commission is not alone in this endeavor. Year after year, this dictum is reawakened by the World Watch Institute in its *State of the World* reports, a main source for ecodevelopers. In these reports, ecology is reduced to a higher form of efficiency, as Wolfgang Sachs (1988) perceptively says.

Although ecologists and ecodevelopmentalists recognize environmental limits to production, a large number do not perceive the cultural character of the commercialization of nature and life integral to the Western economy, nor do they seriously account for the cultural limits which many societies have set on unchecked production. It is not surprising that their policies are restricted to promoting the "rational" management of resources. Environmentalists who accept this presupposition also accept imperatives for capital accumulation, material growth, and the disciplining of labor and nature. In doing so they extrapolate the occidental economic culture to the entire universe. Even the call for a people-centered economy runs the risk of perpetuating the basic assumptions of scarcity and productivism which underlie the dominant economic vision. In sum, by rationalizing the defense of nature in economic terms, advocates of sustainable development contribute to extending the economization of life and history. This effect is most visible in the World Bank approach to sustainable development, an approach based on the belief that, as the President of the Bank put it shortly after the publication of the Brundtland Report, "sound ecology is good economics" (Conable 1987: 6). The establishment in 1987 of a top level Environment Department, and the "Global Environmental Facility" (read: the Earth as a giant market/utility company under Group of Seven and World Bank control) created in 1992, reinforce the managerial attitude towards nature: "Environmental Planning" – said Conable (1987: 3) in the same address – "can make the most of nature's resources so that human resourcefulness can make the most of the future."

Again this involves the further capitalization of nature, the propagation of certain views of nature and society in terms of production and efficiency, not of respect and the common good. This is why Visvanathan (1991) calls the world of Brundtland and the World Bank "a disenchanted cosmos." The Brundtland

53

Report, and much of the sustainable development discourse, is a tale that a disenchanted (modern) world tells itself about its sad condition. As a renewal of the contract between the modern nation-state and modern science, sustainable development seeks not so much to caricature the past, as with early development theory, as to control a future whose vision is highly impoverished. Visvanathan is also concerned with the ascendancy of the sustainable development discourse among ecologists and activists. It is fitting to end this section with his call for resistance to co-option:

> Brundtland seeks a cooptation of the very groups that are creating a new dance of politics, where democracy is not merely order and discipline, where earth is a magic cosmos, where life is still a mystery to be celebrated. ... The experts of the global state would love to coopt them, turning them into a secondary, second-rate bunch of consultants, a lower order of nurses and paramedics still assisting the expert as surgeon and physician. It is this that we seek to resist by creating an explosion of imaginations that this club of experts seeks to destroy with its cries of lack and excess. The world of official science and the nation-state is not only destroying soils and silting up lakes, it is freezing the imagination. ... We have to see the Brundtland report as a form of published illiteracy and say a prayer for the energy depleted and the forests lost in publishing the report. And finally, a little prayer, an apology to the tree that supplied the paper for this document. Thank you, tree.
>
> (1991: 384)

CAPITALIZATION OF NATURE: MODERN AND POSTMODERN FORMS

The sustainable development strategy is the main way of bringing nature into discourse in what still is known as the Third World. The continuous reinvention of nature requires not only bringing nature into new domains of discourse but also bringing it into capital in novel ways. This process takes two general forms, both entailing discursive constructions of different kinds. Let us call these forms the modern and postmodern forms of capital in its ecological phase.

The modern form of capital

The first form capital takes in its ecological phase operates according to the logic of the modern capitalist culture and rationality; it is theorized in terms of what J. O'Connor (1988, 1989, 1991) calls "the second contradiction" of capitalism. Let it be recalled that the starting point of Marxist crisis theory is the contradiction between capitalist productive forces and production relations, or between the production and realization of value and surplus value. It is important to emphasize that from the perspective of traditional Marxist theory capitalism restructures itself through realization crises. But there is a second contradiction of capitalism that has become pressing with the aggravation of the ecological crisis and the social forms of protest this crisis generates. This theorization shows that we need to refocus our attention on the role played by the *conditions of production* in capital accumulation and capitalist restructuring, insufficiently theorized by Marx, yet placed at the center of inquiry by Polanyi's (1957) critique of the self-regulating market. Why? Because capitalist restructuring increasingly takes place at the expense of these conditions. A "condition of production" is everything treated as if it were a commodity, even if it is not produced as a commodity – that is according to the laws of value and the market: labor power, land, nature, urban space, fit this definition. Recall that Polanyi called "land" (that is, nature) and "labor" (that is, human life), "fictitious commodities." The history of modernity and the history of capitalism must be seen as the progressive capitalization of production conditions. Trees produced capitalistically on plantations, privatized land and water rights, genetically altered species sold in the market, and the entire training and professionalization of labor – from its crudest form in slavery to today's Ph.D.s – are all examples of the "capitalization" of nature and human life.

This process is mediated by the state; indeed, the state must be seen as an interface between capital and nature, including human beings and space. As far as human beings are concerned, the disciplining and normalization of labor, the management of poverty, and rise of the social (Donzelot 1979, 1988; Foucault 1979, 1980; Procacci 1991) marked the beginning of the capitalization of life within the modern era, while urban planning normalized and accelerated the capitalization of space (Harvey 1985; Rabinow 1989). This type of capitalization has been central to capitalism since the beginning of the primitive accumulation process and enclosure of the commons. The instrumental tendency of science was crucial in this regard, as discussed by philosophers, feminists, and ecologists (Merchant 1980; Shiva 1989).

In fact, one of the defining features of modernity is the increasing appropriation of "traditional" or pre-modern cultural contents by scientific knowledges, and the subsequent subjection of vast areas of life to regulation by administrative apparatuses based on expert knowledge (Foucault 1979; Giddens 1990; Habermas 1975). The history of capital not only involves exploitation of production conditions; it is also the history of the advance of the scientific discourses of modernity in areas such as health, planning, families, education, economy, and the like, through what Habermas (1987) refers to as the colonization of the lifeworld and Foucault (1980) as the advance of bio-power. The accumulation of capital, in other words, required the accumulation of normalized individuals and the accumulation of knowledge about the processes of capital and populations. This is the primary lesson of the anthropology of modernity of Western societies since the end of the eighteenth century. With this observation we wish to emphasize that the modern form of capital is inevitably mediated by the expert discourses of modernity.

Capital's threatening of its own conditions of production elicits manifold and contradictory attempts at restructuring those conditions in order to reduce costs or defend profits. Conversely, social struggles generated around the defense of production conditions must seek two objectives: to defend life and production conditions against capital's excesses; and to seek control over policies to restructure production conditions, usually via further privatization. In other words, social movements have to face simultaneously the destruction of life, the body, nature, and space, and the crisis-induced restructuring of these conditions (O'Connor 1988, 1991). Often, these struggles pit the poor against the rich as both cultural and economic actors; there is an "environmentalism of the poor" (Guha 1994; Martinez-Alier 1992) which is a type of class struggle and, at the same time, a cultural struggle, to the extent that the poor try to defend their natural environments from material and cultural reconversion by the market. These struggles are often gender struggles, in that many aspects of the destruction of production conditions affect women particularly and contribute to the restructuring of class and gender relations (Mellor 1992; Rao 1989, 1991).

The postmodern form of ecological capital

In the Third World, the continued existence of conventional forms of capitalist exploitation of people and the environment is organized according to the rules of the dominant development discourse of the last forty years, for which nature exists as raw material for economic growth activities (Escobar 1995). While there are areas "sold" to the sustainable development discourse, others remain firmly in the grasp of that crude and reckless developmentalism characterizing the post-Second World War period. As we shall see in the Colombian case, both forms may coexist schizophrenically in the same geographical and cultural region.

M. O'Connor (1993) is right, however, in pointing at a qualitative change in the form capital tends to take today. If with modernity one can speak of a progressive semiotic conquest of social life by expert discourses and economistic conceptions, today this conquest is being extended to the very heart of nature and life. This new conquest takes for granted the normalization already achieved by the modern discourses of science and its administrative apparatuses; not only does it move on to new territories, it also develops new modes of operation, which O'Connor understands particularly in the Baudrillardian sense of the preeminence of the sign. Once modernity is consolidated, once "the economy" becomes a seemingly ineluctable reality (a true descriptor of reality for most), capital and the struggles around it must broach the question of the domestication of all remaining social and symbolic relations, in terms of the code of political economy, that of production. It is no longer capital and labor that are at stake *per se*, but the reproduction of the code. Social reality becomes, to borrow Baudrillard's (1975) phrase, "the mirror of production."

This second form of capital relies not only on the symbolic conquest of nature (in terms of "biodiversity reserves") and local communities (as "stewards" of nature); it also requires the semiotic conquest of local knowledges, to the extent that "saving nature" demands the valuation of local knowledges of sustaining nature. Local, "indigenous" and "traditional" knowledge systems are found to be useful complements to modern biology. However, in these discourses, knowledge is seen as something existing in the "minds" of individual persons (shamans or elders) about external "objects" ("plants," "species"), the medical or economic "utility" of which their bearers are supposed to transmit to us. Local knowledge is seen not as a complex cultural construction, involving movements and events profoundly historical and relational. Moreover, these forms of knowledge usually have entirely different modes of operation and relations to social and cultural fields (Deleuze and Guattari 1987). As they are brought into its politics modern science recodifies them in utilitarian ways.

This triple cultural reconversion of nature, people, and knowledge represents a novel internalization of production conditions. Nature and local people themselves are seen as the source and creators of value – not merely as labor or raw material. The discourse of biodiversity in particular achieves this effect. Species of micro-organisms, flora, and fauna are valuable not so much as "resources" but as reservoirs of value – this value residing in their very genes – that scientific research, along with biotechnology, can release for capital and communities. This is a reason why communities – particularly ethnic and peasant communities in the tropical rainforest areas of the world – are finally recognized as the owners of their territories (or what is left of them), but only to the extent that they accept seeing and treating territory and themselves as reservoirs of capital. Communities in various parts of the world are then enticed by biodiversity projects to become "stewards of the social and natural 'capitals' whose sustainable management is, henceforth, both their responsibility and the business of the world economy" (M. O'Connor 1993: 5). Once the semiotic conquest of nature is completed, the sustainable and rational use of the environment becomes imperative. It is here that the fundamental logic of the discourses of sustainable development and biodiversity must be found.

"Biodiversity conservation" in Colombia

A brief example illustrates the differences between the two forms of capital. The Pacific Coast region of Colombia has one of the highest degrees of biological diversity in the world. Covering about 5.4 million hectares of tropical rainforest, it is populated by about 800,000 Afrocolombians and 40,000 indigenous people belonging to various ethnic groups, particularly Emberas and Waunanas. Since the early 1980s the national and regional governments have increased their development activities in the region, culminating in the elaboration of ambitious development plans (DNP 1983, 1992). The 1992 *Sustainable Development Plan* is a conventional strategy intended to foster the development of capitalism in the region. Since the early 1980s, capital has flowed to various parts of the region, particularly in the form of investments in sectors such as African palm plantations, large-scale shrimp cultivation, gold mining, timber, and tourism. These investments operate, for the most part, in the mode of the first form of capital. All of the activities of this type of capital tend to contribute to ecological degradation and the displacement and proletarianization of local people, who can no longer subsist as farmers and must find precarious jobs in palm oil plantations and shrimp-packing plants.

Parallel to this, the government has launched a more modest, but symbolically ambitious, project for the protection of the region's almost legendary biodiversity, in peril of being destroyed by activities mediated by the development plan. The Biodiversity Project (GEF–PNUD 1993), conceived under the directives of the Global Biodiversity Strategy (WRI/WCU/UNEP 1992) and within the scope of the World Bank's Global Environmental Facility (GEF), purports to effect an alternative strategy for the sustainable and culturally appropriate development of the area. The project is organized along four different axes: "to know" (to gather and systematize modern and traditional knowledge of the region's biodiversity); "to valorize" (to design ecologically sound strategies to create economic value out of biodiversity); "to mobilize" (to foster the organization of the black and indigenous communities so they can take charge of the sustainable development of their environments); and "to formulate and implement" (to modify institutional structures so they can support community-oriented sustainable development strategies).

The Biodiversity Project obeys the global logic of the second form of ecological capital. The Project became possible not only because of international trends, but was formed also by pressure exerted on the state by black and indigenous communities in the context of territorial and cultural rights accorded by the reform of the National Constitution of 1991. The Project designers had to take into account the views of local communities, and had to accept as important interlocutors representatives of the black movement which was growing in the context of the developmentalist onslaught and the reform of the constitution. A few progressive professionals associated with the black movement have been able to insert themselves in the national and regional staff of the project. While these professionals seem aware of the risks involved in participating in a government project of this kind, they also believe that the Project presents a space of struggle they cannot afford to ignore.

Along with new forms of biotechnology, the discourse of biodiversity conservation produced mostly by Northern NGOs and international organizations in the 1990s achieves an important transformation in our consciousness of and practices towards nature. As far as the world rainforests are concerned, this discourse constructs an equation between "knowing" (classifying species), "saving" (protecting from total destruction), and "using" (through the development of commercial applications based on the genetic properties of species). Biodiversity prospectors would roam the rainforest in search of potential uses of rainforest species, and the biotechnological developments that would allegedly ensue from this task would provide the key to rainforest preservation – if appropriately protected, of course, by intellectual property rights so that prospectors and investors have the incentive to invest in the epic enterprise of saving nature (WRI/WCU/UNEP 1992; WRI 1993). Both capitalism and nature would not only survive but would thrive under the new scheme dreamed up by scientists, planners, multinational corporations, and genetic and molecular biology laboratories, among others. Social movements confront a greening of economics, knowledge, and communities more pervasive than ever before.

MAKING NATURE: FROM (MODERN) DEATH TO (POSTMODERN) REINVENTION

It should be clear by now that sustainable development and biodiversity strategies play a crucial role in the discursive production of production conditions. Production conditions are not just transformed by "capital": they have to be transformed in/through discourse. The Brundtland Report, indeed the entire sustainable development movement, is an attempt at resignifying nature, resources, the Earth, human life itself, on a scale perhaps not witnessed since the rise of empirical sciences and their reconstruction of nature – since nature's "death," to use Carolyn Merchant's expression (1980). Sustainable development is the last attempt at articulating nature, modernity, and capitalism before the advent of cyberculture.

The reconversion of nature effected by the discourses of biodiversity and sustainable development may be placed in the broader context of what Donna Haraway (1991) calls "the reinvention of nature." Reinvention is fostered by sciences such as molecular biology and genetics, research strategies such as the Human Genome Project, and biotechnology. For Haraway, however, reinvention began with the languages of systems analysis developed since the early post-Second World War period; it marks the final disappearance of our organic notions of nature. The logic and technologies of command-control have become more central in recent years, particularly with the development of immunological discourses (Martin 1994) and projects such as the mapping of the human genome. The language of this discourse is decidedly postmodern and is not inimical to the post-Fordist regime of accumulation (Harvey 1989), with its new cultural order of "flexible labor," which might also be read symbolically as an attempt to keep dark invaders at a distance or quickly phagocytize them if they come close enough or become numerous enough to threaten contagion and disorder.

Haraway reads in these developments the de-naturalization of the notions of "organism" and "individual," so dear to pre-Second World War science. She sees the emergence of a new entity, the cyborg, which arises to fill the vacuum (1985, 1989b, 1991). Cyborgs are hybrid creatures, composed of organism and machine, "special kinds of machines and special kinds of organisms appropriate to the late twentieth century" (1991: 1). Cyborgs are not organic wholes but strategic assemblages of organic, textual, and technical components. In the language of sustainable development one would say that cyborgs do not belong in/to nature; they belong in/to environment, and the environment belongs in/to systems.

Haraway concludes that we need to develop a different way of thinking about nature, and ourselves in relation to nature. Taking Simone de Beauvoir's declaration that "one is not born a woman" into the postmodern domain of late twentieth-century biology, Haraway adds that "one is not born an organism. Organisms are made; they are constructs of a world-changing kind" (1989b: 10). To be more precise, organisms make themselves and are also made by history. This deeply historicized account of life is difficult to accept if one remains within the modern traditions of realism, rationalism, and organic nature. The histori-cized view assumes that what counts as nature and what counts as culture in the West ceaselessly changes according to complex historical factors. Since at least the end of the eighteenth century, "the themes of race, sexuality, gender, nation, family and class have been written into the body of nature in western life sciences," even if in every case nature "remains a crucially important and deeply contested myth and reality" (1989a: 1). Nature as such (unconstructed) has ceased to exist, if indeed it ever existed.

Nature, bodies, and organisms must thus be seen as "material-semiotic" actors, rather than mere objects of science pre-existing in purity. Nature and organisms emerge from discursive processes involving complex apparatuses of science, capital, and culture. This implies that the boundaries between the organic, the techno-economic, and the textual (or, broadly, cultural) are permeable. While nature, bodies, and organisms certainly have an organic basis, they are increas-ingly produced in conjunction with machines, and this production is always mediated by scientific and cultural narratives. Haraway emphasizes that nature is a co-construction among humans and non-humans. Nature has a certain agency, an "artifactuality" of sorts. We thus have the possibility of engaging in new conversations with/around nature, involving humans and unhumans together in the reconstruction of nature as public culture. Furthermore, "there are great riches for feminists [and others] in explicitly embracing the possibilities inherent in the breakdown of clean distinctions between organism and machine and similar distinctions structuring the Western self" (Haraway 1985: 92).

Haraway's work reflects, and seeks to engage with, the profound transforma-tion brought about by new computer technologies and biotechnology advancing in the core countries of the capitalist system. The advent of the new era – which we can perhaps call cyberculture, as a truly post-industrial and postmodern society (Escobar 1994) – entails a certain cultural promise for more just social configurations. We should have no doubts by now that a fundamental social and cultural transformation is under way, which promises to reshape biological and social life, and which involves both dangers and possibilities. A new regime of bio-sociality is upon us, implying that "nature will be modeled on culture understood as practice. Nature will be known and remade through technique and will finally become artificial, just as culture becomes natural" (Rabinow 1992: 241). This might bring the dissolution of modern society and of the nature/ culture split, marking also the end of the ideologies of naturalism – of an organic nature existing outside history – and even the possibility that the organic might be improved upon by artificial means.

What all this means for the Third World is yet to be examined. This examination has to start with inventing a new language to speak of such issues from Third World perspectives, a language of transformative self-affirmation that allows the Third World to reposition itself in the global conversations and processes that are reshaping the world, without submitting passively to the rules of the game created by them. Sustainable development will not do. Biodiversity, on the contrary, is becoming inextricably linked to other discourses, such as biotechnology, genetics, and intellectual property rights (Shiva 1992). But the implications for the Third World communities placed as "stewards" of organic nature are by no means well understood. The issues are crucial for communities, as the Afrocolombian activists of the Pacific Coast have discovered. Not for nothing are corporations developing aggressive policies of privatizing nature and life. Communities in various parts of the Third World will have to dialogue with each other to face the internationalization of ecological capital. Ecological solidarity (South–South and North–South) must travel this perilous terrain, and perhaps entertain the idea of strategic alliances between the organic and the artificial (in terms of biotechnology applications of rainforests' biodiversity, for instance) against the most destructive forms of capital.

SEMIOTIC RESISTANCE AND ALTERNATIVE PRODUCTIVE RATIONALITY

The role of discourse and culture in organizing and mediating "nature" and "production conditions" is still undeveloped in both the eco-socialist and eco-feminist conceptions. For the most part, the economistic culture of modernity is taken as the norm. Behind this lie the relationships between natural and historical processes. Haraway's work provides valuable elements for examining this relation particularly in the context of rising technoculture. The Mexican ecologist, Enrique Leff, has made a general case for theorizing the mutual inscription of nature, culture, and history in terms useful for thinking about Third World situations. As the ecological becomes part of the accumulation process, Leff argues, the natural is absorbed into history and can thus be studied by historical materialism. Yet he insists that culture remains an important mediating instance. The transformation of nature and ecosystems by capital depends upon the cultural practices of specific societies and the processes of cultural trans-formations that are taking place (Leff 1995; see also Godelier 1986).

Leff's (1986, 1992, 1993, 1995) conceptual effort is linked specifically to the articulation of an alternative, ecologically sustainable productive rationality from an integrated perspective of ecology, culture, and production. For Leff, ecological, technological, and cultural productivity must be woven together to theorize a new view of rationality that generates equitable and sustainable processes. "The environment should be regarded as the articulation of cultural, ecological, technological and economic processes that come together to generate a complex, balanced, and sustained productive system open to a variety of options and development styles" (1993: 60).

At the cultural level, cultural practices should be seen as a principle of productivity for the sustainable use of natural resources. Most clearly in the case of indigenous and ethnic groups, every social group possesses an "ecological culture" that must be seen as forming part of the social relations and forces of production. At the level of production, Leff advocates the development of "a productive paradigm that is not economistic yet pertains to political economy" (1993: 50). The result would be an alternative production paradigm that relates technological innovation, cultural processes, and ecological productivity. Less clear in Leff's work is how concepts such as "production" and "rationality" can be theorized from the perspective of different cultural orders.

Based on this reformed view of the environment, Leff calls on ecology activists and theorists to think in terms of "ecological conditions of production" and a "positive theory of production," in which nature is not seen only as production condition, but actively incorporated into a new productive rationality along with labor and technology. This call parallels J. O'Connor's redefinition of production conditions from the standpoint of the second contradiction, particularly through the actions of social movements. Leff's formulation brings into sharper focus the real need that social movements and communities have for articulating their own views of alternative development and alternative productive schemes specifically from the perspective of ecology. The pressure mounts on social movements and community activists in many parts of the world to engage in this constructive task, as the case of the black and indigenous activists in the Colombian Pacific Coast shows. Leff's ongoing effort at conceptualizing an alternative productive rationality is helpful in this regard.

The creation of a new productive rationality would entail forms of environmental democracy, economic decentralization, and cultural and political pluralism. The creation of spaces in which to foster local alternative productive projects is one concrete way of advancing the strategy. In sum, Leff seeks to redefine and radicalize three basic constructs: production, away from economistic cultural constructions and pure market mechanisms; rationality, away from the dominant reductionistic and utilitarian views; and management, away from bureaucratized practice and towards a participatory approach. A strategy such as this, one might add, implies cultural resistance to the symbolic reconversion of nature; socioeconomic proposals with concrete alternative strategies; and political organizing to ensure a minimum of local control over the entire process. In the landscape of Latin American hybrid cultures (Garcia Canclini 1990), strategies seem to be required combining modern and non-modern, capitalist and non-capitalist forms and practices.

62

One thing is clear in this debate: social movements and communities in the Third World need to articulate alternative productive strategies that are ecologically sustainable, lest they be swept away by a new round of conventional development. The fact that these alternatives must also be culturally defined – from the perspectives of cultures which, although hybrid, nevertheless retain a socially significant difference *vis à vis* Western modernity – necessarily entails that a certain semiotic resistance takes place. The worst would be for communities to opt for conventional development styles. To accede to an era of post-development – in which the hegemonic effect of the constructs of modernity might be held in check (Escobar 1995) – communities will need simultaneously to experiment with alternative productive strategies and to culturally resist capital's and modernity's material and symbolic restructuring of nature. Communities will need to prevent conventional development, green redevelopment via sustainable development discourses, and the greening of communities and local knowledge via discourses of biodiversity.

Is it really possible to imagine an alternative ecological economy based on a different cultural (not only social) order? If one accepts that this has become an essential political task, how could analysts investigate the concrete cultural practices that might serve as a basis for it? What are the macro-economic conditions and political processes that could make its implementation and survival possible? How should this alternative social reality engage with dominant market-driven forces? The importance of such questions will grow as researchers come to realize the increasing complexity of the cultural politics of nature under way in the wake of new forms of capital, technoscience, and globalization.

CONCLUSION: TOWARDS A POSTSTRUCTURAL POLITICAL ECOLOGY

The two socially necessary forms of capital – modern and postmodern – maintain an uneasy articulation depending on local, regional, and transnational conditions. Both forms are mediated by discourse: conventional discourses of development, plus the scientific discourses of modernity, in the case of the first form of ecological capital; discourses of biodiversity and sustainable development (particularly in the Third World), and molecular biology, biotechnology, and cyberculture in the First (and increasingly Third) Worlds, in the case of the second form of ecological capital. The regime of sustainable development in the South, and of bio-sociality and cyberculture in the North show a certain degree of geographical unevenness; yet the connections between them are becoming clearer. While some regions in the Third World join the ranks of the cyberculture, poor communities in the First are affected by the logic of reckless capital and the paradoxes of sustainability. The division between First and Third Worlds is undergoing a fundamental mutation in the wake of post-Fordism, cyberculture, and the ecological phase of capital.

The discursive nature of capital is evident in the case of the production of "production conditions." The resignification of nature as environment; the reinscription of the earth into capital via the gaze of science; the reinterpretation of poverty as an effect of destroyed environments; the destruction of vernacular gender and the concomitant proletarianization and re-articulation of women's subordination under modern principles; and the new interest in management and planning as arbiters between people and nature, all are effects of the discursive construction of sustainable development. As more and more professionals and activists adopt the grammar of sustainable development, the reinvention of production conditions produced by this discourse will be more effective. Institutions will continue to re/produce the world as seen by those who rule it.

Although many people seem to be aware that nature is "socially constructed," many also continue to give a relatively unproblematic rendition of nature. Central to this rendition is the assumption that "nature" exists beyond our constructions. Nature, however, is neither unconstructed nor unconnected. Nature's constructions are effected by history, economics, technology, science, and myths of all kinds as part of the "traffic between nature and culture" (Haraway 1989a). Leff (1986, 1995) emphasizes a similar point in his own way. Capital accumulation, he says, requires the articulation of the sciences to the production process, so that the truths they produce become productive forces. Thus the sustainable development discourse must be seen as part of the creation of knowledge linked to capital, to the extent that the concepts produced participate in reinscribing nature into the law of value. Although the process of transdisciplinarity involved in the sciences of ecology is hopeful, Leff (1986) believes, the lack of epistemological vigilance has resulted in a certain disciplining of environmental themes which has precluded the creation of concepts useful for the formulation of alternative ecological rationalities. The analysis of discourses can serve as a basis for elaborating practical concepts useful to reorient strategies concerning development and the environment.

If nature and other life forms must now be understood as articulations of organic, techno-economic, and cultural elements, does this not imply that we need to theorize this mixture as the appropriate object of biology and ecology, perhaps at the same time – and dialectically – that these sciences seek to theorize the "laws of nature" in and of themselves? As Leff (personal communication) rightly says, one must be cautious in this endeavor, and raise the question of "to what extent by manipulating nature as reality you manipulate the scientific object of biology. By manipulating evolution and genetics, to what extent do we also manipulate and reconstruct the object and the internal laws of biology and genetics?" Perhaps what is needed is a new epistemology of biology, such as that proposed by the phenomenological biology of Humberto Maturana and Francisco Varela (1980, 1987; Varela *et al.* 1991). Works of this kind, attempting to step outside the traditional space of science by taking seriously the continuity between cognizant self and world, between knowledge and the social practices that make that knowledge possible, might contribute important elements to a new biology and ecology. The question of the epistemology of the natural sciences is being broached both from poststructuralist perspectives and reformed phenomenology in Maturana and Varela's case. Should not it be broached from that of political ecology as well?

The worldwide spread of value seems to privilege the new biotechnologies. These further capitalize nature by planting value into it through scientific research and development. Even human genes become conditions of production, an important arena for capitalist restructuring, and so for contestation. The reinvention of nature currently under way, effected by/within webs of meaning and production that link the discourses of science and capital, should be incorporated into a political ecology appropriate to that new age whose dawn we are witnessing. What will count as "organisms" and even "human" for biology, ecology, geography, and biological anthropology will be intimately mediated by these processes.

Nature is now modeled as culture; sooner or later, "nature" will be produced to order. If the production of trees in plantations constituted an important step in the capitalization of nature, for example, the production of genetically produced trees (or the "perfect" tomatoes produced at the University of California at Davis) takes this process to new levels; it takes the tree a step further away from "organic nature." The implications of this are unclear. This is why the rising regime of bio-sociality must find its place at the base of a political ecology and biology as forms of knowledge about material-semiotic objects – organisms and communities – that are historically constituted.

This is to say we need new narratives of life and culture. These narratives will likely be hybrids of sorts; they will arise from the mediations that local cultures are able to effect on the discourses and practices of nature, capital, and modernity. This is a collective task that perhaps only social movements are in a position to advance. The task entails the construction of collective identities, as well as struggles over the redefinition of the boundaries between nature and culture. These boundaries will be reimagined to the extent that the practice of social movements succeeds in reconnecting life and thought by fostering a plural political ecology of knowledge. As the analysis of concrete practices of thinking and doing, discursive approaches have much to contribute to this reimagining. Materialist approaches do not need to exclude this type of analysis.

NOTE

This chapter was originally prepared for the Wenner-Gren Symposium "Political-economic Perspectives in Biological Anthropology: Building a Biocultural Synthesis" 1992. I thank Alan Goodman and Thomas Leathermann, conference organizers, and Richard Peet, James O'Connor, Enrique Leff, and Richard Lichtman for thoughtful comments on previous versions of the manuscript.

REFERENCES
Baudrillard, Jean. 1975. *The Mirror of Production*. St. Louis: Telos Press.

Buttel, Frederick, A. Hawkins, and G. Power. 1990. "From limits to growth to global change: contrasts and contradictions in the evolution of environmental science and ideology," *Global Environmental Change* 1, 1: 57–66.

Clark, William. 1989. "Managing Planet Earth," *Scientific American* 261, 3: 46–57.

Comaroff, Jean and John Comaroff. 1991. *Of Revelation and Revolution*. Chicago: University of Chicago Press.

Conable, Barber. 1987. "Address to the World Resources Institute," Washington, DC: The World Bank.

Deleuze, Gilles and Félix Guattari. 1987. *A Thousand Plateaus*. Minneapolis: University of Minnesota Press.

DNP (Departamento Nacional de Planeación de Colombia). 1983. *Plan de Desarrollo Integral para la Costa Pacífica*. Cali: DNP/CVC.

—— 1992. *Plan Pacífico. Una Estrategia de Desarrollo Sostenible para la Costa Pacífica Colombiana*. Bogotá: DNP.

Donzelot, Jacques. 1979. *The Policing of Families*. New York: Pantheon Books.

—— 1988. "The promotion of the social," *Economy and Society* 17, 3: 217–34.

Escobar, Arturo. 1992. "Planning," in Wolfgang Sachs (ed.) *The Development Dictionary*, London: Zed Books, pp. 132–45.

—— 1994. "Welcome to Cyberia: notes on the anthropology of cyberculture," *Current Anthropology* 35, 3: 211–31.

—— 1995. *Encountering Development: The Making and Unmaking of the Third World*. Princeton, NJ: Princeton University Press.

Foucault, Michel. 1975. *The Birth of the Clinic*. New York: Vintage Books.

—— 1979. *Discipline and Punish*. New York: Vintage Books.

—— 1980. *The History of Sexuality. Volume I. An Introduction*. New York: Vintage Books.

García Canclini, Néstor. 1990. *Culturas Híbridas: Estrategias para Entrar y Salir de la Modernidad*. México, DF: Grijalbo.

GEF–PNUD (Global Environmental Facility–Programa de las Naciones Unidas para el Desarrollo). 1993. *Conservación de la Biodiversidad del Chocó Biogeográfico. Proyecto Biopacífico*. Bogotá: DNP/Biopacífico.

Giddens, Anthony. 1990. *The Consequences of Modernity*. Stanford, CA: Stanford University Press.

Godelier, Maurice. 1986. *The Mental and the Material*. London: Verso.

Green Web. 1989. "Sustainable development: expanded environmental destruction," *The Green Web Bulletin* 16.

Gudeman, Stephen and Alberto Rivera. 1990. *Conversations in Colombia*. Cambridge: Cambridge University Press.

Guha, Ramachandra. 1994. "The environmentalism of the poor," presented at the Conference, "Dissent and Direct Action in the Late Twentieth Century," Otavalo, Ecuador, June.

Habermas, Jürgen. 1975. *Legitimation Crisis*. Boston, MA: Beacon Press.

—— 1987. *The Philosophical Discourse of Modernity*. Cambridge, MA: MIT Press.

Haraway, Donna. 1985. "Manifesto for cyborgs: science, technology, and socialist feminisms in the 1980s," *Socialist Review* 80: 65–107.

—— 1988. "Situated knowledges: the science question in feminism and the privilege of partial perspective," *Feminist Studies* 14, 3: 575–99.

—— 1989a. *Primate Visions*. New York: Routledge.

—— 1989b. "The biopolitics of postmodern bodies: determinations of self in immune system discourse," *Differences* 1, 1: 3–43.

—— 1991. *Simians, Cyborgs and Women. The Reinvention of Nature*. New York: Routledge.

Harvey, David. 1985. *The Urbanization of Capital*. Baltimore, MD: Johns Hopkins University Press.

—— 1989. *The Condition of Postmodernity*. Oxford: Basil Blackwell.

Hvalkof, Søren. 1989. "The nature of development: native and settlers' views in Gran Pajonal, Peruvian Amazon," *Folk* 31: 125–50.

Laclau, Ernesto and Chantal Mouffe. 1985. *Hegemony and Socialist Strategy*. London: Verso.

Leff, Enrique. 1986. "Ambiente y articulación de Ciencias," in Enrique Leff (ed.) *Los Problemas del Conocimiento y la Perspectiva Ambiental del Desarrollo*, México, DF: Siglo XXI, pp. 72–125.

—— 1992. "La dimensión cultural y el manejo integrado, sustentable y sostenido de los recursos naturales," in Enrique Leff and J. Carabias (eds.) *Cultura y Manejo Sustentable de los Recursos Naturales*. Mexico, DF: CIIH/UNAM.

—— 1993. "Marxism and the environmental question: from the critical theory of production to an environmental rationality for sustainable development," *Capitalism, Nature, Socialism* 4, 1: 44–66.

—— 1995. *Green Production: Toward an Environmental Rationality*. New York: Guilford Press.

Marglin, Steve. 1992. "Alternatives to the greening of economics: a research proposal," unpublished proposal.

Martin, Emily. 1994. *Flexible Bodies*. Boston, MA: Beacon Press.

Martinez-Alier, Juan. 1992. *Ecología y Pobreza*. Barcelona: Centre Cultural Bancaixa.

Maturana, Humberto and Francisco Varela. 1980. *Autopoiesis and Cognition: The Realization of the Living*. Boston, MA: D. Reidel Publishing Company.

—— and —— 1987. *The Tree of Knowledge: The Biological Roots of Human Understanding*. Boston: New Science Library/Shambhala.

Mellor, Mary. 1992. *Breaking the Boundaries: Towards a Feminist Green Socialism*. London: Virago.

Merchant, Carolyn. 1980. *The Death of Nature*. New York: Harper & Row.

Norgaard, Richard. 1991a. "Sustainability as intergenerational equity," Internal Discussion Paper No. IDP 97. Washington, DC: World Bank.

—— 1991b. "Sustainability: the paradigmatic challenge to agricultural economics," presented at the 21st Conference of the International Association of Agricultural Economists, Tokyo, 22–9 August.

O'Connor, James. 1988. "Capitalism, nature, socialism: a theoretical introduction," *Capitalism, Nature, Socialism* 1, 1: 11–38.

—— 1989. "Political economy of ecology of socialism and capitalism," *Capitalism, Nature, Socialism* 1, 3: 93–108.

—— 1991. *Conference Papers*, CES/CNS Pamphlet 1, Santa Cruz.

O'Connor, Martin. 1993. "On the misadventures of capitalist nature," *Capitalism, Nature, Socialism* 4, 3: 7–40.

Polanyi, Karl. 1957. *The Great Transformation*. Boston, MA: Beacon Press.

Pred, Alan and Michael Watts. 1992. *Reworking Modernity*. New Brunswick: Rutgers University Press.

Procacci, Giovanna. 1991. "Social economy and the government of poverty," in Graham Burchell, Colin Gordon, and Peter Millers (eds.) *The Foucault Effect*. Chicago: University of Chicago Press, pp. 151–68.

Rabinow, Paul. 1989. *French Modern: Norms and Forms of the Social Environment*. Cambridge, MA: MIT Press.

—— 1992. "Artificiality and enlightenment: from sociobiology to biosociality," in

Jonathan Crary and Sanford Kwinter (eds.) *Incorporations.* New York: Zone Books, pp. 234–52.

Rao, Brinda. 1989. "Struggling for production conditions and producing conditions of emancipation: women and water in rural Maharashtra," *Capitalism, Nature, Socialism* 1, 2: 65–82.

—— 1991. *Dominant Constructions of Women and Nature in Social Science Literature,* CES/CNS Pamphlet 2, Santa Cruz.

Redclift, Michael. 1987. *Sustainable Development: Exploring the Contradictions.* London: Routledge.

Sachs, Wolfgang. 1988. "The gospel of global efficiency," IFDA Dossier No. 68: 33–9.

—— 1992. "Environment," in Wolfgang Sachs (ed.) *The Development Dictionary.* London: Zed Books, pp. 26–37.

Shiva, Vandana. 1989. *Staying Alive. Women, Ecology and Development.* London: Zed Books.

—— 1992. "The seed and the earth: women, ecology and biotechnology," *The Ecologist* 22, 1: 4–8.

Varela, Francisco, Evan Thompson and Eleanor Rosch. 1991. *The Embodied Mind: Cognitive Science and Human Experience.* Cambridge: MIT Press.

Visvanathan, Shiv. 1991. "Mrs. Brundtland's Disenchanted Cosmos," *Alternatives* 16, 3: 377–84.

World Commission on Environment and Development. 1987. *Our Common Future.* New York: Oxford University Press.

World Resources Institute. 1993. *Biodiversity Prospecting.* Washington: WRI.

World Resources Institute, World Conservation Union, and United Nations Environment Program. 1992. *Global Biodiversity Strategy.* Washington: WRI.

Yapa, Lakshman. 1995. "Can postmodern discourse theory help alleviate poverty? Yes!" paper presented at Meeting of the American Association of Geography, Chicago, 17 March.

3

IMPROVED SEEDS AND CONSTRUCTED SCARCITY

Lakshman Yapa

It is widely believed that poverty is caused by economic underdevelopment, and that the problem can be solved through economic development – increasing investment, creating jobs, raising incomes, and improving the general standard of living. I shall refer to this as the "axiom of economic development" – "axiom" because these claims seem so obviously true they have become an integral part of "common sense." The principal purpose of this chapter is to contest the validity of this axiomatic belief. I argue that poverty is not the result of lack of development, poor technology, or scarce resources, but a normal manifestation of the very process of economic development that is supposed to cure it; development causes modern poverty through "socially constructed scarcity." The argument for development-induced scarcity is made by narrating a story about the Green Revolution – how "improved seeds" are implicated in the social construction of scarcity.

Criticisms of "solutions," such as high-yielding seeds, are usually countered by questions like, "Hasn't the Green Revolution staved off massive hunger in the Third World?" It is not possible to provide objective "intelligent" answers to such questions because they presume a certain way of knowing the world. As I show later in the chapter such questions have little meaning outside that epistemology. "If development is not the answer, what is the solution to hunger and poverty?" Over the years, as I became increasingly skeptical about the promise of development, I looked to "other models of development" – socialist, sustainable, authentic, and so on – as possible solutions to poverty. Eventually I concluded that it was impossible to describe such a space in the pages of an article or as a chapter in a book; the search for solutions is a discursive convention stemming from the same epistemology of development that I find so problematic. That still leaves the question unresolved of how one should address issues of hunger, malnutrition, and poverty.

The usual answer, Green Revolution, is not just a matter of producing more food with improved seeds. It embodies a particular epistemology of development, a way of seeing food, technology, nature, culture, and society, an epistemology I wish critically to engage through my narrative of seeds. In this story there is no circumscribed space called "the alternative" to the Green Revolution; in fact there are thousands of site-specific alternatives that emerge in the substantive details of the story. It is not possible to jump ahead for a sneak preview because there is no "solution" to be unveiled at the end of this critique. The "solution" lies within an understanding of the detailed structure of the critique; the critique is the solution.

By "poverty," is meant a situation in which households are unable to satisfy their basic needs for food, clothing, shelter, health care, and functional literacy. "Economic development" refers to efforts to improve "standards of living" through ever-higher levels of production and consumption of material goods and services – that is, through an accelerated growth in GNP. The term "development" is also used to cover all institutions, values, and economic theories that conceive, support, and implement this process. In this chapter I refrain from producing a precise definition of the term "development" because that would not do justice to its many meanings. My plan is to deal with a specific meaning of the term emerging from a story about improved seeds and modern agriculture. A full exposition of the argument of modern poverty as development-induced scarcity would require many other analyses – of nutrition, health care, housing, clothing, literacy, transport, and so on.

There are three principal paradigms within the discourse on economic development: (1) neo-classical economic theories of underdevelopment concerning overpopulation, transfer of technology, and diffusion of innovations; (2) neo-Marxist theories of uneven development concerned with imperialism, dependency, and world systems; and (3) the environmentalist conception of sustainable development. There are two other strands of thought – political ecology and new social movements – that have not wielded much influence in policy circles as yet. Political ecology combines ecological concerns with political economy (Bryant 1992; Peet and Watts 1993; Pickles and Watts 1992). Political ecology in the 1980s has produced, in my judgment, some of the most sophisticated theoretical analyses in development by combining insights from ecology, history, class analysis, the theory of the state, global capital, gender, and local knowledge systems (Blaikie 1985; Carney 1993; Hecht 1985; Richards 1985; Rocheleau 1991; Watts 1983). Despite profound differences in their world views, the major paradigms share the central belief that poverty arises from lack of development, that is underdevelopment, a condition that can be eradicated with more development. The contrary notion, that development creates scarcity, calls for a basic rethinking of the "poverty problem." We need to rethink what we mean when we say "Bangladesh is a basket-case of poverty." We need to examine the suggestive power that geographic boundaries may have on where the poverty problem is located. I argue that the "problem" should no longer be confined to the place where we see tangible, physical evidence of poverty, but should include the very intellect that helped conceptualize poverty in the first place. This leaves us in a serious predicament, because academic tools – that is, the paradigms of development and the epistemology of poverty – pose an obstacle to a solution by distorting understanding of the problem.

New seeds (once called "miracle seeds") are widely understood to be a beneficent technology dramatically increasing agricultural output – a significant technological breakthrough in the fight against hunger in the Third World. That is the official version of the Green Revolution story, inspired by the writings of neo-classical economists. However, these same seeds reveal a considerably more complex and contradictory character when viewed from other vantage points, such as political ecology. Different paradigms, each with its own "way of knowing," direct our attention to attributes of seeds other than high yields, and provide important new meanings to the question, "What are improved seeds?" By moving beyond a view of seeds as material things, to one of seeds as the embodiment of a nexus of interacting relations (social, ecological, academic, and so on), we can see how seeds are one means for dominating people and nature, and how this technology can both create and destroy use values at the same time.

The story of the epistemological transformation of the Green Revolution, from seeds of plenty to seeds of scarcity, is important in another respect. It enables a deeper understanding of the nature of the paradigms themselves – their assumptions, the origins of their language, their strengths, and limitations. Even as paradigms inform us about the nature of seeds, seeds inform us about the nature of paradigms. A seed is an indissoluble nexus of relations. It has been improved, bred, and studied through an epistemology formed by ahistorical, subject-specific disciplines and paradigms. The fragmented nature of that discourse prevents us from seeing the paradox of how improved seeds can give high yields and create scarcity at the same time. Observers of development projects in the Third World may recognize that the story of improved seeds told here is not exceptional.

THE NEXUS OF PRODUCTION RELATIONS

The principal analytical scheme used in this chapter is "the nexus of production relations." Production is an "economic" activity only in the narrowest sense of the word, because it includes far more than technology, goods, and markets. Production is determined at once within a web of relations – technical, social, ecological, cultural, political, and academic – the understanding of which is distorted by the subject-specific views of reductionist science (Figure 3.1). These relations should not be conceived as discrete, analytical categories. Nor is the list meant to be exhaustive. The relations act, interact, and react to maintain a dynamic process of production: analytically, there are no visible seams between them. In a historical sense, the relations are mutually constituted. An entity that appears technological from one perspective, may be academic or social from another. I have borrowed the term "production relations" from Marx (1989 [1869]). However, I have extended the meaning of the term "production relations" beyond the "social," its original usage. I have consciously tried to avoid the problems of the Marxian scheme of associating social relations with the economic base, and matters of culture, knowledge, and ideology with the

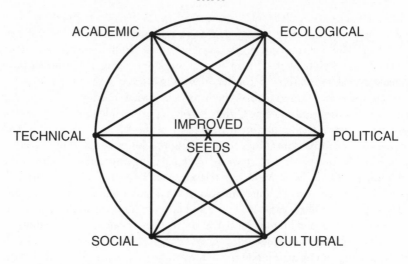

Figure 3.1 Improved seeds: the nexus of production relations

superstructure. I explore the interactions among these relations without any concern for which may be more determinate or "essential." There is no a priori assertion that one kind of relation is more important, determinant, or essential than another. Production exists in an "overdetermined" system where every aspect "is constituted – literally created – as the combined effect of all other aspects" (Wolff and Resnick 1987: 134).

The concept of "overdetermination" stands opposed to essentialism. The latter is the "presumption that complex realities of any sort are ultimately reducible to simpler, or essential, realities" (Graham 1990); under essentialism, some influences producing an outcome can be shown to be non-essential while others can be shown to be the essential causes (Graham 1992). Peet (1992) objects to non-essentialism by insisting that the refusal to separate essential from non-essential leads to indecision and weakens the base of political activism. In my judgment the academic's insistence on discovering the essential (popularly called "finding the root causes") weakens political activism by discounting a large number of "non-essential" sites at which numerous agents engage the circumstances of routine, everyday life. My story of improved seeds testifies that the non-essentialist approach of the nexus of relations can expand the scope of activism in South Asian food politics. While lack of access to land remains an important cause of hunger, the scarcity of food in South Asia is orchestrated through a bewildering array of mechanisms reaching beyond social relations of land ownership into the cultural, ecological, political, and academic realms. Each node in the nexus where scarcity is constructed provides also a site for mounting political resistance, multiplying the scope for activism; however, the agents of such activism, and the choice of strategies, may differ drastically from one node to another.

The nexus of production relations is a discursive materialist formation (Figure 3.1). Adapting a scheme from Foucault (1980) I argue that each node is a site of both discursive and non-discursive practices. Technology, culture, and nature are not only material processes but are constructed and driven by the discourses about them. For example, at the technical node, there are technical "practices" related to the use of land, labor, and capital; but these practices are informed and constituted by various discourses in neo-classical economics, engineering, banking, and so on. Similar discourses/practices occur at all other nodes of the nexus. Improved seeds are not mere material entities, distributed, grown, and eaten; they are constituted by specific social theories (discourses) of technology, culture, and nature. Improved seeds thus exist in a discursive materialist formation.

The phrase "technical relations of production" in Figure 3.1 is similar to Marx's forces of production – the raw materials, resources, labor, and technology used in production (Marx 1989 [1869]). The term "technical relations" calls attention to the determination of attributes of production forces in a larger context (i.e. by other relations). The term "social relations of production" is used in a manner identical to Marxian economics, where it refers to ownership of the means of production, the manner in which the means of production are utilized, and the rules for the social distribution of the final product (Marx 1989 [1869]). Production requires matter and energy as input and a repository to hold waste materials, chemicals, and heat, setting in motion myriad interactions with the biophysical environment – "the ecological relations of production." The phrase "cultural relations of production" refers to the interaction between economy and culture, in particular the interaction of production with "the ways of life" of social groups embodied in shared meaning, beliefs, values, and symbols. Political relations of production include interactions between the state and society in the organization of economic activity. In Third World societies the state plays an all-pervasive role in civil society through its command over development projects. However, I subsume that topic in this chapter under the heading "social relations". Academic relations are of two kinds: internal and external. Internal academic relations arise from the understanding science has of itself in the production of knowledge, that is the view of science as neutral, value-free, and non-political (Proctor 1991). External academic relations refer to the discourses produced at other sites in the nexus: technical, social, cultural, political, and ecological. Academic descriptions of production are not necessarily impartial and neutral, because values, assumptions, objectives, models, and language of representation are all thoroughly imbued by the entire nexus of production relations. This argument has important implications for the academic discourse on poverty.

THE SOCIAL CONSTRUCTION OF SCARCITY

Innumerable mechanisms create scarcity by influencing how "demand end-uses" are matched to "sources of supply." An "end-use" is the "use" to which a good is put in the "end." Scarcity may be created by an expansion of end-uses, as with the creation of demand for new goods through advertising. Even more important is the manipulation of sources of supply. Imagine that a particular end-use is met through several different sources of supply. Through time some of these sources are neglected, "de-developed," and gradually disappear. For example, nitrogenous fertilizer can be provided to a field in a variety of ways, including inorganic commercial fertilizer, animal waste, agricultural residue, crop rotation, the growing of legumes, interplanting, and slurry from methane digesters. However, inorganic commercial fertilizer has become the principal, and often the only, means by which crop nutrients are provided. Increased demand for inorganic fertilizer, and the de-development of other methods, grew partly out of an academic discourse involving universities, national and international research institutes, the state, agribusiness, and international development agencies.

The logic of matching sources to end-uses has application in every area of technology: food, nutrition, agriculture, manufacturing, health care, housing construction, transport, and education. The adoption of this principle in the context of underdevelopment and poverty gives new meaning to the terms "resources," "technology," and "capital." End-use analysis begins with a needed use value and looks for the most direct way of satisfying it, minimizing energy, material, and transport. In fact the entire nexus of production relations can be employed in matching sources to end-uses. Thus the terms "resources," "technology," and "capital" have no universal meaning in the absence of a concrete end-use analysis in a given region. Contrary to the claims of economists "scarcity" has no context-free, universal meaning; indeed, the physical geography of a region, its ecology, cultural values, people's knowledge of plants and animals, class, and power are all essential to determining what constitutes a "resource." Resources are not things and they are not stocks; they are discursive material entities existing in a nexus of relations.

I shall turn next to a detailed consideration of individual sites in the nexus of improved seeds. The nexus has no inner logic ranking different sites according to an order of importance. The nexus is also a useful tool to get a perspective on paradigms of development; by focusing on a class of "essential" causes each paradigm highlights particular relations but fails to see how scarcity is constructed at a large number of other sites (including academia).

TECHNICAL RELATIONS

At its most basic level the Green Revolution represented a new technology to increase the production of cereals, particularly maize, wheat, and rice. It consisted of high-yielding seeds grown in association with chemical fertilizer, pesticides, and irrigation; without these inputs the new seeds yielded poorly (Shiva 1991: 46). Two principal points of this paper follow from the "technical" traits of improved seeds: first, the technical attributes of seeds are not "attributes" as such, but are *relations* that determine, and are determined by, the nature of

74

other relations in the nexus; second, each node of the nexus is a site at which food scarcity is constructed, notwithstanding the evidence of statistical data showing impressive gains in cereal production.

When traditional varieties of cereal are heavily fertilized they lodge (bend over) under the weight of increased yield. Plant breeders solved the problem by developing fertilizer-responsive, thick-stemmed dwarf varieties using hybridization techniques. All modern varieties are the result of hybridization techniques although the term "hybrid" is confined to the first generation progeny obtained from crossing two varieties that have been first inbred through several generations as in the case of corn (Phoehlman and Sleper 1995: 159). A seed grain performs two functions: the endosperm produces food for the germinating plant (also the source of our own food), and the embryo reproduces the plant. In recent years there has been an increasing physical separation of the sites at which these two functions are performed. The reproduction of seeds has moved into the realm of formal science, experimental plots of research institutes, gene banks, commercial seed suppliers, bureaucratic processes of seed certification, and so on. In a word, farmers have lost control over the reproduction of seeds, a tendency that will intensify with gene research in seeds (Kloppenburg 1988; Mooney 1979).

The purchase of inputs (seeds, fertilizer, pesticides, and fuel) is a significant source of expense for small peasant farmers. During the 1970s and early 1980s most governments offered inputs at subsidized rates in an effort to promote the adoption of new varieties. With increasing prices of petroleum and the cutting-back of input subsidies (under economic restructuring), farm costs sky-rocketed. Using 1961 as the base year the index of food production for the developing countries increased at the rate of 0.04 per year over the years 1961–91, with much of the increase coming from cereal production. However, the index of consumption of commercial nitrogenous fertilizer grew at a staggering rate of 0.28 per year over the same period. The quantity of nitrogenous fertilizer used per metric ton of food in the early 1990s was between 3.5 to 4 times the amount used in 1961. The cultivation of pulses, important sources of protein food for poor people and biological nitrogen for the soil showed little or no increase during the entire period (calculations were made from the FAO data diskettes – AGROSTAT-PC 1994). In India all agricultural inputs increased at an aggregate rate of 4.2 per year between the years 1970–71 and 1979–80, but real output increased at only 2.3 per year. As a result the index of productivity declined from 100 in 1970–71 to 75 in 1979–80 (Agarwal and Narain 1985: 160).

A discussion of technical relations of new seeds is a useful place to make a general remark about neo-classical economics and spatial diffusion of innovations. These paradigms assume underdevelopment is caused by an inadequate development of production forces, a situation that can be corrected by the diffusion of inputs: capital, know-how, and technological innovations. Accelerated development of production forces was the answer to underdevelopment. This was the underlying thinking in promoting high-yielding seeds. Regional prosperity would emerge from expanded food production following the adoption of the technical package: seeds, fertilizer, pesticides, and pump-sets for irrigation; this thinking ignored social, ecological, cultural, and academic relations of innovations.

SOCIAL RELATIONS

William Gaud of the United States Agency for International Development first used the term "Green Revolution" in a speech given to the Society for International Development in March 1968 (Spitz 1987). Gaud alluded to the possibility of a green technical revolution in food production as counterposed to a red political revolution. In December 1969, Green Revolution was presented to the U.S. Congress as a major tool of American foreign policy that provided bright market prospects to the pesticide, fertilizer, seed, and tractor industries (Spitz 1987). The very term was thus a political construct. Published documents of the World Bank and the U.S. Agency for International Development during the 1960s and 1970s show the question of land reform was a significant element of the development discourse during that time (World Bank 1974). But, as the productivist logic of the Green Revolution got under way, interest in land reform disappeared from the official agenda of development. Within Third World countries the state was the most important agent promoting Green Revolution technology; the massive project involved national universities, research institutes, ministries, extension services, imports, subsidized inputs, credit and banking, and so on. The entrepreneurial spirit of the state coincided with waning interest in land reform; these state policies were adopted and executed on behalf of powerful class interests of importers and landlords (El-Ghonemy 1990; Herring 1983; Nandy 1992).

The Green Revolution was promoted without a serious consideration of social relations of production. This was a crucial mistake, because interpersonal economic differences and class play important roles in determining who adopts what in rural areas of the Third World (Blaut 1987; Griffin 1974; Yapa and Mayfield 1978). Griffin (1974) has made a persuasive argument to this effect with his model of biased innovations: capital-intensive innovations in the package of high-yielding seeds soon acquired a landlord bias in the fragmented factor markets of India. A large number of studies have shown an increase in income inequality and asset distribution as a result of the Green Revolution (Frankel 1971; Griffin 1974; Harriss 1977; Hewitt de Alcantara 1976; Pearse 1980; Shiva 1991); these claims have been strongly contested in a study of rice growing areas of South India (Hazell and Ramasamy 1991). I shall not intervene in this debate here as it is not possible to evaluate "results" of income and asset distribution studies without detailed attention to specific methodologies employed in each case.

Concern with social relations of Green Revolution technology came out of the academic tradition of political economy. That critique focused on the role of capitalist farming in the exacerbation of class and regional income inequalities through the uneven adoption of high-yielding seeds (Griffin 1974; Hewitt de Alcantara 1976; Yapa and Mayfield 1978). Political economy provides a new answer to the question, "What are improved seeds?" They are a technology that produce higher yields, but they confer benefits unequally to different classes. The state-sponsored productivist logic of the Green Revolution marginalized those concerns. However, missing from the political economy critique were other important questions: Why does technology follow a particular path and not another? What are the ecological and cultural relations of different kinds of technologies?

ECOLOGICAL RELATIONS

Production involves the transformation of materials into use values through the application of information, energy, and labor using the ecosystem as a source of energy and matter and as repository of waste products. This defines myriad interactions within the biophysical environment: these are ecological relations of production. Scarcity is constructed in the ecological realm through two modes: (1) by replacing the "reproductive capacity" of nature with the "productive capacity" of industrial inputs, and (2) by degrading conditions of production (Shiva 1991). I shall illustrate these points by looking at improved seeds from the viewpoint of genetic diversity, required inputs, and so-called economic "externalities."

A serious effect of the introduction of new seeds is the accelerated loss of genetic diversity (Mooney 1979). According to Erlich, a noted biologist, "Aside from nuclear war, there is probably no more serious environmental threat than the continued decay of the genetic variability of crops" (Erlich *et al.* 1977: 344). Genetically uniform varieties of rice, wheat, and corn grown in monocultural stands are more vulnerable to pests and pathogens than older varieties which have co-evolved with the local environment, necessitating the use of pesticides, another example of social construction of scarcity (Bull 1982: 13). Pesticides cause the large-scale destruction of non-target populations, the genetic evolution of pesticide-resistant organisms, the contamination of water and agricultural produce, and the reduction of soil organisms that maintain the quality of humus in the soil (van den Bosch 1978; National Research Council 1989); moreover, they pose health risks to agricultural workers (Wright 1990). Apart from serious environmental hazards, chemicals are also expensive; their use has increased the dependence of Third World farmers on international capital. The pesticide industry – that is, its research, development, and marketing – underdevelops alternative techniques, which include: biological control through prey–predator relationships; cultural methods, such as crop rotation, multiple cropping, and companion planting, that alter the environment by making it less suitable for pests; and crop breeding programs that develop disease-resistant plants. Of all these techniques the chemical ones receive the most support, because they create most exchange value. Even though more than 20,000 serious pests are known, natural enemies are known for less than 10 percent of these. Entomologist Debach (quoted in Nebel 1981: 428–9) believes this research is underfunded because biological methods do not generate profits the way synthetic chemicals do.

The world production of nitrogen fertilizer rose from about 6.5 million metric tons in 1955 to 67.5 million metric tons in 1984. In that year developing countries consumed more than 40 percent of the world's nitrogen fertilizer (Food and Agriculture Organization 1984). Apart from increasing costs the use of chemical nitrogen contributed to increasing scarcity by reducing the supply of naturally available organic nitrogen. It had the effect of underdeveloping knowledge of biological sources of nitrogen related to crop rotation, multiple cropping, incorporation of nitrogen rich legumes in agricultural production, use of agricultural and plant remains, and application of animal manure.

In South Asia, there is widespread evidence that fertilizer and pesticide runoff contaminate groundwater and streams. The fish and crab populations living in rice paddies, an important source of protein for the poor, have declined, or are unsafe to eat (Bull 1982: 63–4). Poor farmers working knee-deep in the rice paddy mud do not wear protective clothing. Moreover, in regions without indoor plumbing or water purification plants, farmers wash themselves in water from the fields, streams, and irrigation channels, which now carry unsafe levels of chemical contaminants. In the context of South Asian farming the very use of the word "environment" can be misleading, because, physically speaking, farmers are an integral part of "the environment"; it is quite harmful to describe the condition of contaminated water as an "externality." One may well ask in what sense is contaminated water an "externality" when farmers have to drink, wash, bathe, and work in it. Thus hybrid seeds are not simply a technique of increasing food production, but represent the emergence of a mode of production that is destroying the productive base of subsistence.

CULTURAL RELATIONS

The term "cultural relations of production" refers to the interaction between culture and economy. Rhodes (1984: 43) describes "culture" as follows:

> Cross-cutting and underlying . . . all anthropological studies is the notion of "culture." A dynamic blueprint or design for living, culture is learned behavior handed down through generations so that each new cohort of babies in a society does not have to start again from scratch. To some degree, what agricultural scientists call tradition is the anthropologist's culture.

Rhodes's remarks about agricultural scientists' conception of culture as tradition are crucial to understanding the cultural relations of the Green Revolution. Improved seeds arrived in the villages of India carrying the authority of science and modernity. The new seeds – sponsored by international aid agencies, developed by crop-breeding science, backed by multinational agribusiness capital, approved by the Government of India, and promoted by an army of trained extension workers – presented a formidable power that confronted peasant farmers living in their "traditional culture of poverty."

Chambers (1983: 76) describes this unequal encounter of modernity and tradition in an essay entitled "Whose knowledge?":

> From rich-country professionals and urban-based professionals in the third world countries right down to the lowliest extension workers it is a common assumption that the modern scientific knowledge of the centre is sophisticated, advanced and valid, and conversely, that whatever rural people may know will be unsystematic, imprecise, superficial and often plain wrong.

To the centuries-long colonial view of peasant agriculture as primitive was added the "modernization" literature of the 1960s and 1970s, which set out to transform "backward" traditional culture, the principal obstacle to adoption of innovations and the diffusion of development (Rogers 1969; Rostow 1960). The concept of traditional culture as backward is an elaborate academic representation of "the other," an intellectual construction which actually reflects the values of sociologists immersed in the dominant world-view of capitalist culture. There is no objective referent in the external world called "backward traditional culture" that is independent of the intellect that constructed it (Said 1979).

Several prominent students of "traditional" agriculture have written persuasively about the complexity and longevity of mixed farming that incorporates animals, manure, and crop rotation – for example, F.H. King (1973 [1911]) of the U.S. Department of Agriculture and Sir Albert Howard (1973 [1940]). Among geographers the pre-eminent student of traditional agriculture was Sauer (1952, 1963 [1938]) who was quite emphatic in his condemnation of the Rockefeller Foundation proposal in the early 1940s to modernize Mexican agriculture (Jennings 1988: 50–5). Based on surveys of traditional farming conducted at several sites in southern Mexico and Central America, Wilken (1987) described traditional resource management techniques in energy supply, soil classification, and the management of soil, water, slope, and space. Wilken adds that traditional tools and techniques are not easily duplicated, because most traditional technology requires understanding local conditions and ways of managing local energy and materials. Other writing describing the importance of indigenous knowledge includes: Altieri (1987); Brokensha et al. (1980); Chambers (1983); Harrison (1987); and Richards (1985.

A good example of the implications of modernization of traditional agriculture comes from the Andean Highlands of Peru. The Andes are the richest gene pool for potatoes, estimated by geneticists to contain 2,000–3,000 varieties. The farmers possess highly developed systems for classifying potatoes that have enabled them to observe, select, and propagate many varieties over large, diverse, mountainous areas. There are also well-developed trading networks for exchange and sale of seed potatoes. Often in a single locality as many as 50–70 varieties may be found, adapted to local conditions. Since 1950, systematic efforts have been made to modernize the potato culture of the Andean Highlands; with the increased adoption of improved varieties, has come an increased demand for external inputs, including seeds. The modernization of the Andean potato culture parallels what we saw in Asia, where there has been a break in the connection in knowledge of local ecology and of the practice of matching native varieties to particular places to minimize losses from frost, drought, and disease (Brush 1986).

The modernization literature on diffusion in the Third World profoundly misrepresented and misinterpreted traditional societies as backward and non-innovative. This cultural bias, abetted in part by academics, affected public policy and the course of diffusion of agricultural innovations. It has also been a bearer of the hegemonic culture of science, capital, and authority that subjugates tradition and the keepers of that knowledge. By doing so the architects of the Green Revolution robbed culture of its power as a problem-solving agency in the everyday life of poor people.

ACADEMIC RELATIONS

The expression "academic relations of production" refers to the work of agricultural scientists who conceived and bred improved seeds, and the work of social scientists who conceived the social theory that facilitated the diffusion of that technology. The story of improved seeds provides an excellent example of the claim made by critical social theorists that science and technology are in fact "social processes" directed by power relations in underlying society, serving to strengthen and reproduce those power relations (Aronowitz 1988; Foucault 1980).

I argued earlier that improved seeds were not just a technology to feed people better by increasing food production, but that they were also an instrument designed to serve the economic interests of particular classes of people. Such claims are usually dismissed as being a naive subscription to conspiracy theory. One resolution to the debate lies in critical social theory which shows how knowledge is constructed to serve the needs of a particular world view, and how this can happen through "internal academic relations," without that intellect being centrally directed by particular agents (Aronowitz 1988; see also *Monthly Review*, July–August 1986). In his critique of "value-free science," Proctor (1991: 268) expressed this in the following way:

> The simplest and perhaps the oldest version of the ideal of neutrality is that science may be used for good or for evil. The problem with this view, though, is that it ignores the fact that science has both social origins and social consequences. Who, one can ask, does science serve, and how? Who has gained from "miracle wheat" and who has lost? . . . Science is not different from other aspects of culture in this sense.

Wheat and corn improvement research conducted by Norman Borlaug is cited as the beginning of the Green Revolution in Mexico. The Mexican story goes further back, however, to the 1930s, when the Ministry of Agriculture in Mexico during the progressive years of President Cardenas (1935–40) initiated a program of scientific research to improve corn, the main staple of the peasantry. The years of Cardenas saw sweeping land reforms, the expropriation of Standard Oil, and the threat of take-over of other U.S. investments in Mexico. With the installation of Camacho as president, the program for the improvement of peasant crops was disbanded. During the 1940s, with help from the Rockefeller Foundation, a new program of agricultural research was started, focusing on hybrid varieties of irrigated wheat for large-scale commercial growers of north-west Mexico. The idea was to reverse the agrarian radicalism of Cardenas and replace it with a model of scientific, industrial agriculture to produce food surpluses for urban areas using industrial inputs. This program later came to be known as the Green Revolution; it did much for agribusiness of pumps, machines, fertilizer, and pesticides and little for the nutrition and welfare of Mexican peasants (Hewitt de Alcantara 1973–4). Drawing on Rockefeller archives, Jennings (1988) reports that Carl Sauer, who was a strong critic of the model of industrial agriculture, believed that the agricultural and nutritional practices of Mexican peasantry were quite sound, but that they needed support and strengthening. Sauer's advice went unheeded at the foundation. Borlaug's work also marginalized the research on rain-fed corn done by Mexican scientists in the Institute of Agricultural Investigations. Hewitt de Alcantara (1973–4: 32) has suggested that Camacho and his advisors specifically reached for a pro-industrial program as a substitute for the agrarian programs of Cardenas.

The International Rice Research Institute (IRRI) was founded in Los Baños, the Philippines in 1960 with funding from the Ford and Rockefeller Foundations. The conception of agricultural research at IRRI was simple, centralized, and productivist; according to Anderson *et al.* (1980: 7–8) it assumed:

> that a single international center would be able to design and breed a small set of new varieties of rice plants that would displace the thousands of locally cultivated plants in the irrigated rice lands of Asia. . . . Although by the commonly accepted standards of academic research the IRRI scientists and technologists were well qualified in their specialties, few had a deep understanding . . . of indigenous practices . . . or the socio-economic contexts in which Asian cultivators operated.

The social theory of the Green Revolution came out of the work of modernization theorists. The promotion of high-yielding varieties spawned a whole new vocabulary that included terms and expressions such as "progressive farmers," "backward farmers," "betting on the fittest," and so on. Capitalist farmers with access to large areas of irrigated land who could purchase the expensive inputs were culturally and linguistically transformed into "progressive farmers." Poor farmers who could not afford to respond, and intelligent farmers who actively rejected the new seeds for ecological reasons, were transformed into "backward farmers," or "laggards," in the language of the sociology of innovation diffusion. In India, the strategy of "betting on the fittest" was a social rationalization of agrarian policies that had nothing to offer the marginal farmers, the landless laborers, or those who cultivated coarse grains in areas of rain-fed agriculture (Frankel 1971).

And so we return again to the question: "What are improved seeds?" The conception of seeds as academic relations shows that what had earlier been called technological, social, ecological, and cultural is in fact constructed through academic processes of research and social theory – "external academic relations." Therefore, it is through academic "deconstruction" that we can begin to understand how improved seeds are actually constituted as a discursive material entity. The question "Hasn't the Green Revolution staved off massive hunger in the Third World?" makes sense only within a particular epistemology; the same inquiry makes little sense when viewed in the context of a nexus.

CONCLUSION

This chapter was partly an exercise in understanding how social problems are defined and solutions are proposed. Since poverty is a serious problem, development was the solution, for example, the Green Revolution. But such solutions are in fact part of the problem, because poverty is a form of scarcity induced by the very process of economic development. That argument was illustrated by narrating a story of improved seeds. In this paper I explored the question "What are improved seeds?" The answer that it is a technology that increased yields, is a reductionist description that ignores how scarcity was constructed in the nexus of relations. Scarcity is not a general condition; it is always socially specific. The Green Revolution created a technology which required poor farmers to buy inputs; it ignored other appropriate technologies of food production such as rain-fed farming, multiple cropping, growing of legumes, and so on. Its productivist logic marginalized political economists' concerns for people's access to land and productive resources. It devalued the "reproductive power" of nature by substituting the "productive power" of industrial inputs. Further, the ecological degradation caused by the use of these inputs reduced the subsistence capacity of land. By marginalizing traditional knowledge it robbed the culture of poor people of its power/agency to address problems of everyday life. It produced an academic discourse that concealed how production can also destroy use values, creating social scarcity at each node of the nexus; thus it disempowered poor people and misled people of goodwill. The problem of poverty must, therefore, be expanded to include not only concrete places and people that experience scarcity, but also the epistemology of how we know scarcity and poverty.

REFERENCES
Agarwal, A. and S. Narain. 1985. *The State of India's Environment, 1984–85.* New Delhi: Centre for Science and Environment.

Altieri, M.A. 1987. *Agroecology: The Scientific Basis of Alternative Agriculture.* Boulder, CO: Westview Press.

Anderson, R.S., P.R. Brass, E. Levy and B.M. Morrison. 1982. *Science, Politics, and the Agricultural Revolution in Asia.* Boulder, CO: Westview Press.

Aronowitz, S. 1988. *Science as Power: Discourse and Ideology in Modern Society.* Minneapolis: University of Minnesota Press.

Blaikie, P. 1985. *The Political Economy of Soil Erosion.* London: Methuen.

Blaut, J.M. 1987. "Diffusionism: a uniformitarian critique," *Annals of the Association of American Geographers* 77: 30–47.

Brokensha, D., D.M. Warren and O. Werner. 1980. *Indigenous Knowledge Systems and Development.* Washington, DC: University Press of America.

Brush, S.B. 1986. "Genetic diversity and conservation in traditional farming systems," *Journal of Ethnobiology* 6: 151–67.

Bryant, R.L. 1992. "Political ecology: an emerging research agenda in Third World studies," *Political Geography* 11: 12–36.

Bull, D. 1982. *A Growing Problem: Pesticides and the Third World Poor.* Oxford: Oxfam.

Carney, J. 1993. "Converting the wetlands, engendering the environment: the intersection of gender with agrarian change in the Gambia," *Economic Geography* 69: 329–48.

Chambers, R. 1983. *Rural Development: Putting the Last First.* New York: John Wiley.

El-Ghonemy, M.R. 1990. *The Political Economy of Rural Poverty: The Case for Land Reform.* New York: Routledge.

Erlich, P.R., A.H. Erlich and J.P. Holdren. 1977. *Ecoscience: Population, Resources, Environment.* San Francisco: W.H. Freeman.

Food and Agriculture Organization. 1984. *FAO Fertilizer Yearbook.* Rome: FAO of the United Nations.

—— 1994. *ARGOSTAT-PC 1994.* Rome: FAO of the United Nations.

Foucault, M. 1980. *Power/Knowledge.* New York: Pantheon Books.

Frankel, F.R. 1971. *India's Green Revolution: Economic Gains and Political Costs.* Princeton, NJ: Princeton University Press.

Graham, J. 1990. "Theory and essentialism in Marxist geography," *Antipode* 22: 53–66.

—— 1992. "Anti-essentialism and overdetermination: a response to Dick Peet," *Antipode* 24: 141–56.

Griffin, K. 1974. *The Political Economy of Agrarian Change: An Essay on the Green Revolution.* Cambridge, MA: Harvard University Press.

Harrison, P. 1987. *The Greening of Africa.* New York: Viking Penguin.

Harriss, J. 1977. "The limitions of HYV technology in North Arcot District: the view from a village," in B.H. Farmer (ed.) *Green Revolution? Technology and Change in Rice Growing Areas of Tamil Nadu and Sri Lanka.* London: Macmillan, pp. 124–42.

Hazell, P.B.R. and C. Ramasamy. 1991. *The Green Revolution Reconsidered.* Baltimore, MD: Johns Hopkins University Press.

Hecht, S.B. 1985. "Environment, development and politics: capital accumulation and the livestock sector in Eastern Amazonia," *World Development* 13: 663–84.

Herring, R.J. 1983. *Land to the Tiller: The Political Economy of Agrarian Reform in South Asia.* New Haven, CT: Yale University Press.

Hewitt de Alcantara, C. 1973–4. "The 'Green Revolution' as history: the Mexican experience," *Development and Change* 5: 25–44.

—— 1976. *Modernizing Mexican Agriculture: Socio-economic Implications of Technical Change, 1940–1970.* Geneva: United Nations Research Institute for Social Development.

Howard, A. [1940] 1973. *An Agricultural Testament.* Emmaus, PA: Rodale Press.

Jennings, B.H. 1988. *Foundations of International Agricultural Research.* Boulder, CO: Westview Press.

King, F.H. [1911] 1973. *Farmers of Forty Centuries.* Emmaus, PA: Rodale Press.

Kloppenburg, J.R. 1988. *First the Seed: The Political Economy of Plant Biotechnology, 1492–2000.* New York: Cambridge University Press.

Marx, K. [1869] 1989. *A Contribution to a Critique of Political Economy.* New York: International Publishers.

Mooney, P.R. 1979. *Seeds of the Earth.* San Francisco: Institute for Food and Development Policy.

Nandy, A. 1992. "State," in W. Sachs (ed.) *The Development Dictionary: A Guide to Knowledge as Power.* London: Zed Books, pp. 264–74.

National Research Council. 1989. *Alternative Agriculture.* Washington, DC: National Academy of Science.

Nebel, B.J. 1981. *Environmental Science.* Englewood Cliffs, NJ: Prentice-Hall.

Pearse, A. 1980. *Seeds of Plenty, Seeds of Want: Social and Economic Implications of the Green Revolution.* Oxford: Clarendon Press.

Peet, R. 1992. "Some critical questions for anti-essentialism," *Antipode* 24: 113–30.

Peet, R. and M. Watts. 1993. "Development theory and environment in an age of market triumphalism," *Economic Geography* 69: 227–53.

Phoehlman, J.M. and D.A. Sleper. 1995. *Breeding Field Crops.* Ames, IO: Iowa State University Press.

Pickles, J. and M. Watts. 1992. "Paradigms for inquiry?" in R.F. Abler, M.G. Marcus, and J.M. Olson (eds.) *Geography's Inner Worlds.* New Brunswick: Rutgers University Press, pp. 301–26.

Proctor, R.N. 1991. *Value-free Science? Purity and Power in Modern Knowledge.* Cambridge, MA: Harvard University Press.

Rhodes, R.E. 1984. *Breaking New Ground: Agricultural Anthropology.* Lima, Peru: International Potato Institute.

Richards, P. 1985. *Indigenous Agricultural Revolution: Ecology and Food Production in West Africa.* London: Hutchinson.

Rocheleau, D. 1991. "Gender, ecology, and the science of survival," *Agriculture and Human Values* 8: 156–65.

Rogers, E.M. 1969. *Modernization Among Peasants: The Impact of Communication.* New York: Holt, Rinehart & Winston.

Rostow, W.W. 1960. *The Stages of Economic Growth: A Non-communist Manifesto.* Cambridge: Cambridge University Press.

Said, E. 1979. *Orientalism.* New York: Vintage Books.

Sauer, C. O. 1952. "The agency of man on Earth," in W.L. Thomas (ed.) *Man's Role in Changing the Face of the Earth.* Chicago: University of Chicago Press, pp. 49–69.

—— [1938] 1963. "Theme of plant and animal destruction in economic history," in J. Leighly (ed.) *Land and Life.* Berkeley: University of California Press, pp. 145–54.

Shiva, V. 1991. *The Violence of the Green Revolution.* London: Zed Books.

Spitz, P. 1987. "The Green Revolution re-examined in India," in B. Glaeser (ed.) *The Green Revolution Revisited.* London: Allen & Unwin, pp. 56–75.

Van den Bosch, R. 1978. *The Pesticide Conspiracy.* Garden City, NY: Doubleday.

Watts, M. 1983. *Silent Violence: Food, Famine and Peasantry in Northern Nigeria.* Berkeley: University of California Press.

Wilken, G.C. 1987. *Good Farmers: Traditional Agricultural Resource Management in Mexico and Central America.* Berkeley: University of California Press.

Wolff, R.D. and S.A. Resnick. 1987. *Economics: Marxian versus Neoclassical.* Baltimore, MD: Johns Hopkins University Press.

World Bank. 1974. *Land Reform*. Washington, DC: World Bank Rural Development Series.

Wright, A. 1990. *The Death of Ramon Gonzalez: The Modern Agricultural Dilemma*. Austin: University of Texas Press.

Yapa, L. and P. Mayfield. 1978. "Non-adoption of innovations: evidence from discriminant analysis," *Economic Geography* 54: 145–56.

4

MOVEMENTS, MODERNIZATIONS, AND MARKETS

Indigenous organizations and agrarian strategies in Ecuador

Anthony Bebbington

This chapter tells a story about popular and non-governmental organizations in the Andes of Ecuador. The tale is related to several of the intellectual and political currents referred to in the opening chapter by Peet and Watts. Of course, the language and concepts of the intellectual currents on which I draw are not those of the actors of the tale. But to relate the story to those concepts helps tell us something about the insights and insufficiencies of the concepts, and about the potentials and limits on the "liberational" possibilities embodied by these organizations.

Telling the story this way also helps us address the challenge of bringing those different language communities closer (Booth 1994; Edwards 1994). Why is this a worthwhile challenge? Because the idea behind writing development stories is not simply to understand the world but rather to change it (Edwards 1994); our understanding and our ability to contribute to such change are the greater if we build bridges across the gulf between the languages of social movements and popular organizations, and those of activists, intellectuals, and "policy makers." To say this is not to collapse into a "people know best" populism taken to task in the opening chapter. Indeed, to understand is not necessarily to agree with. Constructive critique can be a form of support, as long as it has practicable implications.

The story I tell is how federations of indigenous people in highland Ecuador put together strategies for rural social change, and particularly for agricultural development. This empirical story is related to the conceptual discussions of "alternative" development that draw upon concepts of "indigenous technical knowledge," "farmer-first" agricultural development, political ecology, new social movements, and civil society. The assumption of much alternative development writing is that "alternative actors" such as indigenous peoples' organizations will carry forward these "alternative" agendas (Bebbington 1995). This is not always the case. In particular, a commitment to native, traditional, and agro-ecological techniques found in intellectual currents in social science and development activism is often missing among indigenous peoples' organizations: in its place is a commitment to reforming, adapting, and managing modernization. However, at the same time, the principles of local control, democratization, and community-based sustainable development underlying "alternative" development thinking *are* apparent in the strategies of these indigenous federations. Thus while intellectual concepts and popular practice may differ at the level of strategy, they converge at the level of wider political objectives. Looking more carefully at why this happens addresses questions raised by Peet and Watts in their introduction, about "the conditions under which knowledges and practices become part of alternative development strategies." It also problematizes the question by asking *which* knowledges and *which* practices?

In addition to analyzing the internal rationales of the strategies of social movements, questions of effectiveness also must be raised. If, as the Oxfam poster says, "freedom begins with breakfast," then the final measure of the strategies of these organizations is how far they improve the livelihoods of their members – politically and economically. To organize, be innovative, and create "decentered autonomous spaces" (Escobar 1992) is simply not enough. In the final instance, alternative strategies are only worthwhile if they make a concrete difference. Peet and Watts in their introduction are painfully correct in pointing out that the social movements literature says very little about how far, and under what conditions, these movements genuinely contribute to a more robust civil society and (perhaps more importantly) the basics of increased productivity, income, and employment opportunities in the popular sectors. The "alternative visions" of intellectuals and activists are similarly unconvincing in demonstrating how their proposals will create new livelihood possibilities. If we are interested in "liberation ecologies," the proposals of social movements and intellectuals alike fall well short of the practical challenge of liberation.

This chapter first looks at conceptual discussions about indigenous technical knowledge and farmer-first development, and points out – at the risk of caricature – the inherent weaknesses of many such analyses. This leads to a consideration of the political-economic realities within which alternative agricultural strategies are currently pursued in Latin America. This in turn takes us onto the terrain of political ecology, and a brief discussion of social movements and their political and economic context.

This review of conceptual overlaps between political ecology and civil society lays the basis for a short case study that allows us to sustain a conversation between the theoretical concepts of "outside" analysts, and the strategies of civil society organizations. The case study discusses how a group of indigenous (or Indian) peasant federations and non-governmental development organizations (NGOs) in the highland province of Chimborazo in the Central Andes of Ecuador emerged as actors in civil society and composed local development strategies embracing technology, ethnicity, and politics. These strategies have, over time, become increasingly eclectic, pragmatic, and modernizing – and yet the underlying vision of rural development still retains a very "alternative" agenda for an indigenous, grassroots-controlled modernization. This is particularly the case for Indian federations, whose agrarian programs incorporate Green Revolution technologies to promote a form of development that nonetheless aims at reinforcing Indian culture and society.

These agricultural programs differ from the modernization fostered by the state, but also differ from proposals of those intellectuals and activists who suggest that appropriate rural development should build only, or primarily, on farmers' own techniques and innovations. This suggests that what gives a strategy its alternative, indigenous orientation is not its *content* (i.e. that it uses indigenous technologies, etc.) but rather its *goal* (i.e. that it aims to increase local control of processes of social change). Indigenous farmers may well incorporate the modern techniques of those who have long been their dominators, and do so in a way that strengthens an indigenous agenda pitched in some sense against the interests of those dominating groups.

The case suggests also that while it is important to recognize that rural social movements have "agency" (Redclift 1987), it remains important to understand that they are agents *situated* in cultural, economic, agro-ecological, and socio-political contexts which influence how, and why, they manage resources in particular ways. This situatedness should be kept at the forefront of theoretical interpretations. For it is one thing for theoretical analysis and development practice to recover the importance of these long-marginalized actors within civil society. This is important and it is one of the ways in which "voice" can be given to these actors and to the legitimacy of their own strategies and ideas. But that does not mean we have to celebrate all that they do – for much of this may be ineffective, undemocratic, authoritarian, frequently male, and so on. However, if our analyses recover and understand the ways in which actors are situated, and how this affects their rationales, this will draw attention to the limits on their capacities to compose viable and democratizing programs, and to the reasons for the limited impact of these programs. This in turn can be one step in the process of defining potentially more effective strategies that can incorporate "external innovations" but at the same time build from the rationales of the actors involved (rather than from imposed rationales). This *might* perhaps be a more practically oriented version of the theoretical position outlined by Peet and Watts (1993: 249) that

> dialectical analysis . . . provides the possibility of imagining a system of relations that does not consume the autonomy of the particular . . . a dynamic which has pattern, order, and determination without being teleological, a theory of totalities which because it values their unique aspects is not totalizing.

Conceptually this is a step forward in refining the conversation between agrarian populism and political ecology, and the analytical and practical association of political ecology and civil society – two of the themes that Peet and Watts suggest are central to the reconfiguration of the field of political ecology as liberation ecology.

ALTERNATIVE PROPOSALS: INDIGENOUS TECHNOLOGY AND FARMER-FIRST AGRICULTURAL DEVELOPMENT

From the Green Revolution to indigenous technology

The term "Green Revolution" is a shorthand for an agricultural development based on new crop varieties, agrochemicals, and machinery. It is the basic toolkit of most state policies in Latin America, which – via land reforms, rural development programs, and agricultural research and extension systems – seek to modernize the small and middle farm sector (Bebbington and Thiele 1993). Over the years, this approach attracted criticism from which much of the writing on "alternatives" sprang. A large literature accumulated claiming to show that Green Revolution technologies aggravated rural poverty, undermined food security, damaged the biophysical environment, and eroded local cultures (Altieri 1987; Biggs and Farrington 1991; de Janvry 1981; Griffin 1974; Hewitt de Alcantara 1976; Lipton and Longhurst 1989). For these reasons, many cultural and political-ecological geographers oppose the technological modernization of indigenous agriculture (Butzer 1990; Denevan 1989). Their writing is similar to those agro-ecologists who argue that agriculture should be grounded in ecological principles if it is to be sustainable (Altieri 1987; Altieri and Hecht 1990), and to Latin American writing around themes of eco-development, sustainable development, and ecological economics (e.g. see the work of Esteva 1992; Leff 1994; Max-Neef 1991). Some of this writing also shares the more general belief that the transfer of Northern technologies to the South creates unemployment and landlessness, entrenches the power of professional elites who monopolize knowledge, and encourages unrealistic and unsustainable lifestyle aspirations (Lehmann 1990). In that sense it resonates with Nerfin's (1987) call for "another development" which would prioritize needs-orientation, self-reliance, ecological soundness, and popular empowerment.

Out of these critiques have come generalized proposals for another form of agricultural development. Persuasive and powerful proposals argue that viable agricultural development strategies must be based on indigenous peoples' technical knowledge of crops, animals, and the environment (ITK) if they are to be viable (Denevan 1989). They argue that ITK is adapted to peasant production conditions, does not depend on external inputs, and is environmentally sound and culturally appropriate (e.g. Altieri 1987; Brokensha et al. 1980; Chambers et al. 1989; Richards 1985, 1986; Warren et al. 1995). This literature has generated an alternative to orthodox approaches to agricultural development in the so-called "farmer-first" approach (Chambers et al. 1989; Scoones and Thompson

1994). This calls for agricultural development built on farmers' knowledge and participation in agricultural technology development and project planning: a development that rejects the idea that anybody except the farmer is an expert (Chambers 1993).

The case for a "farmer-first" approach is motivated by concerns that are both political and theoretical. The political objectives are clearly to promote farmer participation in agricultural development, to encourage the democratization of agricultural organizations, to support ideas of social equity, and to challenge prevailing "taken-for-granted" power relationships in which the rural poor are always conceived of as "clients," recipients, and the objects of somebody else's development strategy. Theoretically, the concern is to relativize modernist rationality by suggesting that there are equally valid "native" points of view (Geertz 1983; Long and van der Ploeg 1994), to question grand evolutionary theories (Richards 1985), and to suggest that the political economy does not determine quite as many local outcomes as many radical approaches would suggest. These objectives are commendable, and the farmer-first project has achieved a great deal (not least through the work of Chambers). It has helped change attitudes to farmer expertise and indigenous peoples' knowledgeability, and it has undoubtedly helped put rural peoples' agency back into the picture, softening the pessimistic determinism of political economy. But it is also conceptually and practically problematic (Thompson and Scoones 1994).

To begin with, the farmer-first approach is grounded in an exaggerated, over-generalized, and sometimes simplified critique of technological modernization. For there is other research, not written by apologists of the Green Revolution, which suggests that although agrarian modernization has had negative impacts in some cases this need not necessarily always be so. Grossman (1993) cautions against over-hasty generalization about export agriculture. He shows that while export agriculture may have undermined peasant food security in some cases (Grossman 1984), this is not a necessary consequence. In the Windward Islands the relationships between export production and food security are far more complex; in many cases export producers using a "modern" technological package have increased their food security. Similarly, there are cases in which small farmer adoption of Green Revolution crops and varieties has increased food security, offsetting crises that would have occurred without technological change (Goldman 1993; Rigg 1989; Turner et al. 1993). Using the case study in this chapter, I similarly propose that while agrarian modernization led to the erosion of some "indigenous" cultures, this need not be the case: it depends on how the rural poor are able to incorporate and use modernization.

The implication of such cases is that we should treat generalized diagnoses of agrarian crisis with care. We also must be wary of accepting generalized remedies (Richards 1990a). An alternative in one context may not be the appropriate alternative for another. Academic understanding of alternatives may be neither appropriate nor congruent with that of rural people.

Conceptual questions for the farmer-first alternative

The farmer-first approach often constructs an essentialized conception of indigenous agriculture that is homogenized, static, and easily taken out of socio-economic, political, and cultural context (Fairhead 1992; Fairhead and Leach 1994; Scoones and Thompson 1994). Merely by naming something called "ITK," this literature creates the sense that a body of knowledge exists in a coherent form. By discussing ITK with a particular purpose in mind – namely to promote participatory agricultural development strategies building on farmer agronomic knowledge – this literature emphasizes the agricultural dimensions of rural life and the agricultural expertise of the rural poor. But it creates the impression that all rural people are farmers, that agricultural technology is central to solving rural poverty, and that pre-modernized techniques are crucial to any solution. In addition, the emphasis on the "knower" (the farmer), and on the knower's capacity to invent and create, tends to remove agents from structures, and to replace determinism with voluntarism (Giddens 1979; Long 1990). Likewise, an emphasis on what knowers know about technology and ecology diverts attention from the myriad things they do not know about: markets, politics, and the machinations of a world beyond the farm gate.

Recognizing this broader context of peasant livelihoods brings us back to a political-economic (or political-ecological) perspective on agrarian change. Some political-economic formulations may have had excessively deterministic overtones, but they at least kept the impact of wider social, political, and economic processes on farm resource management at the forefront of analysis (Blaikie and Brookfield 1987). They also countered both the populist argument and the dominant argument of the Green Revolution that technological fixes to social problems can be found. Against these arguments, political ecologists stress that the origins of the crisis of peasant agriculture lie in land tenure relations, market dependencies, the organization of the economy, the structure of the state, and the social relations of technological production (Bernstein 1982; Redclift 1987; Watts 1983, 1989; review in Bryant 1992). The implication is that if underlying causes of rural poverty are not addressed, promoting ITK will achieve little – it may not even be an appropriate response.

Furthermore, if ITK is as much *indigenous*, as it is *technical*, knowledge, then it raises issues of ethnic identity and cultural politics. This is especially apparent when we consider social movements which incorporate ideas of indigenous knowledge and practice into their own alternatives. Some "ITKers" do deal with cultural politics (e.g. Richards 1990b), but by and large most writing focuses on the technological rationality of adapted peasant production practices (Brokensha *et al.* 1980; Knapp 1991; Warren *et al.* 1995). Agrarian technology is not merely an instrument for environmental manipulation, but is a symbol speaking to rural people of their social history and relationships, a sign by which they read their identities and their relationships with past, present, and future (Bebbington 1991). Similarly, when peasants incorporate new ideas and material technologies into their practices, this can become a sign that the group is now more distant from a past when they were socially dominated, that their relationship with other social groups is changing, and that they now are claiming rights of access to resources and knowledges previously closed off precisely because of this domination. In short, the incorporation of modern technologies can be a sign of being liberated from a past of domination, even if this may imply new dependencies. It may be that incorporating modern techniques may be politically empowering rather than culturally disempowering.

ITK under neo-liberalism: the problem of sustaining rural livelihoods

Discussions of small farm agricultural development strategies in contemporary Latin America make little sense unless they consider the economic transformations and livelihood crisis faced by poor people in large parts of the Andes and other fragile lands. Currency devaluations lead to rapid price increases in fossil fuel-based agrochemical inputs, making it essential that the use of Green Revolution alternatives be efficient and effective. Trade liberalization and the creation of regional trading blocs are opening agriculture to competitive pressures. These increase pressure on small farmer production to increase productivity, lower costs, increase competitiveness, and use inputs much more efficiently in technical and economic terms.

New proposals for the intensification of agriculture and livelihood possibilities are needed, especially in the Andes (Kaimowitz 1991; Uquillas and Pichón 1995). Although macro-economic indicators of growth and inflation seem healthy, this has yet to feed into any significant growth in the popular economy, especially in rural areas. Rates of rural out-migration are striking. Indeed, "intensify or die" might be a short but to the point development challenge for much of Andes. Unless there is a significant intensification of livelihoods in the region, a combined process of land subdivision, out-migration and continued resource degradation will leave large parts of the Andes, especially the altiplano areas, as little more than labor reserves (cf. the implications of de Janvry 1981).

What can an ITK-based alternative contribute in such a context? At one level, rural people's knowledge of land and crops can make important contributions to technical responses to this challenge, particularly in the identification of lower external input options. Nevertheless, there are few experiences in which low-input agriculture has proven economically viable on a large scale (Ruttan 1991). Furthermore, the economic and technical efficiencies demanded in this new context require capacities for numeracy, economic abstraction, market research (e.g. identifying niche markets), and identification of cost-controlling, productivity-enhancing genetic material that poor people rarely possess (Byerlee 1987). Research in Mexico, Brazil, Paraguay, and Peru in the 1980s found that formal and higher education had positive effects on productivity and income in rural areas precisely because it helped develop skills of abstraction and numeracy required to handle markets (Cotlear 1989; Figueroa and Bolliger 1986).

More generally, the relative weakness of intensification processes in the region suggests that neither indigénous patterns of technical innovation, nor introduced innovations from "modern" science, are sufficient either by themselves or together to trigger intensification. Indeed, it is increasingly argued that livelihood intensification not only requires support for technical change, but also depends on the rural poor having improved access to product and input markets through relationships that allow wealth deriving from natural resource-based and agricultural activities to be captured and reinvested in the Andes. This improved and renegotiated market access is critical for the creation of incentives to the sustainable intensification of natural resource use and livelihood possibilities. Technological innovation alone, however environmentally sound and however grounded in traditional practices, will not achieve this.

The rural poor are firmly integrated into capricious and changing markets (Barsky 1990), and rural livelihoods depend increasingly on non-agricultural, often non-rural income sources (Barsky 1990; Klein 1992; López 1995). Martinez (1991) reports on a region in the Ecuadorean highlands where 40 per cent of rural families have two jobs *within* the countryside. In many areas, and for many people, agriculture is neither the only, nor the main, source of income. An adequate response to the Andean crisis must therefore go beyond the purely agricultural sphere. "Alternative" proposals must consider not only agricultural intensification, but also the expansion and diversification of off-farm rural income and employment opportunities. Indeed, there are many potentially synergistic links between agricultural intensification and expanding off-farm income opportunities. De Janvry and Sadoulet (1988) argue that strategies to alleviate rural poverty should promote rurally based non-agrarian incomes. This involves finding ways of increasing agriculturally derived incomes in order to create a demand for non-agrarian products and services that could be provided locally (see also Klein 1992). The essence of this strategy is to find mechanisms facilitating the retention of surplus within a region. Such mechanisms might include new marketing arrangements and the incorporation of a processing stage to develop new forward and backward linkages within the regional food system.

These observations imply that proposals for alternative agricultural development must go well beyond a focus on ITK in particular, and technology in general. Proposals must begin from the dynamics of the regional economy (cf. the political-ecological perspective).

ALTERNATIVE ACTORS: SOCIAL MOVEMENTS, CULTURAL IDENTITY, AND LIVELIHOODS

The challenge of a specifically indigenous, alternative agenda is to respond in a concrete, income-generating way to an Andean livelihood crisis, and in a way that simultaneously strengthens ethnic identity and politics. The challenge is thus material and cultural, Escobar's (1992) two arenas of struggle of social movements. One especially relevant question, then, is how movements can carry forward the two agendas so that one strengthens the other.

Resistance and identity in rural social movements

Some analyses see rural social movements in Latin America as forms of resistance to domination, exploitation, and subjection (e.g. Redclift 1987). Slater (1985) views them as a protest against traditional politics – in particular against excessive concentration of decision-making power and the incapacity of the state to deliver services. In this sense they are a consequence of the legitimacy problems of the state.

As Peet and Watts's introductory chapter points out, other authors focus more on ways in which movements are expressions of long-dominated and marginalized identities – identities which at the same time are reformulated through the activity of the movement (Evers 1985). Such expression of identity is itself frequently a form of resistance. As Gledhill (1988) has argued for Mexico, the terminology of the "indigenous community" is often used as a way of resisting the all-pervasive intervention of the state in local processes of production and reproduction. Similarly, in Ecuador the recovery and projection of the idea of being Indian is a form of resisting forms of white and mestizo domination, and of regaining a space for the values of being indigenous (Bebbington 1991; Ramón 1988).

These perspectives help us understand the nature, significance, and activities of popular organizations. The thornier question is what it means "to be an Indian" in the context of these new relationships with state and market. The integration of rural areas into the wider economy has brought many lifestyle changes to the Andean countryside. Modernity arrives variously in the form of fertilizers, radios, new textiles, bicycles, vans, school notebooks, school uniforms, and the clothes and vehicles of non-governmental and governmental extension agents. With these come new aspirations, access to which requires increased incomes. Furthermore, with the integration into a national political process and a new set of relationships with the state comes the idea that indigenous people are not only Indians, but are also national citizens with civil rights. Consciously or unconsciously, indigenous movements face two challenges: to reflect the multiple identities of those they represent, and to negotiate a relationship with the state in which they resist its predations and claim autonomous spaces, but at the same time make claims upon it as citizens.

The complexities of this balancing act are apparent in the umbrella organization for Ecuadorean Indians, the Confederation of Indigenous Nationalities of Ecuador (CONAIE) whose sociocultural and political strategy involves "the search for our own identity, or rather, the forging of an identity that continuously adjusts itself to this society and this supposed democracy which does not yet exist," according to one of its leaders, Mario Fares. In June 1990, CONAIE called on Ecuador's Indians to support a national uprising against government apathy toward indigenous peoples' needs and demanding government support for, and recognition of, Indian cultural difference (Macas 1991: 23). This act of protest asserted the values of "traditional" aspects of Indian identity. Yet, at the same time, CONAIE made demands for a full and fair incorporation of Indians into Ecuador's development process as their right as national citizens. These demands, reasonable on their own, do not rest easily together. On the one hand, CONAIE wishes to strengthen a conception of Indian identity largely grounded in past, more autarkic forms of production and social organization. Thus, CONAIE speaks of the recovery of indigenous crops, technologies, crop-environment theories, and cosmologies within larger strategies of ethnic self-determination and cultural revalorization (CONAIE 1989). On the other hand CONAIE demands that Indians be allowed fairer access to markets, credit, research, and extension (Macas 1991: 26). CONAIE supports the perpetuation and recovery of cultural traditions, *and* demands access to the means of rural modernization, the technologies and institutions of the cultural Other. These apparent contradictions point to the difficulty of defining an Indian identity in a modernizing economy. A possible resolution may be found in CONAIE's claim that because indigenous peoples are both Ecuadorean and Indian they are entitled to both community self-determination and rights of access to state resources (Macas 1991: 25–6). The implication is that communities themselves should decide the balance between traditional and modern markers of their ethnic identity.

In the case study, I argue that in some regions such a resolution has taken the form of a "bottom-up" self-management of the modernization process based on indigenous forms of organization.

From identities to livelihoods in rural social movements

Social movements may be expressions of cultural struggles over meaning (Escobar 1992; Peet and Watts, Chapter 1 in this volume), but the meanings over and for which they fight are not always clear. The struggle for meaning is all the more complicated when we consider the material struggles running alongside these cultural contestations. If a material basis for the survival of Andean communities is not assured, a principal element of Indian identity will be eroded away, both metaphorically and actually. Should this happen, then the cultural struggle will have little significance in the longer term, as indigenous people will be unable to secure the material basis on which to sustain a cohesive cultural identity. In the current policy context, this material basis is genuinely threatened. Recovering it demands a rapid intensification of indigenous resource management strategies, more efficient Green Revolution strategies, and the identification of new sources of non-agrarian livelihoods. This is the crucial challenge facing Indian organizations.

Can an ethnically distinct identity be sustained on the basis of transformed and modernized livelihood strategies? If so, how? The experience of the indigenous peoples of Cayambe and Otavalo in Ecuador sheds light on this question (Ramón

1988; Salomon 1981). In a context of severe land subdivision and erosion, local populations followed several strategies to intensify livelihoods. The most renowned is the development of a commercial textile sector, in which production and distribution is controlled by Otavaleño merchants and production is organized through a network of domestic units and small workshops (Salomon 1981). A less remarkable, but therefore more relevant, experience occurred around Cayambe, where farmers developed commercial onion production (Ramón 1988). Another indigenous group, the Chiboleos, became known as producers, purchasers, and distributors of garlic. In all these cases, the intense commercialization of livelihoods and agricultural production is nevertheless associated with the maintenance of strong markers of cultural identity in dress, language, kinship, networks, etc. (Ramón 1988). The opposition traditional/modern is thus not an either/or proposition for the indigenous rural poor.

These groups' responses have involved more than simple adaptive, techno-logical changes. Rather, indigenous people also changed the regional political economy so as to increase the accumulation of capital at the family and regional levels. Indigenous groups gained additional control over relationships of exchange by marketing their own products, enhancing the quality of those goods (e.g. the Chiboleos), and processing more of the materials leaving the region (e.g. the weavers of Otavalo).

A further example from Bolivia is also relevant. This involves a federation of co-operatives, formed in 1977 among farmers settling the high jungle of the Alto Beni (Healy 1988; Trujillo 1993). The development of cocoa, the principal cash crop was constrained because export markets were dissatisfied with the uneven quality of beans supplied, and the farmers' local organization lacked operating capital to buy sufficient volumes of beans up-front from farmers to be able to guarantee quantity (Healy 1988). Responding to this situation, and to facilitate access to technical, management, and financial support, the co-operatives created a formal federation (El Ceibo) to link, service, and represent its member co-ops. Among the federation's early actions was the negotiation of financial support. This removed problems of operating capital, allowing the co-operatives to become more fully involved in marketing activities and processing. Subsequently, Ceibo increased its marketing and processing operations, expanding into export activities. The impact on family income reduced out-migration. It now unites thirty-six separate co-operatives (Trujillo 1993), and in 1988 it sold $1.5 million worth of cacao and cacao derivatives (Healy 1988).

A similar experience is that of Funorsal (the Foundation of Organizations of Salinas) in the central Andes of Ecuador. This federation also elaborated a locally controlled development strategy which, between 1983 and 1992, created some 300 new jobs as well as increased on-farm income. Initially the strategy was mounted around the marketing and transformation of dairy products, and the technical modernization of small farmer dairy activities. Subsequently it moved into textiles, timber, and other forms of local product processing (Bebbington *et al.* 1993).

The successes of El Ceibo and Funorsal owe much to a concerted effort to develop modern business management skills among the administrators of the co-operatives, and to introduce technical innovations of modern cocoa and dairy research into farmer resource management practices. They thus present further cases in which the approach of a peasant organization to a farmer-controlled development was not to reject modernization, but to pursue local and grass-roots control of the modernization process. The experiences suggest also that technological modernization *per se* is not enough to launch successful grassroots-controlled rural development alternatives. There also has to be increased local control over the economic and social relationships traditionally contributing to the transfer of income and wealth from Indians and from the locality to other social groups and other places.

Unfortunately cases such as El Ceibo and Funorsal are as significant for their rarity as they are for their impact in the Andes. But they do suggest one potential path for rural social movements that may allow them to be effective in both their cultural and material struggles. The case study story suggests how other federations are following similar paths – though so far with less impact.

INDIGENOUS AGRARIAN DEVELOPMENT IN CENTRAL ECUADOR

Economic and institutional change in Chimborazo

Located in the Central Andes of Ecuador, the *cantones* of Colta and Guamote in Chimborazo lie in a high altitude area of dominantly *quichua* people (also often referred to as *runas*) that until the 1950s was largely controlled by rural estates, or *haciendas*. In many cases, *runas* were linked to these estates through a land for labor arrangement, often overlaid with debt and other exploitative relationships. Land-use systems were broadly of two types: the hacienda-based production system, which, though low intensity, was the one arena in which aspects of technical modernization were introduced; and the intensively farmed small plots of the *runas*. These *runa* systems were in many respects classic "indigenous" farming systems – diversified, intensive, based on food crops, and organic. Their sustainability, however, depended largely on manure from their animals grazed on the extensive pastures of the haciendas.

The social relationships underlying these labor and human-environment relationships were, however, contested, through more daily forms of resistance (Scott 1985) with occasional land invasions and uprisings. Resistance became more organized and assertive in the 1950s and 1960s, as the national Indian movement became stronger and pressed for land reform. Such pressures, coupled with shifts in policy and political balance, led to land reforms in 1964 and 1973. Because Colta and especially Guamote were conflictive areas, they were defined as priority zones for the implementation of land reform legislation. This led to a series of agrarian, economic, and institutional changes (discussed in more detail in Bebbington 1990).

The textbook agro-ecologically-sound indigenous agriculture began to decline with land redistribution. Hacienda pastures were divided among *runa* farmers and turned to crops. Organic fertilization strategies became increasingly problematic. Also, as population increased, land was further subdivided and fallow periods reduced. No intensification, such as stall feeding of cattle or terracing, occurred, and soils degraded. The use of chemical fertilizers and pesticides has increased with their greater availability, guaranteeing yields from crops weakened by poorer soils.

Such agro-ecological changes were accompanied by socioeconomic trans-formations. Increased market orientation discouraged the cultivation of little-demanded, traditional crops and favored production of marketable food crops (such as potato and broad bean) and new horticultural crops (such as onions, garlic, carrot, and beetroot). Land subdivision in the context of local joblessness (itself an effect of regional underdevelopment and the failure of former estates to reinvest their surpluses productively) led to increased seasonal migration to urban and coastal areas. *Runas* associate periodic migration with mounting social problems and weakened cultural practices in their communities: participation in community activities declined; health problems, petty theft, and violence increased; and manners deteriorated. Post-reform changes suggest an increasing reproduction squeeze on the *runa* economy, with families ever less able to feed themselves and protect the ecological conditions allowing sustained production (Bernstein 1982). This has become increasingly severe since the mid-1980s with rapid inflation and austerity measures aimed at controlling inflation.

Post-reform actors in rural development: the state, NGOs, and the church

Institutionally, land reform marked the beginning of the increased prominence of a modernizing state in local development initiatives. This was reflected in the growth of agricultural extension and integrated rural development programs in the area, which still continue, although with declining resources. These operations, oriented to fostering a modernization of Indian production systems, worked with the basic toolkit of the Green Revolution, introducing new varieties (especially potato), chemical fertilizers, pesticides, etc.

Two other significant institutional changes occurred: an increasingly assertive Church – Catholic and Evangelical – that identifies with indigenous people, and engages in development work; and a growth of non-governmental organizations (NGOs) with a range of social origins, and all pursuing their conceptions of a development alternative. These NGOs are private development agencies, generally supported by European and North American funds, and staffed by people who for whatever reason prefer not to work with state agencies – or are unable to do so.

These different actors have played important roles in the emergence of Indian federations in the area, and in the strategies pursued by these federations. This is so in several regards:

1 Each actor encouraged the creation of federations, often for different reasons, but promoting basically the same organizational structure.
2 Although a number of NGOs and Catholic Church-based organizations were far more cautious than the state in fostering the use of agrochemicals, trying to promote the type of ITK-based alternatives discussed earlier, they have tended with time toward a similar technical packet, sometimes by choice, sometimes as a result of pressure from farmers – a packet based on the scaled-down use of modern varieties and agrochemicals.
3 All tended to use the same model of agricultural development, based on the delivery of inputs and services from an institution (state, NGO, or church), through extension agents working in communities. Similarly they mostly encouraged the idea that project management would later be taken over by the federations.
4 Each of them, even the state, fostered a discussion of the rights of Indian people: rights of equal access to the scriptures, equal citizenship rights as Ecuadoreans, or rights to protect and project their culture.
5 They mostly promoted the idea of grassroots management of development processes, although, somewhat hypocritically they were not always willing to pass on project management to grassroots groups.

These common tendencies are important in several senses. They created the set of ideas and practices on which federations drew as they composed their own agrarian strategies. This alerts us to ways in which social movements reflect the institutional and intellectual environment in which they emerge. Neither movements, nor their strategies, are pristine or entirely self-generated. Not only are they actors "situated" in a political economic context that greatly influences the impact of what they do, they are also "situated" in a mesh of ideas and precedents influencing what they choose to do – the strategies they select to pursue their objectives.

These experiences suggest also that even organizations with a commitment to ITK-based strategies – such as the liberational Catholic Church and several indigenistic and agro-ecological NGOs – have been unable to practice such strategies because farmers are often not especially interested in them, and because their livelihood impact is disappointing.

The agrarian strategies of the Indian federations have a similar focus on technical modernization. But at the same time, this is linked to strategies to which the rural social movements literature draws attention: strategies of resistance, ethnic assertiveness, and questioning the centrality of the state in rural administration – but again in a way influenced by wider economic, political, and ideological relationships.

Indian federations: origins and strategies

Following the struggle for land, the emergence of stronger and more numerous communities as units of territorial administration has been one of the most significant sociopolitical changes in rural Chimborazo. At the same time, a novel form of indigenous organization – the federation of indigenous communities – emerged. Such federations engage in both political and developmental functions, negotiating with public agencies on the one hand, implementing development projects on the other. There are a number of federations in Chimborazo. The more radical of these trace their origins to disputes over land and other matters, often being linked to national indigenous and peasant movements and to the Catholic Church. The more developmentalist of the federations have origins in negotiations over access to resources from the state and donors.

As just noted, the growth and increasingly strong self-management capacity of the federations results in part from support from other organizations and donors. Neither the federations nor their programs are entirely "endogenous" innovations. However, if grassroots control, rather than technological content, gives strategy its "alternative" character, what matters is not that agricultural development strategies are endogenous, but that they are locally controlled. Although this local control is not perfectly democratic – indeed certain groups exercise more influence than others – the federations remain more accessible and accountable to local people than any other development institution.

The emergence of these federations also reflects a further stage in the recovery of Andean space by Indians. Going beyond the recovery of land as means of production, the federations slowly are recovering the administrative control of rural space, taking back terrain once administered by the hacienda and questioning the very control of space by the state and white–mestizo society. As white and mestizo presence declines, rural areas thus return to indigenous people as space in which to practice their culture and agriculture.

The federations' perspectives on relationships between technology and ethnic identity in an indigenous agricultural development can be understood within this increasing Indian control of rural areas. Rationales stemming from this conception do not alone determine the federations' strategies – socioeconomic and ecological processes are equally important factors. But this rationale gives meaning to such strategies. The result is a vision of Indian agricultural development embracing concerns for agrarian technology, a stronger *runa* cultural identity, and control of rural space. The form in which these concerns are combined varies over time and among federations. Nevertheless, the overall objectives are consistent.

In Chimborazo, an important point of debate among, and within, Indian federations is the extent to which they should work with modern agrochemical and crop technologies as opposed to traditional, low-input technologies. The more radical federations emphasized the recovery of traditional culture and technology in their earlier work. Their programs of agricultural development promoted the recovery of Andean crops, use of organic fertilizers, and replacement of pesticides with supposedly traditional methods of pest control. The rationale underlying this strategy was that it constituted a rejection of the agricultural technologies associated with white and capitalist culture, while at the same time affirming and validating indigenous identity (MICH 1989: 199). It also reduced market dependence, costs of production, and environmental pollution. Social and cultural empowerment, in these strategies, was to be based on an agrarian development based on traditional practices. However, it proved difficult to promote this alternative among *runa* farmers already producing for the market, who had little land from which to produce organic manures. Pressures from their members thus led the federations to work with agrochemicals, new varieties, and cash crops.

In making this shift, these radical federations approached the model of the more "developmentalist" federations which endorsed and promoted the use of modern technology through their own research, extension, and input supply programs. They argue that this technology can improve *runa* income. They also deem it a necessary technological response to the grazing crisis and soil degradation in Colta and Guamote. The cultural justification for such strategies is that *modernization, far from being a cause of cultural erosion, is explicitly seen as a means of cultural survival.* Periodic migration, and the problems associated with it, is seen as more of a threat to *runa* cultural coherence than the use of agrochemicals and new crop varieties. Technological modernization, along with the promotion of non-traditional cash crops, is therefore justified as a strategy for increasing local income opportunities and reducing the pressures to migrate. The principle is that indigenous cultural identity hinges on sustained and corporate rural residence, and not so much on retaining traditional technologies. The implication is that indigenous economy and culture must constantly adapt to survive and sustain group cohesion and forms of self-management. In this regard, the federations follow strategies generically similar to those of the Indians of Otavalo, Chiboleo, and Cayambe (Ramón 1988; Salomon 1981).

At the same time, there is a politically radical dimension to this bottom-up modernization. Many *runas* associate "traditional" technologies with the subjugation of the hacienda and wish to distance themselves from the practices associated with Indian life on the hacienda. In this sense, when *runas* reject traditional technology, this is a metaphor for the rejection of the social relations through which they were dominated on the haciendas. By embracing modern technologies they are making a statement – as much to themselves as to others – that they now have the same rights to demand access to resources and benefits (including new technologies) that historically were the preserve of whites and mestizos. This use of modern technology is thus part of a wider discourse on citizenship rights.

In addition, the aspiration is that the hoped-for benefits of modernization –

reduced migration, increased community cohesion – will strengthen indigenous organizations as sociopolitical vehicles for demanding change, access to resources, and a more prominent role for Indians in rural development and government. Such demands are given cogency by the fact that the Indian federations' management of rural modernization reflects an attempt to demonstrate Indian ability to use and manage modern administrative methods in a style similar to state programs. If Indians are able to administer rural modernization through their own organizations, the ethnic exclusivity of state rural development administration is no longer justifiable.

The decision to foster technological modernization thus has clear rationales. Modernization seems a necessary response to the realities of market production, soil degradation, and land subdivision, and is more in tune with farm family concerns than are strategies aimed at recovering traditional practices. It also has politically progressive resonances in that technological modernization need not be interpreted as cultural assimilation. At the same time, it is part of a strategy aimed at offsetting the underlying causes of sociocultural dislocation in communities and strengthening *runa* political organization.

Whether technological modernization alone will be able to achieve all these goals is a moot point, given the many challenges to peasant agriculture. If, as de Janvry and Sadoulet (1988) argue, an increase in farm-level income is a necessary precursor to rural development, additional changes in the marketing sphere are also required, so that the benefits of technological innovation are captured at the farm level.

Recognizing this limitation, several federations have recently initiated marketing and processing activities. With donor support one federation – the Union of Indigenous Communities of Guamote (UCIG) – challenged the position of intermediaries in the marketing chain by establishing its own marketing program, bulking member community produce for sale to other parts of the country. It has also opened a plant to process cereals and the Andean grain, *quinoa*, into flour to capture a higher price. The ownership of the plant is shared between UCIG and the communities providing it with cereals. The program also serves to improve nutritional status in communities by requiring that farmers receive a percentage of their payment in the form of the protein rich *quinoa* flour.

Once again in conjunction with an NGO, the federation is extending the frontiers of modernization. These new programs mark a recognition by the federation that agriculture and technology alone cannot be the basis of *runa* development, and that other income sources are required. The idea that small industries are important in rural development is not new. But the federation is embracing an orthodox idea for quite radical objectives: in the process, an orthodox approach is turned into an "alternative," locally controlled and indigenous strategy for rural development.

CONCLUSION

This chapter opened by suggesting that to improve the conversation between analytical concepts and the rationales of social movements could improve those concepts, and point out weaknesses in the strategies of the movements. The chapter then aimed to do this for the case of Indian federations in highland Ecuador. What has this discussion revealed about the rationales and adequacy of these strategies, the implications for theoretical concepts, and the questions raised in Peet and Watts's introductory chapter?

Understanding indigenous approaches to modernization

Indian people in Chimborazo perceive a close relationship between residing in rural areas and sustaining their identity as indigenous farmers. An "indigenous" agrarian development must therefore allow occupation of traditional Indian spaces. In the current context, productive strategies based on non-modernized technologies do not appear a viable means of ensuring this objective. Ethnic identity will be grounded in other social, cultural, and linguistic practices, and not in traditional technology.

The programs of Indian federations thus encouraged the incorporation of modern technology into local farming practice in an effort to offset migration and improve rural welfare. Contrary to the implications of some critical writings on the Green Revolution, this suggests that technological modernization can be a rational response to crisis in indigenous production and social systems, and yet at the same time have politically and culturally progressive overtones. This challenges how we think about indigenous agricultural revolutions and relationships among culture, technology, and politics.

For an agrarian strategy to carry forward an "alternative, indigenous development" thus depends less on the technological content of that strategy than its social control and objectives. The objective in Chimborazo is to sustain livelihoods to allow the survival of other social practices that continue to mark these people as indigenous.

The limits of technological modernization

To have a political and development coherence is not, however, sufficient. Strategies have to be effective in addressing the causes of the livelihood and sociocultural crisis affecting large parts of the Andes. Recent tendencies suggest that some federations feel that technological modernization alone may be an insufficient response to this crisis. Political economy and regional development theory also suggest that such strategies are insufficient to address the causes of the livelihood crisis. Yet, to recognize this limitation does not have to mean reaffirming the indigenistic and farmer-first proposals for agrarian development. Instead it requires that the case be taken further. A viable indigenous development requires a restructuring of marketing and other social relationships in order to place the production of higher-value and processed products under the control of rural people, thus increasing farm incomes. Only then will some of the underlying forces leading to out-migration begin to be genuinely addressed.

The challenge therefore is not to resist modernization, but to control it, take it further, and increase indigenous peoples' abilities to negotiate market relationships, administer rural enterprises and agro-industry, and compete in a hostile market. In Chimborazo, the case of UCIG suggests that this is the path indigenous federations are beginning to tread. Experiences of other social movements and organizations in Ecuador and Bolivia likewise suggest that a viable indigenous development, which at the same time respects and strengthens ethnic identity, *can* be based on such a modernization strategy. These experiences also suggest, however, that for the organization to carry forward such a strategy takes time and implies that the organization shifts from being largely representative and often informal, to becoming a more formalized, social enterprise.

Conceptual implications

The case study challenges some of the claims made for indigenous technologies, suggesting that much of the writing about ITK takes it out of political economic context. The case endorses the idea that we need to understand ITK as a dynamic response to changing contexts – a response constructed through farmers' practices as active "agents" "situated" within cultural, economic, agro-ecological, and sociopolitical contexts that are products of both local and non-local processes. *Runas'* understanding of their identity and agriculture in Chimborazo responds to wider socioeconomic processes. These processes challenge the viability of indigenous agriculture, most evidently in the declining relevance of traditional practices and in the current pressures deriving from the macro-economic changes. At the same time these wider processes provide resources and ideas that are taken in and reworked by indigenous peoples. The federations' strategies, indeed the very existence of federations, are influenced by the churches, NGOs, and state organizations. Similarly, the ways in which federations and farmers interpret technologies are influenced by local history. Perhaps the most acute illustration of how the wider context can both constrain and enable the strategies of the rural poor is their insertion into the market. While certain forms of insertion can prejudice the sustainability of indigenous farmers' agriculture, if that insertion is renegotiated, as with the current processing and marketing programs of UCIG, the market may be used to strengthen indigenous farmers' strategies, and ultimately, their organizations.

But within this context, indigenous people and their federations are capable of picking and choosing among the different resources, ideas, and technologies that these wider processes make available. They are selectively eclectic, composing strategies that don't necessarily fit the concepts analysts have of them, but which nonetheless are far more coherent and meaningful to *runas* themselves than are those concepts.

The experience in Chimborazo illustrates how analyses of indigenous agrarian strategies can benefit from a more critical look at the rationales behind indigenous strategies and factors underlying them. These strategies are not mere "adaptations to environment." They are influenced also by cultural and political logics and socioeconomic exigencies. Furthermore, they may take a form that on the surface seems counterintuitive – the incorporation of modern technology and administration as part of a strategy of cultural survival. Whether or not these responses are adequate is a secondary question. If we do not understand the reasoning underlying them, we can never make a useful contribution to an "alternative" rural development. Instead we run the risk of imposing our conceptions of what is "alternative" and of what it means to be "indigenous."

What this chapter has not done is to take the discussion of alternatives a step further and ask whether we should be talking about an alternative to development rather than an alternative development (cf. Escobar 1995). The elaboration of such alternative utopias is a valid and important task, as part of a sustained questioning of dominant ideas and policies – and if the social and economic relationships in which the federations are enmeshed were different then it might be the case that other, non-modernizing strategies would be appropriate and feasible. However, these utopias must also be constructed from practice, and grounded in the aspirations of popular sectors. The dilemma is that these aspirations and practices have now incorporated the experience of modernity and development. This may be imperfect, but it cannot be overlooked. Those in the business of alternative utopias must be careful before rejecting popular aspirations for the benefits of modernity as some sort of false consciousness. Furthermore, there is an immediate problem of survival. *Runas* in Chimborazo do not have time to wait for the dawn of new utopias. They demand liberation from where they are now. The challenge then is to build short-term, pragmatic, and realistic responses that work from contemporary contexts, and do so in a way that is coherent with and builds towards longer-term utopias that are already immanent within the strategies and hopes of popular sectors.

The case of the federations challenges theorists to be less idealistic in the ways in which social movements are discussed. These federations open up – and reflect – new ways through which Indians can challenge and question state policies, and new ways through which they can assert and revalorize their identities. Nonetheless, their impacts on the livelihoods of their members remain disappointing. Yet if questions of productivity, income, and employment in the rural popular sectors of Chimborazo – and large parts of the Andes – are not addressed, these organizations will ultimately be a passing phenomenon. Rates of out-migration from communities remain high. With out-migration comes a social disarticulation which undermines communities and the range of social and cultural practices which lie at the heart of what it is to be indigenous.

It is in facing pragmatic and conceptual challenges at this intersection of environment, livelihood, political economy, and effective social movements that political ecology and civil society can be analytically and practically associated.

105

If they are not, neither our concepts nor the strategies of rural social movements will contribute to the construction of a feasible liberation ecology.

NOTE

The paper draws on research conducted at different times over a five-year period, and supported by the Inter-American Foundation, the International Potato Center, La Fundación para el Desarrollo Agropecuario and the Overseas Development Administration. It has also been supported by countless conversations with friends, colleagues, campesinos, and students. My thanks to them all.

REFERENCES

Altieri, M.A. 1987. *Agroecology. The Scientific Basis of Alternative Agriculture*. Boulder, CO: Westview Press.

Altieri, M.A. and S.B. Hecht (eds.). 1990. *Agroecology and Small Farm Development*. Boston: CRC Press.

Barsky, A. 1990. *Políticas agrarias en América Latina*. Santiago: Imago Mundial.

Bebbington, A.J. 1990. "Indigenous agriculture in the central Ecuadorian Andes. The cultural ecology and institutional conditions of its construction and its change." Ph.D. dissertation, Graduate School of Geography, Clark University, Worcester, MA.

—— 1991. "Indigenous agricultural knowledge, human interests and critical analysis," *Agriculture and Human Values* 8, 1/2: 14–24.

—— 1995. "Rural development: policies, programmes and actors," in D. Preston (ed.) *Latin American Development: Geographical Perspectives*, 2nd edn. Harlow: Longman, pp. 116–45.

Bebbington, A.J. and G. Thiele. 1993. *NGOs and the State in Latin America. Rethinking Roles in Sustainable Agricultural Development*. London: Routledge.

Bebbington, A.J., H. Carrasco, L. Ramón Peralvo, V.H. Torres, and J. Trujillo. 1993. "Fragile lands, fragile organisations: Indian organisations and the politics of sustainability in Ecuador," *Transactions of the Institute of British Geographers* 18: 179–96.

Bernstein, H. 1982. "Notes on capital and peasantry," in J. Harris (ed.) *Rural Development. Theories of Peasant Economy*, London: Hutchinson, pp. 160–77.

Biggs, S. and J. Farrington. 1991. *Agricultural Research and the Rural Poor: A Review of Social Science Analysis*. Ottawa: International Development Research Centre.

Blaikie, P. and H. Brookfield. 1987. *Land Degradation and Society*. London: Methuen.

Booth, D. (ed.). 1994. *Rethinking Social Development: Theory, Research and Practice*. Harlow: Longman.

Brokensha, D., D.M. Warren, and O. Werner (eds.). 1980. *Indigenous Knowledge Systems and Development*. Lanham, MD: University of America Press.

Bryant, R. 1992. "Political ecology," *Political Geography* 11: 12–36.

Butzer, K. 1990. "Cultural ecology," in C.J. Wilmott and G.L. Gaile (eds.) *Geography in America*. Washington, DC: Association of American Geographers and National Geographic Society.

Byerlee, D. 1987. *Maintaining the Momentum in Post-Green Revolution Agriculture: A Micro-level Perspective from Asia*. Michigan State University International Development Paper 10. Department of Agricultural Economics, Michigan State University, East Lansing.

Chambers, R. 1993. *Challenging the Professions*. London: Intermediate Technology Publications.

Chambers, R., A. Pacey, and L.A. Thrupp (eds.). 1989. *Farmer First. Farmer Innovation and Agricultural Research*. London: Intermediate Technology Publications.

Confederación de Nacionalidades Indígenas del Ecuador (CONAIE). 1989. *Nuestro proceso organizativo*. Quito: CONAIE.

Cotlear, D. 1989. *El desarrollo campesino en los Andes. Cambio tecnológico y transformación social en las comunidades de la Sierra del Perú.* Lima: Instituto de Estudios Peruanos.

de Janvry, A. 1981. *Land Reform and the Agrarian Question in Latin America.* Baltimore, MD: Johns Hopkins University Press.

de Janvry, A. and E. Sadoulet. 1988. *Investment Strategies to Combat Rural Poverty: A Proposal for Latin America.* Mimeo. Department of Agricultural and Resource Economics, University of California, Berkeley.

Denevan, W.M. 1989. "The geography of fragile lands in Latin America," in J. Browder (ed.) *Fragile Lands of Latin America,* Boulder CO: Westview Press, pp. 11–25.

Edwards, M. 1994. "Rethinking social development: the search for relevance," in D. Booth (ed.) *Rethinking Social Development: Theory, Research and Practice.* Harlow: Longman, pp. 279–97.

Escobar, A. 1992. "Imagining a post-development era? Critical thought, development and social movements," *Social Text* 31–2.

—— 1995. *Encountering Development: The Making and Unmaking of the Third World.* Princeton, NJ: Princeton University Press.

Esteva, G. 1992. "Development," in W. Sachs (ed.) *The Development Dictionary: A Guide to Knowledge and Power.* London: Zed Books.

Evers, T. 1985. "Identity: the hidden side of new social movements in Latin America," in D. Slater (ed.) *New Social Movements and the State in Latin America.* Amsterdam: CEDLA, pp. 43–77.

Fairhead, J. 1992. "Representing knowledge: the 'new farmer' in research fashions," in J. Pottier (ed.) *Practising Development: Social Science Perspectives.* London: Routledge, pp. 187–204.

Fairhead, J. and M. Leach. 1994. "Declarations of difference," in I. Scoones and J. Thompson (eds.) *Beyond Farmer First: Rural People's Knowledge, Agricultural Research and Extension Practice.* London: Intermediate Technology Publications, pp. 75–9.

Figueroa, A. and Bolliger, F. 1986. *Productividad y aprendizaje en el medio ambiente rural. Informe comparativo.* Rio de Janeiro: ECIEL.

Geertz, C. 1983. *Local Knowledge. Further Essays in Interpretive Anthropology.* New York: Basic Books.

Giddens, A. 1979. *Central Problems in Social Theory.* London: Macmillan.

Gledhill, J. 1988. "Agrarian social movements and forms of consciousness," *Bulletin of Latin American Research* 7: 257–76.

Goldman, A. 1993. "Population growth and agricultural change in Imo State, Southeastern Nigeria," in B.L. Turner, R. Kates, and G. Hyden (eds.) *Population Growth and Agriculture Intensification Studies for Densely Settled Areas of Africa.* Gainesville: University of Florida Press.

Griffin, K. 1974. *The Political Economy of Agrarian Change,* London: Macmillan.

Grossman, L. 1984. *Peasants, Subsistence Ecology and Development in the Highlands of Papua New Guinea.* Princeton, NJ: Princeton University Press.

—— 1993. "The political ecology of banana exports and local food production in St. Vincent, Eastern Caribbean," *Annals of the Association of American Geographers* 83, 2: 347–67.

Healy, K. 1988. "From field to factory: vertical integration in Bolivia," *Grassroots Development* 11, 2: 2–11.

Hewitt de Alcantara, C. 1976. *Modernizing Mexican Agriculture.* Geneva: United Nations Research for Social Development.

Kaimowitz, D. 1991. *The Role of NGOs in the Latin American Research and Extension System.* Mimeo. Instituto Interamericano de Cooperación Agropecuaria: San Jose.

Klein, E. 1992. "El empleo rural no agricola en America Latina," paper presented at conference "La sociedad rural Latinoamericana hacia el siglo XXI," 15–17 July 1992. Centro de Planificación y Estudios Sociales.

Knapp, G. 1991. *Andean Ecology: Adaptive Dynamics in Ecuador.* Boulder, CO: Westview Press.

Leff, E. 1994. *Ecologia Política y Capital: hacia una perspectiva ambiental del desarrollo.* Mexico City: Siglo XXI.

Lehmann, A.D. 1990. *Development and Democracy in Latin America. Economics, Politics and Religion in the Postwar Period.* Cambridge: Polity Press.

Lipton, M. and R. Longhurst. 1989. *New Seeds and Poor People.* London: Unwin Hyman.

Long, N. 1990. "From paradigm lost to paradigm regained? The case for an actor-oriented sociology of development," *European Review of Latin American and Caribbean Studies* 49: 3–24.

Long, N. and J.D. van der Ploeg. 1994. "Heterogeneity, actor and structure: towards a reconstitution of the concept of structure," in D. Booth (ed.) *Rethinking Social Development: Theory, Research and Practice.* Harlow: Longman, pp. 62–89.

López, R. 1995. *Determinants of Rural Poverty in Chile: A Quantitative Analysis for Chile,* Technical Department, Rural Poverty and Natural Resources, Latin America Region. Washington: World Bank.

Macas, L. 1991. "El levantamiento indígena visto por sus protagonistas," in Instituto Latinoamericano de Investigaciones Sociales (ed.) *Indios.* Quito: ILDIS, El Duende, Abya-Yala, pp. 17–36.

Martinez, L. 1991. "Situación de los campesinos artesanos en la Sierra Central del Ecuador: Provincia de Tungurahua," manuscript. Quito.

Max-Neef, M. 1991. *Human-scale Development. Conception, Application and Further Reflections.* New York: Apex Press.

Movimiento Indígena de Chimborazo (MICH). 1989. "Movimiento Indígena de Chimborazo, MICH," in *Nuestro Proceso Organizativo.* Quito: CONAIE, pp. 195–202.

Nerfin, M. 1987. "Neither prince nor merchant: citizen – an introduction to the Third System," *Development Dialogue* 1: 170–95.

Peet, R. and M. Watts. 1993. "Introduction: development theory and environment in an age of market transformation," *Economic Geography* 69, 3: 227–53.

Ramón, V.G. 1988. *Indios, crisis y proyecto alternativo.* Quito: Centro Andino de Acción Popular.

Redclift, M. 1987. "Introduction: agrarian social movements in contemporary Mexico," *Bulletin of Latin American Research* 7, 2: 249–55.

Richards, P. 1985. *Indigenous Agricultural Revolution. Ecology and Food Production in West Africa.* London: Hutchinson.

—— 1986. *Coping with Hunger. Hazard and Experiment in a West African Rice Farming System.* London: Allen & Unwin.

—— 1990a. "Local strategies for coping with hunger: Central Sierra Leone and northern Nigeria compared," *African Affairs* 89: 265–75.

—— 1990b. "Indigenous approaches to rural development: the agrarian populist tradition in West Africa," in M. Altieri and S. Hecht (eds.) *Agroecology and Small Farm Development.* Boston: CRC Press, pp. 105–11.

Rigg, J. 1989. "The new rice technology and agrarian change: guilt by association?" *Progress in Human Geography* 13, 2: 374–99.

Ruttan, V. 1991. "Challenges to agricultural research in the 21st century," in P.G. Pardy, J. Roseboom, and J.R. Anderson (eds.) *Agricultural Research Policy: International Quantitative Perspectives.* Cambridge: Cambridge University Press.

Salomon, F. 1981. "The weavers of Otavalo," in N. Whitten (ed.) *Cultural Transformations and Ethnicity in Modern Ecuador.* Urbana. University of Illinois Press, pp. 420–49.

Scoones, I. and J. Thompson (eds.). 1994. *Beyond Farmer First: Rural People's Knowledge, Agricultural Research and Extension Practice.* London: Intermediate Technology Publications.

Scott, J.C. 1985. *Weapons of the Weak. Everday Forms of Peasant Resistance.* Yale: Yale University Press.

Slater, D. (ed.). 1985. *New Social Movements and the State in Latin America.* Amsterdam: CEDLA.

Thompson, J. and I. Scoones. 1994. "Challenging the populist perspective: rural peoples' knowledge, agricultural research and extension practice," *Agriculture and Human Values* 11, 2–3: 58–76.

Trujillo, G. 1993. "El Ceibo," in A.J. Bebbington and G. Thiele (eds.) *Non-Governmental Organizations and the State in Latin America: Rethinking Roles in Sustainable Agricultural Development.* London: Routledge.

Turner, B.L., R.W. Kates, and G. Hyden (eds.) 1993. *Population Growth and Agricultural Intensification: Studies from the Densely Settled Areas of Africa.* Gainesville, FL: University of Florida Press.

Uquillas, J. and F. Pichón. 1995. *Rural Poverty Alleviation and Improved Natural Resource Management Through Participatory Technology Development in Latin America's Risk-prone Areas.* World Bank Technical Environmental Unit, Latin America and the Caribbean Region. Washington: World Bank.

Warren, D.M., D. Brokensha and J. Slikkerveer (eds.). 1995. *The Cultural Dimension of Development. Indigenous Knowledge Systems.* London: Intermediate Technology Publications.

Watts, M. J. 1983. "Populism and the politics of African land use," *African Studies Review* 26: 73–83.

—— 1989. "The agrarian crisis in Africa: debating the crisis," *Progress in Human Geography* 13, 1: 1–41.

5

DISCOURSES ON SOIL LOSS IN BOLIVIA

Sustainability and the search for socioenvironmental "middle ground"

Karl S. Zimmerer

THE EROSION PROBLEM IN BOLIVIA

By the late 1970s various reports were sounding the alarm about worsening soil erosion in Bolivia, a landlocked and mostly mountainous republic of over 4 million people in central South America. Books such as *Bolivia: The Despoiled Country* by Walter Terrazas Urquidi (1974) and *The Wasted Country: The Ecological Crisis in Bolivia* by Mariano Baptista Gumucio (1977) alerted many Bolivians and Latin Americans to the country's grave dilemma. The widely read *Losing Ground* by North American Erik Eckholm (1976) introduced it to a still larger audience in the United States and Western Europe. Academic and governmental studies spelled out some of the serious consequences of Bolivia's erosion crisis (Grover 1974; LeBaron *et al.* 1979; Preston 1969). Accelerating erosion was degrading farm and range-land, forcing floods downstream, and leading to destructive desertification and dust storms. Recent estimates in Bolivia's major newspapers surmise that between 35 and 41 percent of the country now suffers moderate to extreme loss of soils (*Los Tiempos* 1991; *Presencia* 1990).

Widespread alarm about soil erosion cannot be attributed solely to the problem's new gravity. A variety of historical sources from the 1570s to the 1920s allude to severe soil loss resulting from grazing and farming and to deforestation carried out in the absence of conservation measures (Larson 1988; Zimmerer 1993b). The predominant base of erosion-prone sedimentary rock, steep terrain, and semi-arid climate and vegetation in the Bolivian Andes have long rendered its mountainous landscape vulnerable to soil loss (de Morales 1990; Montes de Oca 1989). While the Inca overlords probably enforced soil conservation in the fourteenth and fifteenth centuries, the colonial rulers (1530s–1825) and republican governments (1825–present) did not instill soil conservation or grant official notice to the erosion problem. The countryside's environmental destruction almost disappeared from view for the broader Bolivian public between the

110

1530s and 1970s. Even advisors from the United States Department of Agriculture, present from 1943, failed to bring attention to the country's catastrophic erosion (USDA 1962).

A new awareness of erosion in Bolivia during the 1970s was inspired by international conferences and publications coming from other Latin American countries, the United States, and the United Nations. Their environmental ideas, together with the downright worsening of erosion, led agronomists Walter Terrazas Urquidi and Mariano Baptista Gumucio to publicize their country's environmental crisis. Yet the widely held concern about soil loss in Bolivia has not cemented a public consensus about the nature of the erosion problem or the prospects for conservation. In fact individuals and institutions living and working in Bolivia hold a variety of divergent views on the causes, as well as the preferred cures, for the problem. My interest in the predicament of Bolivian soil erosion and conservation prospects led me to study the diverse perceptions of these issues in relation to the political ecology of the changing environment (Zimmerer 1993a, 1993b, 1994). I chose to focus on the period of the recent past (1950s–present) in the Cochabamba region, a geographical area of stark contrasts and abrupt ecological transitions between the mountainous Andes and the rainforest lowlands of the upper Amazon. This focus enabled me to study people's perceptions and the political ecology of the changing environment in a well-grounded historical and regional setting.

Often claimed to be the "heartland" or "breadbasket" of Bolivia, the productivity of Cochabamba has been slipping dangerously during recent decades. Estimates show that 64 per cent or 790 square kilometers of the region's land surface is at least moderately eroded. Approximations of annual erosion range from 50 to 150 tons per hectare, rates far exceeding those of soil formation (CORDECO 1980; Zimmerer 1991). Severity of erosion in Cochabamba surpasses even the debilitating national average. Its threat has become a pressing issue for many institutions and inhabitants – including more than 100,000 peasant farmers in the region whose livelihoods depend on small-scale cropping, livestock raising, and a wide variety of non-farm work. Adding to their voices, development institutions of the government and non-governmental organizations (NGOs) have voiced alarm and launched analyses and programs to address the erosion problem. In all, three primary perspectives have taken shape on Cochabamba's erosion crisis and proposals for conservation-with-development (i.e. sustainable development): (1) government and non-government institutions; (2) peasants in their personal perspectives; and (3) rural trade unions.

SOIL LOSS AND DISCOURSE IN POLITICAL ECOLOGY

To assess the perspectives on soil erosion and sustainable development I combine a framework of concepts from political ecology and the analysis of articulated perceptions or discourse (Blaikie 1985; Blaikie and Brookfield 1987; Eagleton 1983; Emel and Peet 1989; Peet and Watts, Chapter 1 in this volume; Watts 1983,

1985; Zimmerer 1993b, 1994, 1995). From political ecology I adopt a socio-economic and political analysis of environmental change. A process of uneven economic development has in effect cornered peasant farmers in the Cochabamba countryside, pressuring them to modify land use in erosion-inducing ways – such as curtailing once common conservation measures as a consequence of labor–time constraints – while not offering economic alternatives sufficiently remunerative to permit land-use alteration in a more environmentally sound manner (Zimmerer 1993b; see also de Janvry *et al.* 1989; Storper 1991). The analysis shows that the so-called "scissors effect" has led peasants to intensify production without the use of conservation techniques.

I also pursue a pair of other themes recently being brought to bear in the political ecology approach. These themes are the analysis of civil society, especially so-called "new social movements" distinct from government institutions and the analysis of environmental discourses (Peet and Watts, Chapter 1 in this volume; Zimmerer 1993a). Rural trade unions in Cochabamba resemble other "new social movements" that seek to revitalize existing institutions at the grassroots or popular level. The rural trade unions gained a growing prominence in public debate and discussion on environmental issues, among which soil erosion looms large. Interestingly, the recent renewal of Cochabamba trade unions as social movements – as well as the region's proliferation of NGOs – returns us to a main theme of political ecology: how government policies in many developing countries enforce the extremely biased processes of uneven development. Amid the biases of distorted development policies, non-state institutions spread pell-mell in certain regions and economic sectors, including those of peasants (Slater 1985).

Applying a political ecology approach to the language-rich realm of articulated perceptions about erosion requires a focus on discourses representing the ideas and ideologies held by groups of individuals and institutions (Peet and Watts, Chapter 1 in this volume; Zimmerer 1993a). The discourses on soil erosion of groups in the Cochabamba region differ in significant ways. Through my fieldwork I realized that people and institutions there did not form their environmental discourses in absolute isolation or, figuratively speaking, as discursive islands of self-contained dialogue (cf. the narrowly poststructuralist interpretation of Orlove 1991). Instead, they expressed and gave distinctiveness to their viewpoints through interaction within and among the groups. Processes of resistance and contestation as well as accommodation and agreement guided their elaboration of environmental ideas. A historical approach and sensitivity to discursive alterations over time were crucial to my gaining insights on their views of changing nature and efforts to conserve it.

I seek also to renew consideration of local knowledges and peasants' personal or "everyday" perspectives on the soil erosion dilemma as part of an effort to invigorate political ecology through the analysis of discourse: i.e. as part of what Peet and Watts call liberation ecology in this book. Such projects include the study of conservation-related knowledges and institutions (Zimmerer 1993b,

1994). Focusing on the social situation of erosion-related knowledge expressed in peasants' discourses, the present study offers an example of a broadened political ecology and the impetus for a more open "liberation ecology" (Peet and Watts, Chapter 1 in this volume). Indeed the local knowledge apparent in the everyday discourses of Cochabamba peasants about erosion has rarely been manifest as a strictly self-contained and self-referential dialogue. Their common-place observations suggest strongly that such expressions have influenced, and been influenced by, ideas on erosion of other social groups in the region.

SOIL EROSION DISCOURSES IN COCHABAMBA

In undertaking the present study I relied on written and sometimes published materials and open-ended interviews with thirty-four Cochabamba peasants taped and transcribed with their consent. Our conversations usually combined Quechua, the first language of Cochabamba peasants, and Spanish, a widely used second language.

Development institutions: blaming the peasants

the land users have not developed any awareness about the problems of soil erosion. . . . Overgrazing and trampling by livestock, together with the removal of shrub cover for fuel in the Altiplano and the Mesothermic Valleys, are the most important causes of soil erosion.

(IIDE and USAID 1986)

Government institutions in Bolivia paid little heed to soil erosion and failed to support any sizeable effort at conservation despite the accumulating accounts of a dramatic erosion dilemma. Throughout the 1980s national governments refused to establish a policy or program on soil conservation (IIDE and USAID 1986). When agencies in the Bolivian government did address the erosion problem, they placed the blame squarely on the shoulders of peasant farmers and herders. A 1977 report on "Renewable natural resources" by its Ministry of Peasant Agriculture and Ranching, for instance, claimed that the primary cause of soil erosion could be found in the land-users' failure to employ modern techniques (MACA 1977). Such reports reasoned that the transfer of proper tools and techniques to ill-equipped peasants would stem erosion. Market signals and the articulation of the peasants' economy with agricultural businesses would induce the necessary innovations and transfer of land-use techniques including modern soil conservation (Adams 1980).

But the capacity of Cochabamba's peasant sector to generate market demand for modern technologies fell during the 1980s under an onslaught of sectoral, social, and spatial inequalities enforced by national economic policies. Agribusiness integration, meanwhile, was confined to small segments of the total peasant economy (Weil 1983). Restructuring the national economy since August

1985 in accord with a neo-liberal model imposed by the International Monetary Fund and the World Bank dashed most remaining hopes that market-induced technological change would reduce erosion. To be sure, some agencies in the Bolivian government have recently urged assistance programs to aid in transferring modern farm tools and techniques for conservation goals. In a 1987 "National Meeting on Natural Renewable Resources," governmental institutions, together with major international aid agencies, proposed establishing a national soil conservation program (MACA 1987). Yet such programs faced scant chance of government financing in the aftermath of 1985 restructuring and Decree 21060 that officially abdicated the nation's responsibility for rural development in highland regions (Pérez Crespo 1991).

As the fiscal and administrative capacity of the Bolivian government stagnated, soil conservation became the hallmark mission of a proliferating pool of international aid agencies and NGOs. Beginning in the 1970s, these institutions advocated technical assistance to peasant producers. USAID, for instance, sponsored a number of soil erosion and conservation studies in peasant communities in the guise of larger projects aimed at modernizing Bolivian agriculture (LeBaron et al. 1979; Wennergren and Whittaker 1975). Another international aid organization, the Swiss Technical Corporation, supported a number of soil conservation programs, most notably in the form of small-scale forestry projects. Overall NGOs accounted for the bulk of the new erosion prevention and conservation programs. By the late 1980s, more than 300 NGOs had initiated assistance programs in Bolivia. At least eighty clustered in the Cochabamba region, where many sponsored studies and small projects designed to abate soil erosion.

The statements on erosion causes by international aid agencies and most NGOs coincided in large part with assessments by the government agencies. Most non-Bolivian development institutions concurred with the mainstream belief that peasant ignorance was culpable for the erosion crisis. Consider for example the "Environmental profile of Bolivia" in which NGO and USAID authors allege that "land users were not at all aware of the soil erosion problem" (IIDE and USAID 1986: 99). Other USAID reports went further in concluding that worsened erosion originated in the "cultural backwardness" of rural inhabitants (LeBaron et al. 1979; Wennergren and Whittaker 1975). The director of the Center for Forestry Development (CDF) in Cochabamba held that "men cause soil erosion where they do not know better" (Estrada 1991). Rooting their logic in the perceived bane of cultural backwardness, the modernizers deduced that the techniques and technologies of peasant land use were the chief inducers of erosion (de Morales 1990: 52; Estrada 1991; IIDE and USAID 1986; MACA 1977).

Curiously an indictment of ill-suited techniques and inadequate knowledge similar to this modernizing logic was shared by an anti-modernizing group of mostly NGOs. They held that environmentally damaging land use and the lack of necessary knowledge are consequences of cultural degradation, in effect too

114

much modernity rather than too little (Eckholm 1976; van den Berg 1991). The erosion-inducing peasant farmers of Cochabamba and other Bolivian regions are thought to have fallen from an earlier, near-Edenic state of advanced culture and sophisticated environmental knowledge. Current residents represent, it is said, the culturally deprived descendants of the Inca and Aymara empires that ruled before the onset of Spanish colonialism. Despite the radical difference in interpretation, the anti-modernizing analysts arrived at a conclusion not much different to their mainstream counterparts: cultural inadequacy and lack of knowledge cause the current erosion crisis.

Hence, many soil conservation projects in Cochabamba were designed to address the perceived ignorance of peasant farmers (e.g. IIDE and USAID 1986). Characteristically the projects disseminated educational pamphlets and organized conservation seminars for local farmers. They funded posters, supported demonstration plots, set up farmer-to-farmer forms of knowledge transfer, and arranged for speakers at meetings of peasant communities or unions known as *sindicatos*. Though the project methods were not fundamentally wrong, their analysis was. The projects frequently prescribed new conservation strategies without really assessing present practices and the rationales behind them. Circumstances in peasant farmers' livelihoods shaping their land use were largely ignored. A number of soil conservation projects found themselves promoting measures such as the large-scale construction of contour terraces that were impractical given the reality of peasant resources (Zimmerer 1993b).

The peasants: diverging perspectives

It was not like this before, the hills weren't barren nor were there many erosion gullies. Look, I'm only 27 years old but I've seen it deteriorate bit by bit . . . the soil has lost its productive force, each year it no longer produces as before. Soil from the slopes is being swept downwards – leaving bare rock, subsoil, and gullies – due to the heavy rains . . . the development institutions claim that they know the solutions, but when we look at it, we recognize that we know as a result of our experience, we know how to take care of the earth.

(Interview, Ubaldina Mejía, Aiquile, Cochabamba, 14 Oct. 1991)

Peasant farmers in Cochabamba blamed themselves for soil erosion. But their viewpoints are distinct from the development institutions in two principal ways: (1) most peasant farmers express a sophisticated knowledge of soil erosion, utilizing a complex lexicon from Quechua and Spanish to discuss diverse erosional landforms and their management, while relating erosion to soil types and farming practices (Zimmerer 1994); (2) they invoke the supernatural world of religious beliefs and customs in explaining the causes of erosion (see also Staedel 1989). In both regards, a distinguishing characteristic of everyday peasant perceptions of soil erosion is the sense – at times vivid and quite personal – of prolonged historical time and a close relation to place.

115

Historical illustrations of peasants' environmental discourse are revealed in their judicial depositions lodged in efforts to defend the right to land, water, and forest resources. Records filed during the nineteenth century, currently housed in the Municipal Archive of Cochabamba (*Archivo Municipal de Cochabamba*, AMC) reveal the twin features of prolonged time and a close relation to place. For example, in 1832 Isidoro Ayllita who, as the Cacique of Colcapirgua, was a local indigenous authority in the Cochabamba countryside, defended his right to irrigate with waters of the Collpa River on the basis of traditional and hence long-term historical use: "we have possessed these waters of the Collpa since time immemorial . . . and we have used them continuously . . . since the creation of the world" (AMC 1832). Similar in form to concepts embraced by many present-day environmentalists, Ayllita evoked the rights conferred by sustained use over the long-term past. By referring to various local places, he offered a detailed as well as highly personal and familiar knowledge of resources and landscapes.

A personalized and long-term view of resources continues to infuse the everyday accounts of Cochabamba peasants. In their commonplace discourse on soil erosion most attribute the worsening problem to an increased frequency and intensity of torrential downpours referred to as "crazy rains" or *loco paras*, an amalgam of Spanish (*loco* for "crazy") and Quechua (*para* for "rain"). This may seem to blame nature. But the ultimate responsibility for "crazy rains" is seen as personal. A neglect of ritual obligations toward the main non-Christian deity – the climate-controlling "Earth Mother" – has brought on recent worsening of "crazy rains." As Leocardia González said:

> When I was a child my parents made offerings to the "Earth Mother" [*Pachamama*]. They cooked special foods which they buried in the soil, along with maize beer [*aqha*]. They did all this so that they would be looked on favorably by her. But today these practices aren't common although we still make offerings on Carnival and on Saint John's Day, and when we start to plant. But it's less than before; perhaps for this reason she's angry with us and maybe that's why there are so many "crazy rains" [*loco paras*].
>
> (Leocardia González, Tiraque, Cochabamba, 2 March 1991)

This account attests to how most peasants envision personalized reciprocity as the basis for obligations to the "Earth Mother." Such a customary reciprocity may form the basis of a peasant ethic on environmental conservation according to some indigenist anthropologists in Cochabamba (Rocha 1990; van den Berg 1991). Yet, while soil erosion is often attributed to ritual neglect, the region's peasants do not see this transgression as solely responsible for the divine thought to wreak dire environmental consequences. Many people reason that transgressions in the realm of social reciprocity also incite punishment from the earth deity who orders the heightened onset of "crazy rains" and ensuing soil erosion. Following their style of cause–effect thinking, erosion is born in the breakdown of customary social rights and obligations. In our conversations, numerous persons commented on a social world being undermined by disrespect, animosity, inequality, and violence.

Although the ideal of social reciprocity continues to infuse life and livelihood in the Cochabamba countryside, this belief often veils outright domination and growing differentiation of groups defined by wealth, status, age, and gender (Mallon 1983; Orlove 1974; Weismantel 1988). Among the Cochabamba peasantry, the elderly and the young adults differ in amount of schooling and non-farm work experience. Discourses on soil erosion correspond in an indirect, albeit important, way to this rift in experience. Elderly peasants are most likely to voice the explanations of soil erosion similar to those outlined above. Some elderly peasants find young people in their communities inviting divine wrath saying, for example, that "some people rebel against their parents," "children do not respect us," and, in one conversation, "parents are being killed." Numerous young peasants in Cochabamba, by contrast, cast soil erosion less in terms of the earth deity's wrath and more in terms of human-induced causes. The young people commonly blame their elders, those from whom they are inheriting degraded fields and pasture of obviously diminished value. More schooled, more likely to speak Spanish as well as Quechua, and more experienced in off-farm work, many young peasants admit skepticism about the earth deity, although few deny her existence outright. Conversational accounts of two 23-year-olds, a man and a woman, illustrate the generational shift:

It's true that "crazy rains" have increased, the thunder too is greater than before. They have increased the problem of erosion, but the problem of erosion is due also to the fact that the ground is "naked." It no longer has grass or trees. These were depleted by our parents and the others [elders].
(Casimiro Vargas, Tarata, Cochabamba,
15 June 1991)

Due to erosion the fields that we [young adults] inherit are infertile. Seeing this some of us migrate to the Chapare. Furthermore, there's not much land left, and all of it is pure rock or at least rocky. There aren't good agricultural lands available for inheritance or partitioning. Look up there, for example, it's bedrock, along with some other rock-filled fields. It looks as though the rain or perhaps the wind has removed the soil.
(Ninfa Salazar, Tarata, Cochabamba,
11 Oct. 1991)

During the 1980s, the discontent of young Cochabamba peasants about economic, political, and environmental dilemmas led to their increased involvement in peasant unions or trade unions. An increasingly common perspective on soil erosion formed as several rural trade unions initiated critiques that combined local perspectives with a broader consideration of related national and international issues. By the decade's end, young peasant voices were mingling with, and eventually adding to, prior explanations on environment and development in a revitalized branch of the traditional trade unions.

Rural trade unions

The peasants will no longer tolerate . . . the exploitation of our natural resources by the oligarchy and the imperialists . . .

(Resolutions of the Third Congress, FSUTCC
["The Sole Trade Union Federation of the
Peasant Workers of Cochabamba"], Cochabamba, 1986)

Rural trade unions were first organized in Bolivia after national defeat in the Chaco War with Paraguay (1932–5) and the ensuing weakening of its small governing elite. Through the mid-1980s, the rural trade unions – which belonged initially to the main miner-led Confederation of Bolivian Workers (COB) – did not state official positions on soil erosion or other degradation of rural environments, although they did advance criticisms and social analyses of water pollution in mining centers. The apparent absence of critique persisted even as leadership in the COB shifted in 1977 from mining centers to an axis combining city with countryside under the growing ethnic and social movement known as *katarismo. Katarismo* quickly gained much popularity among Bolivia's indigenous peasants for its commitment to cultural autonomy, social justice, and economic betterment. But during its early years the popular *katarismo* movement did not address the erosion dilemma. Even after 1979, when *katarismo* activists founded the first national trade union for peasants, known as "The Sole Trade Union Confederation for the Peasant Workers of Bolivia" or CSUTCB, the problem of erosion could not be counted among its concerns (Albó 1987; Flores 1984; Healy 1989).

The absence of a discourse on soil loss became more conspicuous and perplexing since the *katarismo* leaders and their rural followers rose to power amid a growing awareness of worsening erosion in the late 1970s. That discursive silence did not result from mere coincidence, nor simple oversight, for at the same time the trade union movement advanced its critique of other environmental problems, especially water pollution and lowland deforestation. In fact, the official resolutions drafted at the national and regional meetings of rural trade unions and other union groups through the mid-1980s repeatedly detailed these environmental dilemmas (Calla *et al.* 1989; COB 1985; CSUTCB 1989; FSUTCC 1986). Yet the notable absence of erosion in this early environmental discourse coincided with the epistemology implicit in the trade union analysis of resource-related problems. Economic and political domination by transnational corporations and imperialist countries could be held accountable for water pollution in the mining centers and large-scale deforestation. Soil erosion, on the other hand, was not obviously extra-local in origin. Instead, for most trade union members, the causes of this problem were contained in local settings, among local people, including themselves.

Prevailing accounts in other social sectors and settings reinforced the otherwise perplexing omission of soil erosion from syndicalist discourse. Neither the vocal discourse of government and non-governmental institutions that blamed peasants' land use nor the personal perspectives that singled out ritual neglect offered much of a model that could be absorbed into a union position. But as erosion in Cochabamba worsened and a clamor began about its effects, the rural trade unions began to articulate their concern for the first time. Their efforts coincided with many young adult peasants gaining leadership posts in the late 1970s and the 1980s. The new trade union discourse on soil erosion stressed land-use practices as the main cause, extending to unfavorable economic policies affecting the peasant farm sector. One leader in the Campero province of Cochabamba said the following:

> the national government maintains a contradictory position ["thinks two times"]; on the one hand, they want us [peasants] to conserve the environment but on the other hand they pressure us to exploit the environment because we keep having to produce more to earn a livelihood.

The success of soil conservation projects, he continued, hinges on favorable policies for peasant farming.

This leader and others in Campero, who became especially involved in discussing soil erosion and conservation, referred to a lengthy historical past and personalized views of the environment in their trade union discourse. With a broad base of popular support and participation, Campero unions resembled other social movements in Latin America (Slater 1985). The Campero *sindicatos* selectively adopted several views and beliefs about resources from earlier generations of peasants, such as the Cacique Isidoro Ayllita mentioned above, which they used to defend their resource rights. Yet the environmental traditions were reinvented under the inspiration of a growing politics of cultural revindication launched by the ethnically charged *katarismo* movement. Reinvention of earlier ideas led one Campero peasant leader to thread environmental deterioration into the much-debated 1992 quincentential of the Spanish invasion:

> Throughout the last 500 years we peasants have been stepped on by the wealthy, the mestizos, and the Spaniards; the trees and animals similarly have been abused and are being extinguished, and thus we share much suffering along with the environment.
>
> (Victor Flores, Aiquile, Cochabamba, 30 March 1991)

SUSTAINABILITY AND THE SEARCH FOR A "MIDDLE GROUND"

Discourses on soil loss in the Bolivian "heartland" of Cochabamba conjure distinct visions not only of the causes of degradation but also of scenarios for conservation and development. In fact, each of the region's three main discourses are interwoven with a particular perspective on conservation-with-development or sustainability. Most aid agencies – national, international, and NGO – used their backward-peasant argument to justify conservation programs emphasizing on-farm technical assistance and training. By contrast, recent analyses by rural trade union leaders combined assessments of local and extra-local conditions. Some union officials in Cochabamba's Campero province, for instance, pressed the NGOs to design intermediate-scale technologies – such as small dams for irrigation that would improve existing land-use patterns – while also urging regional and national leaders to protest against unfavorable farm policies that worsen erosion. Clearly these discourses on soil loss were not simply mirrors of experience. They also constituted differing efforts to shape conservation and development (cf. Giddens 1979).

Notwithstanding contrasting prescriptions for sustainable development, the major groups nonetheless found themselves motivated by a similar concern: the impacts of soil loss. In a general way the groups came close to a broadly defined "middle ground." This concept has recently been applied by ethnohistorians to places and issues where different groups, cultural, social, political, economic, with quite distinct self-interests, were able to negotiate shared understandings and solutions to everyday problems (e.g. Merrell 1989; White 1991). Characterizing the erosion crisis in Cochabamba as a broadly defined "middle ground" resembles other environmental dilemmas where diverse groups pursue linked goals of sustainability. Indeed the idea of a "common future" was claimed by now classic statements on sustainability (Mathews 1989: World Commission on Environment and Development 1987). Political analysis and individual case studies indicate that such assumptions need to be tempered by the realization that environmental dilemmas may divide as much as they unite (Denevan 1973; Hecht and Cockburn 1990; Schmink and Wood 1992; White 1966).

The dilemma of soil erosion in Cochabamba did in fact divide people. There was little evidence of a neatly defined "common future" or "middle ground" at work in the major discourses on soil erosion. Yet, taking a broader view, it is possible to find some semblances of "middle ground." First, as individuals and groups altered standpoints through contending or accommodating other discourses, they created at least partially commensurate realms of meaning. Consider the changing perspective on erosion of many young Cochabamba peasants and the leaders of their unions' social movement. Reminded daily of their degraded landscape, and in many cases frustrated by the abysmal failure of solely technical solutions to soil conservation, they sought sustainability in terms different than their parents and the region's development agencies. Yet even as the young people's discourse opposed others, it nonetheless drew them closer together in debate. Similarly some Cochabamba NGOs recently reacted to peasant discourse by advocating political and economic reform, thus extending beyond the counsel of technical assistance alone (Rist and San Martín 1991). Though their dialogues did not self-consciously seek "middle ground," an awareness and engagement with other positions enacted a preliminary sort of negotiation on the soil erosion issue.

Another semblance of "middle ground," albeit small, was evident in the processes of justifying, or challenging, the government's role in conservation. At first the rural trade unions, development institutions, and peasants reinforced one another by indirectly affirming the Bolivian state's denial of an erosion problem. Beginning in the early 1980s, when this shared account of soil loss first started to fall apart in Cochabamba, the contestors' goal was not just accurate diagnosis but rather challenging and changing the government's policy. For its part, the Bolivian government moved to marginalize peasant and union discourses on erosion, typically by omitting them. Even after national officials acknowledged the erosion problem, they undermined the prospect of a mediated "middle ground" by excluding peasants and their organizations. But when the national government hastened its withdrawal from rural aid under neo-liberal reforms in the mid-1980s, the prospect for "middle ground" processes improved. The Cochabamba peasants and their local unions stood a better chance of negotiating conservation measures with the new development institutions.

The small semblances of a "middle ground" do not of course grant an arena in which sustainability can be easily attained. The activity of the new development agencies, dominated by NGOs, is rife with potential difficulties and possible cross-purposes. The NGOs are not held directly responsible to a citizenry or in some cases even to the Bolivian government. Thus far they offer few models for facilitating the sort of public debate and democratic participation necessary for a consolidation of "middle ground" approaches toward sustainability. Though the role of the Bolivian government in discourses on soil loss diminished in the late 1980s and early 1990s it is likely that national government will continue to shape whether a fairly negotiated "middle ground" can be approached. Without a duly and democratically established arena for debate on sustainability, Cochabamba and other complex, developing societies are unlikely to find a sufficient area of "middle ground," leaving little hope for resolving environmental dilemmas.

NOTE

A Post-Doctoral Fellowship from the Social Science Research Council and a research grant from the National Science Foundation funded the field project. The Graduate School at the University of Wisconsin – Madison supported the analysis and interpretation of fieldwork findings. I am grateful for the collaboration and co-operation of numerous peasant farmers, rural trade unions, government agencies, and NGOs in Cochabamba, Bolivia. I have chosen to use pseudonyms for living farmers in order to maintain their anonymity. Their lack of secure civil and human rights makes this precaution a necessity since even a seemingly innocuous discussion might conceivably jeopardize them at some later day.

REFERENCES

Adams, K. 1980. "Agribusiness integration as an alternative small farm strategy," Report to the Consortium for International Development (CID), Cochabamba.
Albó, X. 1987. "From MNRistas to kataristas to katari," in S.J. Stern (ed.) *Resistance,*

Rebellion, and Consciousness in the Andean Peasant World, Eighteenth to Twentieth Centuries. Madison: University of Wisconsin Press, pp. 379–420.

Archivo Municipal de Cochabamba (AMC). 1832. "Ysidoro Ayllita, por mi y mis compartes," Legajo 5, Provincias. Cochabamba.

Baptista Gumucio, M. 1977. *El país erial: la crisis ecológica boliviana.* La Paz: Los Amigos del Libro.

Blaikie, P. 1985. *The Political Economy of Soil Erosion in Developing Countries.* Essex: Longman.

Blaikie, P. and H. Brookfield. 1987. *Land Degradation and Society.* London: Methuen.

Calla, R., N. Pinelo and M. Urioste. 1989. *CSUTCB: debate sobre documentos políticos y asamblea de nacionalidades.* Talleres CEDLA Number 8. La Paz: EDOBOL.

Central Obrera Boliviana (COB). 1985. *VI Congreso de la COB: protocolos y tésis de la discusión política.* La Paz: HISBOL.

Confederacion Sindical Unica de Trabajadores Campesinos de Bolivia (CSUTCB). 1989. *IV Congreso Nacional Ordinario: Comision de recursos naturales y tenencia de tierras.* Tarija: CSUTCB.

Corporación de Desarrollo de Cochabamba (CORDECO). 1980. *Análisis de problemas y potenciales en el desarrollo regional de Cochabamba.* Cochabamba: CORDECO.

de Janvry, E. Sadoulet and L.W. Young. 1989. "Land and labour in Latin American agriculture from the 1950s to the 1980s," *Journal of Peasant Studies* 16, 3: 396–424.

de Morales, C.B. 1990. *Bolivia: medio ambiente y ecología aplicada.* La Paz: Universidad Nacional Mayor de San Andrés.

Denevan, W.M. 1973. "Development and the imminent demise of the Amazon rain forest," *Professional Geographer* 25, 2: 130–5.

Eagleton, T. 1983. *Literary Theory.* Minneapolis: University of Minnesota Press.

Eckholm, E. 1976. *Losing Ground: Environmental Stress and World Food Prospects.* New York: Norton.

Emel, J. and R. Peet. 1989. "Resource management and natural hazards," in R. Peet and N. Thrift (eds.) *New Models in Geography.* London: Unwin Hyman, pp. 49–76.

Estrada, V.J.G. 1991. "El fenómeno erosivo en Bolivia," *Pro Campo* (Cochabamba) 22: 29–30.

Federación Sindical Unica de Trabajadores Campesinos de Cochabamba (FSUTCC). 1986. *Resoluciones del III Congreso Campesino, 17–19 de julio 1986.* Cochabamba: FSUTCC.

Flores, G. 1984. "Estado, políticas agrarias y luchas campesinas: revision de una década en Bolivia," in F. Calderón and J. Dandler (eds.) *Bolivia: la fuerza histórica del campesinado.* Cochabamba: CERES, pp. 445–545.

Giddens, A. 1979. *Central Problems in Social Theory: Action, Structure, and Contradiction in Social Analysis.* Berkeley: University of California Press.

Grover, B. 1974. *Erosion and Bolivia's Future.* Utah State University Series 31/74, USAID Contract 511–56T. USAID, La Paz.

Healy, K. 1989. *Sindicatos campesinos y desarrollo rural, 1978–1985.* La Paz: HISBOL.

Hecht, S. and A. Cockburn. 1990. *The Fate of the Forest: Developers, Destroyers, and Defenders of the Amazon.* New York: Harper Collins.

IIDE and USAID. 1986. *Perfil ambiental de Bolivia.* La Paz: USAID.

Larson, B. 1988. *Colonialism and Agrarian Transformation in Bolivia: Cochabamba, 1550–1900.* Princeton, NJ: Princeton University Press.

LeBaron, A., L.K. Bond, P.S. Aitken, and L. Michaelsen. 1979. "Explanation of the Bolivian highlands grazing-erosion syndrome," *Journal of Range Management* 32, 3: 201–8.

Los Tiempos (Cochabamba). 1991. "Más del 35% del territorio nacional está sufriendo un proceso de erosión," 12 May 1991.

Mallon, F.E. 1983. *The Defense of Community in Peru's Central Highlands: Peasant Struggle and Capitalist Transition, 1860–1940*. Princeton, NJ: Princeton University Press.

Mathews, J.T. 1989. "Redefining security," *Foreign Affairs* 68, 2: 162–77.

Merrell, J. 1989. *The Indians' New World*. Chapel Hill: University of North Carolina Press.

Ministerio de Asuntos Campesinos y Agropecuarios (MACA). 1977. "Recursos naturales renovables," in M. Baptista Gamucio (ed.) *El país erial: la crisis ecológica boliviana*. La Paz: Los Amigos del Libro, pp. 43–56.

—— 1987. *National Meeting on Natural Renewable Resources*. Tarija: MACA.

Montes de Oca, I. 1989. *Geografía y recursos naturales de Bolivia*. La Paz: Editorial Educacional.

Orlove, B.S. 1974. "Reciprocidad, desigualdad, y dominación," in G. Alberti and E. Mayer (eds.) *Reciprocidad e intercambio en los Andes peruanos*. Lima: Instituto de Estudios Peruanos, pp. 290–321.

—— 1991. "Mapping reeds and reading maps: the politics of representation in Lake Titicaca," *American Ethnologist* 18: 3–38.

Pérez Crespo, C. 1991. "Why do people migrate? Internal migration and the pattern of capital accumulation in Bolivia," IDA Working Paper No. 74. Binghamton, NY: IDA.

Presencia (La Paz). 1990. "El problema de la erosion de suelos," 16 May.

Preston, D.A. 1969. "The revolutionary landscape of highland Bolivia," *Geographical Journal* 135, 1: 1–16.

Rist, S. and J. San Martín. 1991. *Agroecología y saber campesino en la conservación de suelos*. Cochabamba: Runa.

Rocha, J.A. 1990. *Sociedad agraria y religion: cambio social e identidad en los valles de Cochabamba*. La Paz: HISBOL.

Schmink, M. and C.H. Wood. 1992. *Contested Frontiers in Amazonia*. New York: Columbia University.

Slater, D. 1985. "Social movements and a recasting of the political," in D. Slater (ed.) *New Social Movements and the State in Latin America*. Dordrecht: CEDLA, pp. 1–26.

Staedel, C. 1989. "The perception of stress by campesinos: a profile from the Ecuadorian sierra," *Mountain Research and Development* 6, 1: 35–49.

Storper, M. 1991. *Industrialization, Economic Development, and the Regional Question in the Third World: From Import Substitution to Flexible Production*. London: Pion.

Terrazas Urquidi, W. 1974. *Bolivia: país saqueado*. La Paz: Ediciones Camarlinghi.

Universidad Mayor de San Simón (UMSS). 1963. *Mesa redonda sobre desarrollo agroeconómico del valle de Cochabamba, 19–21 Marzo 1963*. Cochabamba: Imprenta Universitaria.

Urioste Fernández de Córdova, M. 1984. *El estado anticampesino*. Cochabamba: El Buitre.

USDA Mission to Bolivia (USDA). 1962. *Bolivian Agriculture: Its Problems, Programs, Priorities, and Possibilities*. Washington, DC: USDA.

van den Berg, H. 1991. "Conviven con la tierra," *Cuarto Intermedio* (Cochabamba) 18: 64–83.

Watts, M. 1983. "On the poverty of theory: natural hazards research in context," in K. Hewitt (ed.) *Interpretations of Calamity: From the Viewpoint of Human Ecology*. Boston: Allen & Unwin, pp. 231–62.

—— 1985. "Social theory and environmental degradation," in Y. Gradis (ed.) *Desert Development: Man and Technology in Sparse Lands*. Dordrecht: Reidel, pp. 14–23.

Weil, C. 1983. "Migration among landholdings by Bolivian campesinos," *Geographical Review* 73: 182–97.

Weismantel, M.J. 1988. *Food, Gender, and Poverty in the Ecuadorian Andes.* Philadelphia: University of Pennsylvania Press.

Wennergren, E.B. and M.D. Whittaker. 1975. *The Status of Bolivian Agriculture.* New York: Praeger

White, G.F. 1966. "Formation and role of public attitudes," in G.F. White (ed.) *Environmental Quality in a Growing Economy: Essays from the Sixth RFF Forum.* Baltimore: Johns Hopkins University Press, pp. 105–27.

White, R. 1991. *The Middle Ground: Indians, Empires, and Republics in the Great Lakes Region, 1650–1815.* Cambridge: Cambridge University Press.

World Commission on Environment and Development. 1987. *Our Common Future.* Oxford: Oxford University Press.

Zimmerer, K.S. 1991. *Informe diagnóstico: uso de la tierra y la erosion de suelos en la cuenca del Río Calicanto.* Report, Cochabamba: Centro para la Investigación de Desarrollo Rural (CIDRE).

—— 1993a. "Soil erosion and social discourses: perceiving the nature of environmental degradation," *Economic Geography* 69, 3: 312–27.

—— 1993b. "Soil erosion and labor shortages in the Andes with special reference to Bolivia, 1953–1991: implications for conservation-with-development," *World Development* 21, 10: 1659–75.

—— 1994. "Local soil knowledge and development: answers to some basic questions from highland Bolivia," *Journal of Soil and Water Conservation* 49, 1: 29–34.

6

MARXISM, CULTURE, AND POLITICAL ECOLOGY
Environmental struggles in Zimbabwe's Eastern Highlands

Donald S. Moore

INTRODUCTION

Concern with the politics of environmental resource conflicts in the Third World has grown steadily over the past decade. Early studies emphasized the changing dialectic between society and environment, underscoring the interrelation of political economy and resource management (Blaikie and Brookfield 1987; Dove 1985; Schmink and Wood 1987). In its broadest conceptual reaches "political ecology" has addressed issues of resource access and control, the political processes influencing land rights, and the ethics of technology and development (Hecht and Cockburn 1992). While the term "political ecology" lacks a coherent theoretical core, its Marxian versions direct attention to history, social relations of production, and the embeddedness of local land-use practices in regional and global political economies (see Bassett 1988; Bryant 1992; Neumann 1992; Sheridan 1988; Zimmerer 1991). Recent regional and local studies situate agrarian and pastoral struggles within historical patterns of access to critical environmental resources across Africa (Bassett and Crummey 1993; Fairhead and Leach 1994; Little 1992; Peters 1994), Latin America (Faber 1993; Hecht and Cockburn 1990; Painter and Durham 1995; Schmink and Wood 1992; Stonich 1993), and South Asia (Agarwal 1994a; Gadgil and Guha 1992; Guha 1989; Peluso 1992).

Political ecology came from a realization that ecological processes could not be understood outside the contexts of local productive relations and wider economic systems. Influential critiques of neo-Malthusian explanations of environmental degradation, food insecurity and famine, and land conflicts looked to political and social factors, not the "natural" mechanism of population pressure, to explain resource struggles (Durham 1979; Harvey 1977; Richards 1983; Watts 1983a). Growing dissatisfaction with cultural ecology's metaphors of "adaptation" and "homeostasis," and their resulting neglect of power and history, encouraged greater attention to the social relations of production influencing land use (Durham 1995; Watts 1983b; Zimmerer 1994). Marxian political economy

125

offered a vision of causal explanations for ecological transformations that stressed the social and historical factors shaping relations between land users and their environments. Recently, tenets of a "new ecology" that stress complexity and heterogeneity rather than homeostatic systems have focused greater attention on how socioeconomic relations are part and parcel of environmental history (Demeritt 1994b; Fairhead and Leach forthcoming; Zimmerer 1994).

Much political ecology remains within a macro-structural framework that emphasizes the determining influence of broad economic forces. Global capitalism, from this perspective, not only shapes but also exactly determines heterogeneous local histories, cultures, and societies. Cultural practices and beliefs usually enter this discussion derivatively, as exotic trappings to the nuts and bolts of "underlying" structures. Although it would be wrong to minimize the force of political economy, too much emphasis on structural determination elides other factors shaping conflicts over Third World environmental resources: (1) the micro-politics of peasant struggles over access to productive resources; and (2) the symbolic contestations that *constitute* those struggles. Similarly, the neglect of local politics may produce a misleadingly monolithic model of the "state," conceiving it as an actor with a unified intentionality, internally consistent in its agenda, structurally and automatically opposed to local interests. Macro-structural accounts miss also local differentiation among resource users, particularly those mediated by class, gender, ethnicity, and age. In Peet and Watts's chapter's terms, "*political* ecology has very little politics." By fusing together an understanding of the mutual constitution of micro-politics, symbolic practices, and structural forces, it may be possible to unravel how competing claims to resources are articulated through cultural idioms in the charged contests of local politics. It is this new tendency that Peet and Watts refer to as a move to liberation ecology.

In this chapter, I examine conflicts over access to environmental resources and land in the Kaerezi area of Nyanga District, eastern Zimbabwe. I employ a Gramscian metaphor for environmental resources – viewing Nyanga's landscape as materially and symbolically contested terrain – to analyze conflicts among a differentiated peasantry and a state made up of multiple actors sometimes with disparate interests. While recognizing plural axes of differentiation among both state and peasantry, I focus an ethnographic perspective on: (1) state agencies whose representatives sought to influence land use in Kaerezi and (2) gendered relationships to the local landscape fundamentally shaped by male wage labor migration in the regional political economy. As Peet and Watts stress, "cultural politics are rarely visible in conventional analyses" of Third World social movements. Ethnography provides a critical medium for exploring the dynamics of cultural politics which animate environmental conflicts. My analysis thus tries to avoid tired orthodoxies by placing Marxism, political ecology, and cultural theory in a more productive conversation, echoing recent attempts to incorporate "place, space, and environment into cultural and social theory" (Harvey 1995: 96).

126

A GRAMSCIAN PERSPECTIVE ON ENVIRONMENTAL RESOURCE STRUGGLES

Gramsci's work has been central to critiques of "economism" in Marxist theory, that is the assumption of a "single, productivist logic" privileging economy and class relations (Peet and Watts, Chapter 1 in this volume). Gramsci's fragmentary writings, many penned in an Italian prison, offer an "analytical palimpsest" (Bove 1994: x) from which multiple meanings may be drawn. In particular, Gramsci drew attention to the dynamic interplay of culture, power, and history. Efforts to theorize a less economistic "post-Marxism" (Chakrabarty 1993; Laclau and Mouffe 1985, 1987; McRobbie 1994) bear his imprint. Interpretations of Gramsci have recently explored how gender and cultural identity, as well as productive relations and class, shape material interests (Hall 1986; Holub 1992; Landy 1994).

I am especially interested in Gramsci's thoughts on cultural forms and their efficacy within social relations. Concerned with the "cultural 'deployment' of Marxism," Gramsci (1983 [1957]: 85) argued that the working class must "think about organizing itself culturally" just as "it has thought to organize itself politically and economically" (1991 [1920]: 42). For Gramsci, "'popular beliefs' are ... themselves material forces" (1971: 165) and "ideologies ... organize human masses, and create the terrain on which men [*sic*] move, acquire consciousness of their position, struggle, etc." (1971: 377). He was, of course, building on Marx's (1986 [1859]: 187) famous passage: "a distinction should always be made between the material transformation of the economic conditions of production ... and the legal, political, religious, aesthetic, or philosophic – in short, ideological forms in which men [*sic*] become conscious of this conflict and fight it out."

By emphasizing the mutual interdependence of culture and politics, Gramsci underscored how symbolic struggles effect material transformation. Values and beliefs mobilize action, shape social identities, and condition understandings of collective interests. In this sense, cultural meanings are *constitutive* forces, that is shapers of history, and are not simply reflections of a material base (Donham 1990; Williams 1991). Ideologies contribute to the formation of productive relations and do not derive, mechanically, from them. Struggles over symbolic processes are conflicts over material relations of production, the distribution of resources, and ultimately power. Gramsci's notion of hegemony – that is, the process through which dominant representations color, yet never determine, practical consciousness and everyday lived experiences – reminds us that dominant meanings are continually contested, never totalizing, and are always unstable (Williams 1977). This insight warns against easy divisions between economy and culture, the material and the symbolic, or structure and agency in social analysis.

In turning to Gramsci for this chapter, I borrow from work which shows how struggles over land and environmental resources are simultaneously struggles over cultural meanings. Berry (1988, 1993), Peters (1984, 1994), Carney and Watts (1990, 1991), Alonso (1992), Agarwal (1994a, 1994b), and Nugent and Alonso (1994) explore connections between meanings and control over resources in diverse Third World agrarian contexts. A key insight is that "people may invest in meanings as well as in the means of production – and struggles over meaning are as much a part of the process of resource allocation as are struggles over surplus or the labor process" (Berry 1988: 66). Historical patterns of access to resources and exclusion from them mold cultural understandings of rights, property relations, and entitlements; in turn, these competing meanings influence people's land and resource use (see Goheen and Shipton 1992; Shipton 1994; Watts 1991). Struggles over cultural idioms and access to resources reveal salient gendered differences along with those of class (Agarwal 1992, 1994a, 1994b; Carney 1993; Leach 1994; Mackenzie 1993, 1995; Moore and Vaughan 1994).

These theoretical issues inform an investigation of particular resource conflicts in Kaerezi: the siting and delayed operation of a community cattle dip near a state-defined "protected area"; the demarcation of a national park boundary; and women's cultivation of prohibited crops on mountain slopes. In each case, I emphasize the importance of attending simultaneously to the material and the meaningful in these struggles. My focus on the situated practices of women and men offers a much needed grounding to recent efforts to blend cultural studies and environmental politics, which I engage after an analysis of conflicts in this particular Third World locale. Cultural politics offers a key site to link broader theoretical concerns with the specific contexts and cadences unfolding across a terrain of memories, peasant livelihoods, and state interventions. Rather than relying on these small cases to bear the weight of generalizing theoretical pronouncements, I hope to use them to provoke critical reflection on the conceptual categories and theoretical potential of an emerging environmental cultural studies.

To understand how resource conflicts play themselves out in post-colonial Kaerezi, I first turn to a long history of competing claims to landscape in this Zimbabwean river valley of lush, rolling hills. A historical perspective reveals myriad struggles over the cultural categories through which access to critical environmental resources are contested. Historical transformations in social relations of production and state policy during the colonial period shaped the post-independence terrain of Kaerezi resource struggles. An ethnographic approach helps us understand how peasant social memories of colonial experiences also animated environmental politics since Independence. Differing cultural understandings of the meanings of land were central to these resource conflicts.

Peasants and bureaucrats also assigned competing meanings to the cultural forms of state institutions and practices (cf. Abrams 1988; Corrigan and Sayer 1985). The Kaerezi case demonstrates how the post-independence state, inheriting colonial legacies, is itself an internally differentiated entity with branches often at cross-purposes. Finally, turning from peasant–state conflicts to the local salience of gendered patterns of resource use, differing women's and men's relationship to Kaerezi's landscape cautions against positing homogeneous communities in which cultural idioms for environmental resources are univer- sally shared. In contrast to macro-structural approaches, my analysis of the colonial political economy, state policies and practices, and gender relations in Kaerezi seeks to understand what Hall (1990: 225) aptly terms the "continuous 'play' of history, culture and power."

RHODES'S LEGACY IN NYANGA: ALIENATING LAND AND LABOR

In 1890, a "pioneer column" of white settlers from South Africa arrived in present-day Zimbabwe, forming a new colony in the name of Cecil Rhodes's British South Africa Company (BSAC). Superior firepower put down violent African resistance to colonial rule in 1896–7. By the turn of the century, Rhodes was among the first group of whites to acquire legal rights to land in what is now Nyanga District, in the Eastern Highlands bordering Mozambique. He purchased a little under 40,000 hectares in the district, now constituting the core of Nyanga National Park's 47,000 hectare estate (Figure 6.1).

In 1902, the year of Rhodes's death, a Barwe chief fled political violence in Portuguese East Africa and settled at the site of the current Kaerezi Resettlement Scheme, just north of Rhodes's estate. Chief Dzeka Tangwena claimed his ancestral territory included Kaerezi. He joined lineages that had inhabited the river valley and its surrounding hills for generations. Shortly after his arrival, a Johannesburg-based company purchased the block of land encompassing Kaerezi from the BSAC, fitting the familiar pattern of "paper alienation" in colonial Rhodesia, where land expropriation represented less a physical European presence – since the white "owners" did not occupy the land – and more of a requirement for Africans living on alienated land to enter the labor market to procure cash for taxes and rents (see Palmer 1977: 227).

On the land alienated by these means in Nyanga District, most adult African men either worked under labor tenancy agreements with white landowners, or as migrant laborers in the colony's urban centers and in the Union of South Africa. Europeans seized much more land than they had the capacity to farm or pasture (Roth and Bruce 1994; Weiner 1991). Today, many Zimbabweans remember these seizures through the cultural idiom of "greed" (*mbayo*). In a mixture of Shona and English, one Kaerezi Resettlement Scheme farmer recalled a European commercial farmer who controlled almost 100,000 hectares in the district, including Gaeresi Ranch, core of the present-day Kaerezi Resettlement Scheme. With a sweeping gesture of his hand to sketch the vast expanse, he proclaimed: "*Nzvimbo yese, yese* [This whole place, all of it]. It's too much fucking land for one man."

Rhodes's estate was converted into a National Park in 1947 in the wake of growing government concern over natural resources and conservation (Ranger

129

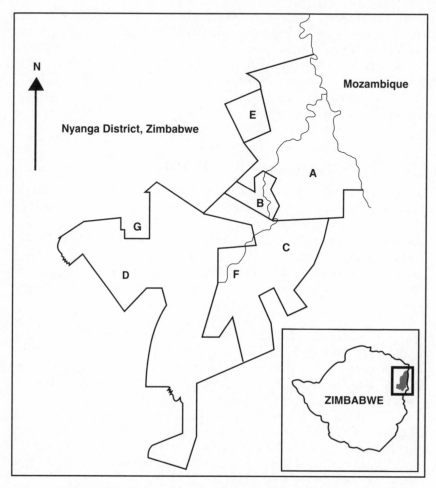

Figure 6.1 Kaerezi Resettlement Scheme and Nyanga National Park, Nyanga District, Zimbabwe. (A) Kaerezi Resettlement Scheme (including Kaerezi extension); (B) proposed Kaerezi River Protected Area; (C) Eastern Extension of Nyanga National Park, comprising properties annexed to the park estate in 1987; (D) Nyanga National Park Estate prior to extension in 1987; (E) Tsanga Valley Wheat Estate; (F) Kaerezi River; (G) Nyanga District Administrative Center.

1989). Like many white farms in the district, the estate had faced the problem of men migrating to distant urban centers to meet growing cash demands, often sending taxes back through the mail. In 1963, the government evicted families from Rhodes Inyanga National Park enforcing a strict policy of racial segregation by denying Africans the right to reside in areas designated "European." In the same year, a state land inspector cited "excessive damage to natural resources" on the commercial Gaeresi Ranch, allegedly caused by tenants and "squatters" living on the property. When residents refused to renew a "labor agreement"

[handwritten margin note: racial segrega]

130

stipulating conditions of their tenancy, government enforcement of the Land Apportionment Act led 'to forced evictions of followers of Chief Rekayi Tangwena, who had been living on the ranch without tenancy contracts.

Between 1967 and 1972, the chief and his followers fought evictions from Gaeresi Ranch in highly publicized court cases. They also protested on the land, arguing that their ancestral rights pre-empted the alienation of commercial land in the area, suffering arrest and brutal beatings at the hands of the Rhodesian authorities. Rekayi, Dzeka's son, emphasized the pre-colonial chieftainship that spanned the international border: "We have lived on this land for many generations, since long before the whiteman came . . . it is where we believe we have a right to live. . . . I inherited it from my forefathers who were found there by Rhodes and the early Europeans" (quoted in Clutton-Brock 1969: 3). Arguing that his ancestral rights to land were inalienable, Rekayi became a national symbol of defiance to the Rhodesian state during a liberation war (1966–79) in which African grievances against settler colonialism focused on land expropriation (Moyana 1984; Ranger 1985).

The government and the white landowners, however, had a different understanding of Tangwena's claims to land on Gaeresi Ranch. In 1972, after years of thwarted attempts to oust Tangwena and his followers, Rhodesian police and security forces burned huts and crops, seized cattle, and set police dogs on unarmed civilians. Some 300 families were dispersed, a great number seeking refuge in the forests bordering Kaerezi, now located within Nyanga National Park's boundary. Some hid there for almost two years. The majority eventually migrated to Mozambique, remaining there until Zimbabwean Independence in 1980. During this forced migration, the Tangwena people aided current Zimbabwean President Robert Mugabe's escape from Rhodesian security forces into Mozambique, ensuring Chief Rekayi Tangwena's place in nationalist legend and accumulating future political capital for Tangwena's followers.

INDEPENDENCE, RESETTLEMENT, AND AN ENCROACHING NATIONAL PARK

Kaerezi Resettlement Scheme began in 1980, after the liberation war, as part of the post-colonial state's National Resettlement Program. It is now home to some 1,000 peasant families (5,500 official residents) spread across 18,500 hectares of rugged terrain in Zimbabwe's Eastern Highlands. All land in Kaerezi is officially designated state property and administered by a government official, the scheme's resettlement officer. Household heads – married men, widows, and divorcees – are given permits to occupy a residential stand, cultivate a designated 3.5 hectare arable field, and pasture livestock on a grazing commons. The functional and spatial divisions of land uses (arable, grazing, residential) follow closely many of the principles of colonial land-use planning actively resisted during the liberation struggle (Drinkwater 1991; Ranger 1985).

Colonial racial land policies shaped a dual agricultural economy in

Zimbabwe. Large-scale European commercial farmers benefited from protective legislation, while small-scale African peasants produced on the poor soils of over-crowded labor reserves (Moyo 1987). At independence in 1980, the Zimbabwe government's National Resettlement Program sought to alleviate this "inherited uneven qualitative and quantitative distribution of agricultural land" (DERUDE 1991: 1). The program entailed purchasing land from the white commercial farming sector and redistributing it to the rural poor. In 1980, the government proposed resettling 18,000 families nationally on 1.1 million hectares of land over a three-year period; it revised its target to 162,000 families in 1982; as of July 1989 only 54,000 families had been allocated resettlement land (Roth and Bruce 1994: 21). Government plans conceive of an eventual resettlement sector of 8.3 million hectares, roughly 21 percent of Zimbabwe's total land area (DERUDE 1991).

Kaerezi differs significantly from other national schemes in three important respects. First, many of its current residents lived on the former commercial farm prior to the liberation war (1966–79), claiming ancestral rights, while nationally most resettled African farmers move into schemes where they had no prior ties to the land. Some Kaerezi residents view the scheme as a "reward" for the late Chief Rekayi Tangwena, a state senator buried as a national hero in 1984, and his followers' contributions to the liberation struggle. Second, unlike most schemes, Kaerezi is in Zimbabwe's most-preferred "agro-ecological zone," a high-rainfall belt running along the mountainous Eastern Highlands. Its location has provoked debate on the most viable land uses in an area perceived to have high agricultural and tourist potential. Third, Kaerezi shares a border with Nyanga National Park, a major tourist attraction whose clear headwaters spill out of Mount Inyangani (2,592 meters), Zimbabwe's highest peak, and flow northward through the scheme via the Kaerezi River.

COMPLEXITIES OF THE POST-COLONIAL STATE

Such resettlement schemes tested the post-Independence state's administrative capacity and political will during the 1980s. The Ministry of Lands identified and purchased commercial farms or properties for resettlement; the planning branch of Agricultural and Technical Extension (AGRITEX) in the Ministry of Agri-culture developed a land-use plan; the Department of Rural Development (DERUDE), in the Ministry of Local Government, Rural and Urban Develop-ment, implemented the plan. Plans were supposed to be approved by an inter-ministerial committee, described by many as a battleground for opposing ideas. A senior official involved with the Inter-ministerial Committee on Resettlement described it as rife with "conflicts over invading other ministries' territory."

In Kaerezi, post-colonial partitions of land were layered on a century of conflicts, revealing a state highly differentiated by ministries and departments often pursuing cross-cutting agendas in their claims to administer the local landscape and its inhabitants. In the 1980s, the Department of National Parks and Wildlife Management sought to protect an "ecologically sensitive watershed" on its border, seen as a potential tourist attraction capable of generating foreign currency. Meanwhile, the Ministry of Lands attempted to acquire land from commercial farm owners to extend the National Park estate and for resettlement, even as DERUDE tried to administer a resettlement scheme.

While underscoring the "difference in perceptions of land users and their government[s]" (Blaikie 1989: 21), the Nyanga case suggests the need to move beyond an opposition between what Bernstein (1990: 69) only half-facetiously calls "virtuous peasants and vicious states." If the state is internally differentiated, itself a site of struggle, then "the state is no longer to be taken as essentially an actor, with the coherence, agency, and subjectivity that term presumes" (Mitchell 1991: 90). Rather, the "state can be opened as a theatre in which resources, property rights, and authority are struggled over" (Watts 1989: 4). An ethnographic approach situates the interventions of state officials in Kaerezi, revealing the "concrete procedures by which social actors simultaneously borrow from a range of discursive genres, intermix them and, as a result are able to invent original cultures of the State" (Bayart 1993: 249). By focusing on the actions of particular bureaucrats and peasants on the ground in Kaerezi, I want to show how "state power *creates*, through its administrative and bureaucratic practices, a world of meanings" (Mbembe 1992: 2). Yet these meanings, far from stable and secure, are contested by local peasants through cultural idioms forged within their historical relationships to the local landscape. Hence the "state" can best be described as "those aspects of the governing, administrative, and coercive apparatus that are experienced as external yet hegemonic" (Tsing 1993: 26).

The cross-purposes of the post-colonial state manifested themselves in conflicts over a cattle dip, a critical community resource protecting herds from tick-borne disease, constructed near the river by the administrative branch of the ministry overseeing resettlement (DERUDE) in 1988. State law required stock owners to dip cattle weekly during rainy season and bi-weekly during dry season. For those living near the disputed dip, it represented a significant decrease in the regular labor required to herd their cattle to distant dips. But when National Parks discovered that the dip was built within a proposed "Protected Area" – a corridor of approximately 500 meters bordering the Kaerezi River in which grazing and cultivation would be prohibited – it strongly objected, lobbying against its operation. A private white trout-fishing club complicated the situation by writing letters to various ministries inquiring about the dip's fate. While the entire proposed protected area fell within the resettlement scheme, it bordered on the park, and neither peasants nor administrators were clear on who, exactly, had state-sanctioned authority over its utilization. In October 1990, the government officer administering the scheme met with Kaerezi resettlement farmers to explain why the cattle dip was not yet functional: "The National Park wants to use the river. We together with our Member of Parliament want to use the river. Peter and other whites [members of the trout-fishing group] want to use the river. These are the factors stopping the diptank from working."

The local headman was quick to pick up on the white fishing club's claim, using a racial idiom to demand state action in defense of local residents' rights. Sitting in the tall grass amidst a circle of farmers, he pointed a bony finger at the resettlement officer: "Then *you* are the one who must go and fight these people. Why are invaders (*vatorwa*) coming into an area bought for people to settle on? You must want the whites to take the river again." When the resettlement officer countered that it was not the white club that controlled the land, but the government, the headman tactically concluded: "So *you* want to kill the cattle? It's better to kill me and leave the cattle." The state official then produced a letter voicing concern over pollutants from the cattle dip seeping into the river: "This shows that the National Park was trying to take over the river, since the dip has already been approved by Veterinary Services," a department within yet another ministry, that of agriculture. The headman pushed this logic: "Then the National Park is making a mistake." Suddenly defending the National Park, the resettlement officer shot back: "The National Park is a branch of the government." This only opened up the space for the headman's closing salvo: "Yes, but it's a bad branch."

In this pattern of "challenge and riposte" (Bourdieu 1977: 12), the administrator unraveled and reconstituted complex alliances and conflicts among different state agencies, sometimes represented as a unified "we," other times depicted as "us against them"; the headman watched competing claims *within* the Zimbabwean state unfold in his area, preventing him from dipping cattle near his home. The headman responded by turning the old colonial "divide and rule" strategy, this time as tactic, back on the state officials. I stress *tactic* precisely because peasants in Kaerezi, in a style of opposition described by de Certeau (1984: xix), have had to "constantly manipulate events in order to turn them into opportunities'" and "turn to their own ends forces alien to them" from a subordinate position. By insinuating a non-state, white actor into the theatre of struggle, the headman drew on symbolic capital garnered from a legacy of racial exclusion. The failure of post-colonial state officials to deliver the cattle dip, an important community resource, became their moral failure to redress a history of racial inequality. Without over-romanticizing their efficacy, these tactics pushed the resettlement officer to promise publicly to lobby on the peasants' behalf. He dispatched a letter to the district administrator the morning after the meeting, and a permit slowly emerged from the bureaucratic machinery. In mid-1991, after nearly three years lying idle, the diptank opened.

LANDSCAPES OF MEMORY, CULTURE, AND POWER

Since its establishment in 1988, the National Park's expanded border has caused deep local resentment. The precise location of the boundary was in dispute when families thought to be living in the park were evicted in 1991, precipitating a meeting attended by the district administrator, the senior state official in Nyanga, and the local Member of Parliament (MP). Chief Tangwena, Rekayi's successor, drew applause from his constituents when he invoked the memory of the 1970s evictions from Gaeresi Ranch: "The National Park wants to burn huts in my area. We thought the whites had returned." The MP responded by stressing that a "farm and a chieftainship are not the same thing." Today, he continued, the government recognizes property boundaries as they are written in title deeds and demarcated with beacons placed by the office of the surveyor general, not as they are remembered in oral traditions.

The chief invoked his ancestral right to "rule the land," describing his "traditional" boundaries by reciting prominent features of the landscape. The MP chastised chief and followers alike: "You don't know your boundary. Without boundaries there would be only war." Invoking the colonial separation of cordoned spaces sanctioned by property deeds, the MP scolded his constituents for not having a map revealing the "true boundary." Maps, as many have noted, have been critical to state control of subject populations (Bassett 1994; Harley 1988; Kain and Baigent 1992). The particular cultural practice of dutifully recording property boundaries on a map underpinned the MP's belief in Kaerezi's status as a commodity owned and administered by its present owner, the state. The chief invoked an alternate cultural vision of place, property, and territory by claiming ancestral inheritance to a chieftainship. Particular cultural practices – propitiating ancestral spirits, recognizing sacred features of the land-scape, and enforcing "respect for the land" (*kuremekedza nyika*) – constructed Kaerezi as an inalienable birthright. The chief, who received a government salary, saw no contradiction between his opposition to state administration of his "inheritance" (*nhaka*) and his being a civil servant.

Appadurai (1990: 7) has stressed that metaphors of landscape portray "not objectively given relations which look the same from every angle of vision, but rather . . . deeply perspectival constructs, inflected very much by the historical, linguistic and political situatedness of different sorts of actors" (see also Demeritt 1994a; Duncan 1994; Duncan and Ley 1993). Kaerezi peasants, as well as their chief, see the surrounding landscape as saturated with power, meanings, and historical struggles for land rights, a "'fractal' world of overlapping boundaries" (Guyer 1994: 215). Most Kaerezi farmers use the term *zvisikwa*, literally "things created," to refer to features of the local landscape. *Zvisikwa* do not posit a "natural" essence outside of history, but rather suggest symbolic and material interactions among spiritual forces, humans, and a particular landscape. Moreover, *zvisikwa* are understood through a range of cosmologies and religious idioms: Christianity, ancestral spirits of lineage and family (*vadzimu*), and guardian spirits for a particular territory (*mhondoro*). *Zvisikwa* are imbued with use-values and take on meanings through people's daily livelihood struggles: the collection of firewood, pasturing of cattle, drawing water from springs and streams, hunting and fishing. As one elder told me: "People must use *zvisikwa*. . . . When we see an animal, we say we have seen meat and when we see trees we now have firewood."

For Kaerezi peasants, social memories as well as daily and seasonal resource uses animate features of the local landscape. None of these cultural idioms for land and resources, however, are beyond the reach of historical incorporation into a southern African regional political economy and state intervention in rural settlement patterns, agricultural practices, and prohibitions on the utilization of the environment. Peasants have cultural categories for commercial farms and the townships near Nyanga's administrative center – as well as for myriad demarcations of the landscape introduced since the advent of colonial rule. These historical sedimentations underscore what Massey (1994: 156) calls the "accumulated history of a place, with that history itself imagined as the product of layer upon layer of different sets of linkages, both local and to the wider world" (see also Crush 1994). As competing uses and memories of place vie for features of the landscape, claimants contest access to critical environmental resources.

While many Kaerezi peasants assert rights of access to *zvisikwa* through patterns of resource use recognized as legitimate by state policy, others defy state law prohibiting all hunting of wildlife. A man evicted from Gaeresi Ranch in the 1970s asserted hunting rights by invoking what Marx (1975: 232, 234) called, in the context of resource conflicts in nineteenth-century England, "the customary rights of the poor" to harvest the "alms of nature." Poetically, the Kaerezi man mused: "*Mhuka dze sango inhaka ye povo*" (The wildlife in the forest is the poor's inheritance). Another hunter produced a lively mixture to defend his customary access to park resources:

> The Bible says cattle is meat, goats are meat, and a buck is meat. That's why people should eat all different kinds of animals, except those prohibited in the Bible like pigs. Meat, I want it, but the National Park doesn't allow us to hunt. . . . But it's *our* wildlife, the animals are *ours*.

ENGENDERING ENVIRONMENTAL RESOURCE STRUGGLES

Local idioms dealing with claims to environmental resources – *nhaka* as just inheritance, *zvisikwa* as resources to be utilized – reflect what Bourdieu (1977) calls "habitus," a set of historically developed dispositions and cultural habits. People's historical relationship to the Kaerezi landscape and migration within the region has not been uniform, leading to the "coexistence of multiple historicities within . . . [a] particular locality" (Feierman 1990: 29). Different historical experiences among Kaerezi peasants caution against assuming that all people share a common habitus. Cultural idioms may be pervasive, but they are not always harnessed for the same purposes, nor experienced in the same way. Class, gender, age, and ethnicity are important in shaping not only the experience of resource struggles in Kaerezi, but also social actors' participation in those fields of conflict.

Let me focus on the question of gender. In Zimbabwe, gendered divisions of labor have arisen from men's incorporation into wage labor and women's responsibilities for remaining on the land and managing it on a daily and seasonal basis

(Schmidt 1992). In turn, these have filtered perceptions of the local landscape and resources. In the 1990s, Kaerezi women were particularly concerned about the potential impact of the National Park on their access to two critical resources: firewood for daily use and reeds woven into handicrafts marketed in Nyanga's tourist centers. Because weaving could be done by firelight after the evening meal, it did not compete with the domestic labor expected of most women. In poorer households, these handicrafts represented an important source of income. Women also gathered wild fruits and herbs within the park estate, supplementing the local diet and selecting plants efficacious for healing.

In Kaerezi, married women contend both with patrilineal notions of land rights and a state policy of allocating permits to male household heads. While widows and divorcees were eligible for arable plots, the permit system in resettlement schemes channels married women's *de jure* land rights through a male household head (Gaidzanwa 1994; Jacobs 1989). Women cultivated one crop in particular, *tsenza* (*Coleus esculentus*), outside fields allocated to "households" in the scheme's planned villages and those allocated by local headmen, or claimed through patrilineal inheritance. While *tsenza* cultivation doesn't reveal a "women's world" separate and distinct from that of men's, women's cultivation unfolded through gendered spaces and practices (see Guyer 1991; Leach 1992).

Groups of women collectively selected and allocated themselves isolated plots on steep mountain slopes far from arable fields, since *tsenza* was widely believed to "poison" the soil by robbing it of nutrients. Most groups resolved disputes arising in *tsenza* plots among themselves, outside state and patrilineal control. Groups of women usually pooled labor for burning and clearing a common plot, dividing the cleared area into a complex grid of small patches (*huma*). This process carefully interspersed any one woman's plots across a cleared hillside, evenly distributing the risk of a poor harvest due to soil, slope, and drainage. If a husband or older son of one of the women plowed initially to break the soil, his labor was not seen as laying claim to the land's produce; rather, it was "helping" (*kubatsira*). Women performed all land preparation, planting, and harvesting on their individual plots by hoe and hand, sometimes utilizing their children's labor.

Cash from selling *tsenza*, more than from any other marketable produce, was most likely to be subject to women's discretion within their household budgets. *Tsenza* cultivation was an important source of relatively autonomous cash income in an area where few women engaged in formal wage labor, improving their bargaining position within the "conjugal contract" through which household economic decisions and provisioning responsibilities are negotiated (Jackson 1995; Whitehead 1981). Significantly, *tsenza* production hinged on access to cultivable plots mediated neither by patriarchal nor state conceptions of land rights in Kaerezi. While officially prohibited in the resettlement scheme – ostensibly because *tsenza* poisoned the soil and eroded steep slopes, but also because it challenged the government's ordered pattern of regulated land use – it was widely tolerated as a benign practice. Women feared that the protected area would encourage greater attention to land-use prohibitions elsewhere in

Kaerezi, endangering their *tsenza* fields lying far beyond the river corridor's boundaries.

Women retooled the patrilineal discourse of *nhaka*, or rightful inheritance, to defend *tsenza* cultivation by arguing for the birthright of "freely cultivating" (*kurima madiro*) without intervention from state officials or conservation regulations. Many recalled "suffering for the land" (*kutambudzikira nyika*) during the 1972 evictions from Gaeresi Ranch when women were beaten by police and defiantly stripped off their clothes. They marched more than 30 kilometers to the district offices and protested, bare-breasted, making national headlines and getting mentions in international human rights reports. Many women invoked these memories, refashioning *nhaka* as a claim to land rights validated through historical struggle rather than patrilineal inheritance. They did so through particular cultural *and* cultivation *practices*. In this sense, Richards' (1989) emphasis on viewing farming operations as "embedded in a social context" and agriculture as a particular kind of cultural "performance" helps us understand the significance of *tsenza* cultivation.

In Kaerezi, a "gender analysis from a micro political economy perspective" (Leach 1991: 17) reveals how women's use of park and resettlement scheme resources grow out a particular history. Colonial land alienation and the intro-duction of taxes forced men into migrant labor to meet cash demands in the changing regional economy. These structural transformations in the wider political economy, however, were given texture through cultural idioms and the everyday practices of women and men. Local patterns of resource use arise from the complex interplay of gender relations, history, and culture and articulate with, but are not determined by, regional and global economic forces (Katz and Monk 1993; Leach *et al.* 1995). The historical and local specificity of the Kaerezi case, however, points to the more general advantage of weaving attention to cultural forms into a retooled political ecology.

REWORKING POLITICAL ECOLOGY, MARXISMS, AND CULTURAL THEORY

> The proper names "Marx" and or "Marxism" have always already been plural nouns.
>
> (Magnus and Cullenberg 1995: x)

The history of Marxism, as Watts (1994: 462) reminds us, "is a long succession of crises and transformations." There is a lack of theoretical unity among diverse approaches. As a means of countering economic reductionism in Marxism, critics have long stressed "struggle in the cultural dimension" (Hall 1978: 10). Well before the relatively recent explosion of poststructuralism and cultural studies, Thompson (1966 [1963]): 13) emphasized that "class is a cultural as much as an economic formation." Similarly, Hall's (1960: 1) inaugural editorial in the *New Left Review* argued that the "humanist strengths of socialism . . .

must be developed in cultural and social terms, as well as in economic and political." Thompson (1995: 303) later stressed that paying attention to culture should not mean neglecting the "controlling context of power."

The insight that "culture is laced with power and power is shaped by culture" (Rosaldo 1994: 525) remains at the core of cultural studies, an interdisciplinary array of academic and associated practices spanning the humanities and social sciences. Marxian influences, particularly from Gramsci, continue to shape the field and its keyword *culture*: "the actual, grounded terrain of practices, representations, languages and customs of any specific historical society" (Hall 1986: 26). Practitioners of cultural studies have, simultaneously, criticized Marx's "Eurocentrism" and "the great evasions of Marxism . . . culture, ideology, language, the symbolic" (Hall 1992: 279).

The complexity of Third World resource struggles offers an opportunity for rethinking Marxism, political ecology, and their relationships to cultural theory. In Kaerezi, Zimbabwe, gendered relationships to environmental resources are embedded within a history of myriad symbolic and material struggles on a contested terrain. Some twenty years ago, Harvey (1977: 226) stressed how environmental "'resources' can be defined only in relationship to the mode of production which seeks to make use of them and which simultaneously 'produces' them through both the physical and mental activity of the users." Political ecology follows suit by appreciating the importance of social relations of production and labor for understanding livelihood struggles. As Collins (1992: 186) suggests, however, political ecology needs to be more grounded in the "lived practices of production," giving life to abstract analytical categories. We need to move still further beyond the narrow confines of the labor process, however, and situate resource struggles within the *cultural* production of landscape and resources.

Seen as a surface of semiotics as well as soil, Kaerezi's landscape joins issues of symbol and livelihood, culture and environment, the meaningful and the material. Environmental politics thus offers a much needed groundedness to cultural studies where concrete practices are often effaced in the pursuit of grand theory. Peasants' historical understandings of "suffering for the land" fused the material experience of forced colonial evictions, including cattle seizures and crop and hut burnings with a symbolic understanding of ancestral claims and entitlements shaped by a century's struggles over access to grazing, hunting, and arable land. The simultaneity of symbolic and material conflicts is made manifest through particular interventions in the landscape: the demarcation of a national park boundary; gathering firewood; pasturing and dipping cattle; and the cultivation of *tsenza*. These cultural practices have produced meanings as well as harvests.

A Gramscian approach opens up space for a "dialectical tacking between political economy and representational forms required to unravel the dense meanings encoded in landscapes" (Watts 1992a: 122). The micro-politics of resource struggles are animated by local history, mediated by cultural idioms, and gendered through the different practices men and women have pursued in defense of local livelihoods. The competing agendas of state functionaries pursuing their ministries' agendas in Kaerezi layer over this contested terrain, and warn against any simple structural opposition between a monolithic state and an undifferentiated peasantry. Rather, conflicts between bureaucrats and peasants reveal the "polysemic, ambiguous, contradictory quality of . . . putative state forms" (Sayer 1994: 369). These forms unfold through symbolic contestation as well as material conflict, producing meanings through the concrete interventions of state functionaries in time and space. The "state" is not "outside" cultural politics, but rather a constellation of practices and institutions constituted through struggles over meanings of rights, legitimacy, and authority. The importance of the *practices* of historical actors brings us back, also, to the relationship between cultural theory and environmental politics.

Commentators have recently stressed a "more imaginative role for the social sciences in environmental debate," advocating the need to explore "the ways in which patterns of social relationships, cultural forms, political practices, and economic institutions are all implicated in the production of environmental change" (Redclift and Benton 1994: 1). Recent initiatives have sought to explore the "culture of nature" (Simmons 1993; Wilson 1991) and the cultural meanings pervading environmentalism (Berland and Slack 1994; Escobar 1995: 192–211; Harvey 1993; Milton 1993). The 1990s have seen increasing critical reflection on the relevance of cultural theory for environmental struggles (Gelder and Jacobs 1995; Haila and Heininen 1995; Jacobs 1994; Whitt and Slack 1994). Anthropologists contributed detailed analyses of the cultural understandings of environment and categories of "nature" (Croll and Parkin 1992; Descola 1994; Leach 1994; Richards 1993; Strathern 1992). At the same time, environmental historians seek "to synthesize and integrate environmental phenomena with cultural and socio-economic change" (Williams 1994: 3). This convergence among diverse disciplinary perspectives underscores the need to integrate approaches formerly considered distinct and irreconcilable. Rather than viewing cultural forms as derivative of, or "outside," structural entities such as "the state," or transformations in "the economy," the challenge becomes to explore how symbol and meaning give form and content to material transformation. It is not a question of attending to either culture or power, political economy or symbolic forms, but the interrelations among them.

The legacies of colonialism and the often conflicting mandates of a post-colonial state remind us that women and men in Kaerezi, Zimbabwe, as elsewhere in the Third World, manoeuver in material and symbolic terrain not entirely of their own choosing (cf. Marx 1961 [1852]). "Local" conflicts reflect historical shifts in the regional political economy, interventions by state officials, and what has become an increasingly global discourse of environmentalism. However, people make sense of these transformations through cultural *practices* that take place in particular social and historical contexts. The simultaneity of symbolic and material struggles is played out across landscapes rich in meanings

140

and productive resources. Contestations over these meanings, in turn, shape the outcome of environmental politics. The challenge of a retooled political ecology is to conceptualize how external forces "are articulated with internal agency, with locally shared knowledges and practices, with shared but socially differentiated meanings and experiences" (Watts 1992b: 15) in the context of environmental conflicts. The cultural practices of women and men, within constraining yet not determining "structures," remain critical to the form, texture, and ultimate outcome of those struggles.

NOTE

I would like to thank Orin Starn, Suzana Sawyer, Michael Watts, and Dick Peet for their constructively critical comments. Some of the ideas presented here appear in "Contesting terrain in Zimbabwe's Eastern Highlands," *Economic Geography* 69, 4: 380–401.

REFERENCES

Abrams, Philip. 1988. "Notes on the difficulty of studying the state," *Journal of Historical Sociology* 1, 1: 58–89.

Agarwal, Bina. 1992. 'The gender and environment debate: lessons from India," *Feminist Studies* 18, 1: 119–58.

—— 1994a. "Gender, resistance and land: interlinked struggles over resources and meanings in South Asia," *Journal of Peasant Studies* 22, 1: 81–125.

—— 1994b. *A Field of One's Own: Gender and Land Rights in South Asia.* Cambridge: Cambridge University Press.

Alonso, Ana Maria. 1992. "Gender, power, and historical memory: discourses of Serrano resistance," in Judith Butler and Joan W. Scott (eds.) *Feminists Theorize the Political.* New York: Routledge, pp. 404–26.

Appadurai, Arjun. 1990. "Disjuncture and difference in the global cultural economy," *Public Culture* 2, 2: 1–24.

Bassett, Thomas. 1988. "The political ecology of peasant–herder conflicts in the northern Ivory Coast," *Annals of the Association of American Geographers* 78, 3: 453–72.

—— 1994. "Cartography and empire building in nineteenth-century West Africa," *Geographical Review* 84, 3: 316–35.

Bassett, Thomas and Donald Crummey (eds.). 1993. *Land in African Agrarian Systems.* Madison: University of Wisconsin Press.

Bayart, Jean-François. 1993. *The African State: The Politics of the Belly.* London: Longman.

Berland, Jody and Jennifer Daryl Slack. 1994. "On environmental matters," *Cultural Studies* 8, 1: 1–4.

Bernstein, Henry. 1990. "Taking the part of peasants?" in Henry Bernstein, Ben Crow, Maureen Mackintosh, and Charlotte Martin (eds.) *The Food Question.* London: Earthscan, pp. 69–79.

Berry, Sara. 1988. "Concentration without privatization? Some consequences of changing patterns of rural land control in Africa," in S.P. Reyna and R.E. Downs (eds.) *Land and Society in Contemporary Africa.* Hanover, NH: University Press of New England, pp. 53–75.

—— 1993. *No Condition is Permanent: The Social Dynamics of Agrarian Change in Sub-Saharan Africa.* Madison: University of Wisconsin Press.

Blaikie, Piers. 1989. "Environment and access to resources in Africa," *Africa* 59, 1: 18–40.

141

Blaikie, Piers and Harold Brookfield. 1987. *Land Degradation and Society*. London: Methuen.

Bourdieu, Pierre. 1977. *Outline of a Theory of Practice*, translated by Richard Nice. Cambridge: Cambridge University Press.

Bove, Paul. 1994. "Foreword," in Marcia Landy, *Film, Politics, Gramsci*. Minneapolis: University of Minnesota Press, pp. ix–xxv.

Bryant, Raymond. 1992. "Political ecology: an emerging research agenda in Third World studies," *Political Geography* 11, 1: 12–36.

Carney, Judith. 1993. "Converting the wetlands, engendering the environment: the intersection of gender with agrarian change in the Gambia," *Economic Geography* 69, 3: 329–48.

Carney, Judith and Michael Watts. 1990. "Manufacturing dissent: work, gender, and the politics of meaning in a peasant society," *Africa* 60, 2: 207–41.

—— and —— 1991. "Disciplining women? Rice, mechanization, and the evolution of Mandinka gender relations in Senegambia," *Signs* 16, 4: 651–81.

Chakrabarty, Dipesh. 1993. "Marx after Marxism: subaltern histories and the question of difference," *Polygraph* 6/7: 10–16.

Clutton-Brock, Guy. 1969. *Rekayi Tangwena: Let Tangwena Be*. Salisbury: Cold Comfort Society.

Collins, Jane. 1992. "Marxism confronts the environment: labor, ecology and environmental change," in Sutti Ortiz and Susan Lees (eds.) *Understanding Economic Process*. Lanham, MD: University Press of America, pp. 179–88.

Corrigan, Philip and Derek Sayer. 1985. *The Great Arch: English State Formation as Cultural Revolution*. Oxford: Basil Blackwell.

Croll, Elisabeth and David Parkin (eds.). 1992. *Bush Base: Forest Farm. Culture, Environment and Development*. New York: Routledge.

Crush, Jonathan. 1994. "Scripting the compound: power and space in the South African mining industry," *Environment and Planning D: Society and Space* 12: 301–24.

de Certeau, Michel. 1984. *The Practice of Everyday Life*. Berkeley: University of California Press.

Demeritt, David. 1994a. "The nature of metaphors in cultural geography and environmental history," *Progress in Human Geography* 18, 2: 163–85.

—— 1994b. "Ecology, objectivity, and critique in writings on nature and human societies," *Journal of Historical Geography* 20, 1: 22–37.

DERUDE. 1991. *Resettlement Progress Report: 1991*. Harare: Ministry of Local Government, Rural and Urban Development.

Descola, Philippe. 1994. *In the Society of Nature: A Native Ecology of Amazonia*. Cambridge: Cambridge University Press.

Donham, Donald. 1990. *History, Power, and Ideology: Essays in Marxism and Social Anthropology*. Cambridge: Cambridge University Press.

Dove, Michael. 1985. *Swidden Agriculture in Indonesia: The Subsistence Strategies of the Kalimantan Kantu*. New York: Mouton.

Drinkwater, Michael. 1991. *The State and Agrarian Change in Zimbabwe's Communal Areas*. London: Macmillan.

Duncan, James. 1994. "The politics of landscape and nature, 1992–1993," *Progress in Human Geography* 18, 3: 361–70.

Duncan, James and David Ley (eds.). 1993. *Place/Culture/Representation*. New York: Routledge.

Durham, William. 1979. *Scarcity and Survival in Central America: Ecological Origins of the Soccer War*. Stanford, CA: Stanford University Press.

—— 1995 "Political ecology and environmental destruction in Latin America," in Michael Painter and William Durham (eds.) *The Social Causes of Environmental*

Destruction in Latin America. Ann Arbor: University of Michigan Press, pp. 249–64.

Escobar, Arturo. 1995. *Encountering Development: The Making and Unmaking of the Third World*. Princeton, NJ: Princeton University Press.

Faber, Daniel. 1993. *Environment Under Fire: Imperialism and the Ecological Crisis in Central America*. New York: Monthly Review Press.

Fairhead, James and Melissa Leach. 1994. "Contested forests: modern conservation and historical land use in Guinea's Ziama Reserve," *African Affairs* 93, 373: 481–513.

—— and —— forthcoming. *Reversing Landscape History: Power, Policy and Socialised Ecology in East Africa's Forest–Savanna Mosaic*. Cambridge: Cambridge University Press.

Feierman, Steven. 1990. *Peasant Intellectuals: Anthropology and History in Tanzania*. Madison: University of Wisconsin Press.

Gadgil, Madhav and Ramachandra Guha. 1992. *This Fissured Land: An Ecological History of India*. Berkeley: University of California Press.

Gaidzanwa, Rudo. 1994. "Women's land rights in Zimbabwe," *Issue* 22, 2: 12–16.

Gelder, Ken and Jane Jacobs. 1995. "'Talking out of place': authorizing the Aboriginal sacred in postcolonial Australia," *Cultural Studies* 9, 1: 150–60.

Goheen, Mitzi and Parker Shipton (eds.). 1992. "Rights over land: categories and controversies," Special Issue of *Africa* 62, 3.

Gramsci, Antonio. 1971. *Selections from the Prison Notebooks*, edited by Quentin Hoare and Geoffrey Nowell-Smith. London: Laurence & Wishart.

—— 1983 [1957]. "Marxism and modern culture," translated by Louis Marks. *The Modern Prince and Other Writings*. New York: International Publishers, pp. 82–9.

—— 1991 [1920]. "Questions of culture," in David Forgacs and Geoffrey Nowell-Smith (eds.) *Selections from Cultural Writings*. Cambridge, MA: Harvard University Press, pp. 41–3.

Guha, Ramachandra. 1989. *The Unquiet Woods: Ecological Change and Peasant Resistance in the Himalaya*. Berkeley: University of California Press.

Guyer, Jane. 1991. "Female farming in anthropology and African history," in Micaela di Leonardo (ed.) *Gender at the Crossroads of Knowledge: Feminist Anthropology in the Postmodern Era*. Berkeley: University of California Press, pp. 257–77.

—— 1994. "The spatial dimensions of civil society in Africa: an anthropologist looks at Nigeria," in John Harbeson, Donald Rothchild, and Naomi Chazan (eds.) *Civil Society and the State in Africa*. Boulder, CO: Lynne Reinner, pp. 215–29.

Haila, Yrjo and Lassi Heininen. 1995. "Ecology: a new discipline for disciplining?" *Social Text* 42: 153–71.

Hall, Stuart. 1960. "Editorial," *New Left Review* 1: 1–4.

—— 1978. "Marxism and culture," *Radical History Review* 18, Fall: 5–14.

—— 1986. "Gramsci's relevance for the study of race and ethnicity," *Journal of Communication Inquiry* 10, 2: 5–27.

—— 1990. "Cultural identity and diaspora," in Jonathan Rutherford (ed.) *Identity: Community, Culture, Difference*. London: Lawrence & Wishart, pp. 222–37.

—— 1992. "Cultural studies and its theoretical legacies," in Lawrence Grossberg, Cary Nelson, and Paula Treichler (eds.) *Cultural Studies*. New York: Routledge, pp. 277–94.

Harley, J.B. 1988. "Maps, knowledge, and power," in Denis Cosgrove and Stephen Daniels (eds.) *The Iconography of Landscape*. Cambridge: Cambridge University Press, pp. 277–312.

Harvey, David. 1977. "Population, resources, and the ideology of science," in Richard Peet (ed.) *Radical Geography*. Chicago: Maaroufa, pp. 213–42.

—— 1993. "The nature of environment: dialectics of social and environmental change," in Ralph Milliband and Leo Panitch (eds.) *Socialist Register 1993*. London: Merlin Press, pp. 1–51.

—— 1995. "Militant particularism and global ambition: the conceptual politics of place, space, and environment in the work of Raymond Williams," *Social Text* 42: 69–98.

Hecht, Susanna and Alexander Cockburn. 1990. *The Fate of the Forest: Developers, Destroyers and Defenders of the Amazon.* New York: HarperCollins.

—— 1992. "Realpolitik, reality and rhetoric in Rio," *Environment and Planning D: Society and Space*, 10, 4: 367–75.

Holub, Renate. 1992. *Antonio Gramsci: Beyond Marxism and Postmodernism.* New York: Routledge.

Jacobs, Jane. 1994. "Earth honoring: western desires and indigenous knowledges," in Alison Blunt and Gillian Rose (eds.) *Writing Women and Space: Colonial and Postcolonial Geographies.* New York: Guilford Press, pp. 169–96.

Jacobs, Susan. 1989. "Zimbabwe: state, class, and gendered models of land resettlement," in Jane Parpart and Kathleen Staudt (eds.) *Women and the State in Africa.* Boulder, CO: Westview Press, pp. 161–84.

Kain, Roger and Elizabeth Baigent. 1992. *The Cadastral Map in the Service of the State: A History of Property Mapping.* Chicago: University of Chicago Press.

Katz, Cindi and Janice Monk (eds.). 1993. *Full Circles: Geographies of Women Over the Life Course.* New York: Routledge.

Laclau, Ernesto and Chantal Mouffe. 1985. *Hegemony and Socialist Strategy: Towards a Radical Democratic Politics.* London: Verso.

—— 1987. "Post-Marxism without apologies," *New Left Review* 166: 79–106.

Landy, Marcia. 1994. *Film, Politics, and Gramsci.* Minneapolis: University of Minnesota Press.

Leach, Melissa. 1991. "Engendered environments: understanding natural resource management in the West African forest zone," *IDS Bulletin* 22, 4: 14–24.

—— 1992 "Women's crops in women's spaces: gender relations in Mende rice farming," in Elisabeth Croll and David Parkin (eds.) *Bush Base: Forest Farm.* New York: Routledge, pp. 76–96.

—— 1994. *Rainforest Relations: Gender and Resource Use Among the Mende of Gola, Sierra Leone.* Washington, DC: Smithsonian Institution Press.

Leach, Melissa, Susan Joekes and Cathy Green. 1995. "Gender relations and environmental change," *IDS Bulletin* 26, 1: 1–8.

Little, Peter. 1992. *The Elusive Granary: Herder, Farmer, and State in Northern Kenya.* Cambridge: Cambridge University Press.

Mackenzie, Fiona. 1993. "'A piece of land never shrinks': reconceptualizing land tenure in a smallholding district, Kenya," in Thomas Bassett and Donald Crummey (eds.) *Land in African Agrarian Systems.* Madison: University of Wisconsin Press, pp. 194–221.

—— 1995. "A farm is like a child who cannot be left unguarded: gender, land and labour in Central Province, Kenya," *IDS Bulletin* 26, 1: 17–23.

Magnus, Bernd and Stephen Cullenberg (eds.). 1995. *Wither Marxism? Global Crisis in International Perspective.* New York: Routledge.

Marx, Karl. 1961 [1852]. *The Eighteenth Brumaire of Louis Bonaparte.* New York: International Publishers.

—— 1975. "Debates on the law on thefts of wood," in Karl Marx and Friedrich Engels, *Karl Marx and Friedrich Engels: Collected Works*, Volume 1. New York: International Publishers, pp. 224–63.

—— 1986 [1859]. "Preface to the *Critique of Political Economy*," in John Elster (ed.) *Karl Marx: A Reader.* Cambridge: Cambridge University Press, pp. 187–8.

Massey, Doreen. 1994. *Space, Place, and Gender.* Minneapolis: University of Minnesota Press.

Mbembe, Achille. 1992. "The banality of power and the aesthetics of vulgarity in the postcolony," *Public Culture* 4, 2: 1–30.

McRobbie, Angela. 1994. "Post-Marxism and cultural studies," in Angela McRobbie, *Postmodernism and Popular Culture*. New York: Routledge, pp. 44–60.

Milton, Kay (ed.). 1993. *Environmentalism: The View from Anthropology*. New York: Routledge.

Mitchell, Tim. 1991. "The limits of the state: beyond statist approaches and their critics," *American Political Science Review* 85, 1: 77–96.

Moore, Henrietta and Megan Vaughan. 1994. *Cutting Down Trees: Gender, Nutrition, and Agricultural Change in the Northern Province of Zambia, 1890–1990*. London: James Currey.

Moyana, Henry. 1984. *The Political Economy of Land in Zimbabwe*. Gweru, Zimbabwe: Mambo Press.

Moyo, Sam. 1987. "The land question," in Ibbo Mandaza (ed.) *Zimbabwe: The Political Economy of Transition, 1980–1986*. Harare: Jongwe Press.

Neumann, Rod. 1992. "Political ecology of wildlife conservation in the Mt. Meru area of northeast Tanzania," *Land Degradation and Rehabilitation* 3: 85–98.

Nugent, Daniel and Ana Maria Alonso. 1994. "Multiple selective traditions in agrarian reform and agrarian struggle: popular culture and state formation in the Ejido of Namiquipa, Chihuahua," in Gilbert Joseph and Daniel Nugent (eds.) *Everyday Forms of State Formation: Revolution and the Negotiation of Rule in Modern Mexico*. Durham, NC: Duke University Press, pp. 209–46.

Painter, Michael and William Durham (eds.). 1995. *The Social Causes of Environmental Destruction in Latin America*. Ann Arbor: University of Michigan Press.

Palmer, Robin. 1977. "The agricultural history of Rhodesia," in Robin Palmer and Neil Parsons (eds.) *The Roots of Rural Poverty in Central and Southern Africa*. Berkeley: University of California Press, pp. 221–54.

Peluso, Nancy. 1992. *Rich Forests, Poor People: Resource Control and Resistance in Java*. Berkeley: University of California Press.

Peters, Pauline. 1984. "Struggles over water, struggles over meaning: cattle, water and the state in Botswana," *Africa* 54, 3: 29–49.

—— 1994. *Dividing the Commons: Politics, Policy and Culture in Botswana*. Charlottesville: University Press of Virginia.

Ranger, Terence. 1985. *Peasant Consciousness and Guerrilla War in Zimbabwe*. Harare: Zimbabwe Publishing House.

—— 1989. "Whose heritage? The case of the Matobo National Park," *Journal of Southern African Studies* 15, 2: 217–49.

Redclift, Michael and Ted Benton (eds.) 1994. *Social Theory and the Global Environment*. New York: Routledge.

Richards, Paul. 1983. "Ecological change and the politics of African land use," *African Studies Review* 26, 2: 1–72.

—— 1989. "Agriculture as performance," in Robert Chambers, Arnold Pacey, and Lori Ann Thrupp (eds.) *Farmer First: Farmer Innovation and Agricultural Research*. London: Intermediate Technology Publications, pp. 39–43.

—— 1993. "Natural symbols and natural history: chimpanzees, elephants and experiments in Mende thought," in Kay Milton (ed.) *Environmentalism: The View from Anthropology*. New York: Routledge, pp. 144–59.

Rosaldo, Renato. 1994. "Whose cultural studies?" *American Anthropologist* 96, 3: 524–9.

Roth, Michael and John Bruce. 1994. *Land Tenure, Agrarian Structure, and Comparative Land Use Efficiency in Zimbabwe: Options for Land Tenure Reform and Land Redistribution*. Madison: Land Tenure Center.

Sayer, Derek. 1994. "Everyday forms of state formation: some dissident remarks on hegemony," in Gilbert Joseph and Daniel Nugent (eds.) *Everyday Forms of State Formation: Revolution and the Negotiation of Rule in Modern Mexico.* Durham, NC: Duke University Press, pp. 367–77.

Schmidt, Elizabeth. 1992. *Peasants, Traders, and Wives: Shona Women in the History of Zimbabwe, 1870–1939.* London: James Currey.

Schmink, Marianne and Charles Wood. 1987. "The 'Political Ecology' of Amazonia," in P.D. Little, M.M. Horowitz, and A.E. Nyerges (eds.) *Lands at Risk in the Third World.* Boulder, CO: Westview Press, pp. 38–57.

—— 1992. *Contested Frontiers in Amazonia.* New York: Columbia University Press.

Sheridan, Thomas. 1988. *Where the Dove Calls: The Political Ecology of a Peasant Corporate Community in Northwestern Mexico.* Tuscon: University of Arizona Press.

Shipton, Parker. 1994. "Land and culture in tropical Africa: soils, symbols, and the metaphysics of the mundane," *Annual Review of Anthropology* 23: 347–77.

Simmons, I.G. 1993. *Interpreting Nature: Cultural Constructions of the Environment.* New York: Routledge.

Stonich, Susan. 1993. *"I Am Destroying the Land": The Political Ecology of Poverty and Environmental Destruction in Honduras,* Boulder, CO: Westview Press.

Strathern, Marilyn. 1992. *After Nature: English Kinship in the Late Twentieth Century.* Cambridge: Cambridge University Press.

Thompson, E.P. 1966 [1963]. *The Making of the English Working Class.* New York: Vintage Books.

—— 1995 [1985]. "Agenda for radical history," *Critical Inquiry* 21, 2: 299–305.

Thompson, J.B. 1984. *Studies in the Theory of Ideology.* Berkeley: University of California Press.

Tsing, Anna. 1993. *The Realm of the Diamond Queen.* Princeton, NJ: Princeton University Press.

Watts, Michael. 1983a. *Silent Violence: Food, Famine, and Peasantry in Northern Nigeria.* Berkeley: University of California Press.

—— 1983b. "On the poverty of theory: hazards research in context," in K. Hewitt (ed.) *Interpretations of Calamity.* Boston: Allen & Unwin, pp. 231–62.

—— 1989. "The agrarian question in Africa: debating the crisis," *Progress in Human Geography* 13, 1: 1–41.

—— 1991. "Entitlement or empowerment? Famine and starvation in Africa," *Review of African Political Economy* 51: 9–26.

—— 1992a. "Space for everything (a commentary)," *Cultural Anthropology* 7, 1: 115–29.

—— 1992b "Capitalisms, crises, and cultures I: notes toward a totality of fragments," in Allan Pred and Michael Watts. *Reworking Modernity: Capitalisms and Symbolic Discontent.* New Brunswick: Rutgers University Press, pp. 1–19.

—— 1994. "Post-Marxism," in R.J. Johnston, Derek Gregory, and David Smith (eds.) *The Dictionary of Human Geography.* Oxford: Blackwell, pp. 461–3.

Weiner, Daniel. 1991. "Socialist transition in the capitalist periphery: a case study of agriculture in Zimbabwe," *Political Geography Quarterly* 10, 1: 54–75.

Whitehead, Ann. 1981. "'I'm hungry Mum': the politics of domestic budgeting," in Kate Young *et al.* (eds.) *Of Marriage and the Market.* London: Routledge and Kegan Paul, pp. 93–116.

Whitt, Laurie Anne and Jennifer Daryl Slack. 1994. "Communities, environments and cultural studies," *Cultural Studies* 8, 1: 5–31.

Williams, Michael. 1994. "The relations of environmental history and historical geography," *Journal of Historical Geography* 20, 1: 3–21.

Williams, Raymond. 1977. *Marxism and Literature.* Oxford: Oxford University Press.

—— 1991 [1980]. "Base and superstructure in Marxist cultural theory," in Chandra

Mukerji and Michael Schudson (eds.) *Rethinking Popular Culture: Contemporary Perspectives in Cultural Studies.* Berkeley: University of California Press, pp. 407–23.

Wilson, Alexander. 1992. *The Culture of Nature: North American Landscape from Disney to the Exxon Valdez.* Oxford: Blackwell.

Zimmerer, Karl. 1994. "Human geography and the 'new ecology': the prospect and promise of integration," *Annals of the Association of American Geographers* 84, 1: 108–25.

7

DEFINING DEFORESTATION IN MADAGASCAR

Lucy Jarosz

World-scale environmental problems such as tropical deforestation, global warming, ozone depletion, and the erosion of cultural and ecological diversity have recently garnered public concern, fuelling intellectual debates on development and environment. This concern was tangibly manifest at the 1992 United Nations Conference on Environment and Development (UNCED), held in Rio de Janeiro. In his opening remarks, UNCED Secretary-General Maurice Strong indicated that overpopulation in the South and overconsumption in the North were root causes of environmental degradation (*Multinational Monitor* 1992). In the case of tropical deforestation, poverty among shifting cultivators and rapid national population growth rates are common explanations. Such cause–effect relationships exemplify the most commonly employed Western discourse concerning tropical deforestation in the South. Neo-Malthusian theory links population growth to shifting cultivation and this, in turn, to tropical deforestation. Multilateral organizations, mainstream environmentalists, and the Western media deploy and disseminate versions of this explanation throughout the world. This particular definition makes "the problem" amenable to technical solutions of modernization and birth control.

The political ecology approach to deforestation can provide a counter-discourse. It positions people, places, and practices in relation to broader processes of social and economic change at the global and local levels. Increasingly, it also considers the cultural politics of changing land-use practices, revealing the meanings and practices which constitute places. Together, political ecology of tropical deforestation and discourse analyses situate environmental change in relation to the politics and ideology of development and local responses or resistances to the existing discourse of development. In general, this materialist and discursive analysis reveals how explanations and definitions of deforestation are socially and politically constructed to the advantage of powerful people.

CONTRIBUTIONS FROM RECONSTRUCTED REGIONAL GEOGRAPHY AND POLITICAL ECOLOGY

Reconstructed regional geography argues that regions and their transformations are inextricably social processes (Pred 1984; Pudup 1988; Rosenblatt 1992; Thrift 1983; Urry 1987). Human activities shape, and are shaped by, place and history; human identities and activities constitute the economic, political, and ideological processes which form and transform regions. In turn, the particular contextual details of place shape human activities. Rather than offering a cause–effect explanation of tropical deforestation *per se*, reconstructed regional geography reveals how resource extraction, control, and distribution are social processes which shape, and are shaped by, particular regional contexts, contingencies, and activities. In this view, regional transformation is constituted from, and constitutive of, multiple overlapping processes occurring at varying geographic scales, from the local to the global (Murphy 1991). This approach is especially useful in situating regional transformations in relation to external, global processes.

Similarly, the political ecology perspective advocates placing human activities and regional change in concrete spatial and historical contexts (Blaikie and Brookfield 1987; Zimmerer, Chapter 5 in this volume). Regional political ecology complements, and in some sense parallels, a reconstructed regional geography. Both adopt a political economy framework that emphasizes the social relations of production as central to geographic understanding. Much regional political ecology is inspired by Marxist analysis, which concerns itself with resource access and control, relations of surplus extraction, and capitalist intrusion, set within African, Asian, and Latin American contexts (Agarwal 1988; Bassett 1988; Blaikie and Brookfield 1987; Hurst 1990). Considerations of ecological change and land degradation must be taken into account in theorizing regional transformation. Regional political ecology typically focuses on analyses of regional transformation and rural development in the so-called Third World (see Bryant 1992; Pickles and Watts 1992), whereas reconstructed regional geographies customarily focus on urban industrial development in North America and Europe (see Sayer 1989 for representative examples). Nonetheless, as I seek to demonstrate, concepts derived from both regional political ecology and reconstructed regional geography may be employed to reveal how regional transformation and environmental change are embedded in social processes. By synthesizing these approaches, this chapter attempts to integrate two areas of traditional geographic concern – those of society–space relations and society–nature relations – in an attempt to contest and problematize deforestation in the humid tropics.

Discussions about the incorporation of ideology, consciousness, and meaning have emerged within the reconstructed regional geography literature (Pred 1986; Sayer 1989; Watts 1991) and have recently been included within political ecology (see Bryant 1992; Carney and Watts 1991). I shall contribute to this discussion by focusing on the way some aspects of Euro-American culture and ideology shape knowledge and beliefs about deforestation. Shifting cultivation and population growth are conventionally seen as ecological villains destroying the forested landscape. Growing numbers of poor people using a nomadic style of cultivation mean shorter fallow periods and increasing deterioration of tropical forest cover. Solutions typically are defined in terms of population control and the introduction of sedentarized, intensive agriculture. The key issue here is how various groups construct definitions of deforestation as they establish the "reality" most conducive to their interests. In particular, this shows how the colonial state defined the population-shifting cultivation–deforestation nexus and how (little) this definition changed in the post-colonial period.

DEFINING TROPICAL DEFORESTATION: THE GLOBAL LEVEL

Since at least the Neolithic period shifting cultivation has been a widespread form of land use that varies in character through space and time (Conklin 1961). Shifting cultivation is "a continuous system of cultivation in which temporary fields are cleared, usually burned, and subsequently cropped for fewer years than they are fallowed" (Peters and Neuenschwander 1988: 1). The practice is now largely concentrated in Southeast Asia, sub-Saharan Africa, and Latin America, embracing regions of great geographic, demographic, ethnic, and ecological diversity. Conventionally, shifting cultivation has been interpreted as inefficient, destructive, and primitive – the province of ignorant and marginalized subsistence producers, isolated from global and national economic systems (see Dove 1983: 93). More recently, shifting cultivation has been viewed as a rather inflexible static system ill-suited for adapting to changes brought about by modernity. This latter view holds that it slows agricultural production and increases ecological degradation. It is well illustrated in a 1991 World Bank study, which I use as an example. This position argues that

> Shifting cultivation and grazing have been appropriate traditional responses to abundant land, scarce capital, and limited technology. . . . This slowly evolving system has, however, proved unable to adapt to sharply accelerated population growth over the past four decades. Traditional uses of land and fuel have depleted soil and forests and contributed to agricultural stagnation.
>
> (World Bank 1992: 27)

Shifting cultivation is defined as environmentally harmful, because too many people employ this traditional land-use practice. Traditional land-use practices are thus constructed as environmentally detrimental when too many people use them.

Roughly 250–300 million people practice shifting cultivation on approximately half the land area of the tropical and mountain ecosystems which span the globe (Dove 1983: 85). All too frequently, shifting cultivators are represented as exploding hordes of faceless, poverty-stricken peasants felling forests for fields and food. They are depicted as either ignorant of the irreversible destruction of primary forest cover or compelled by grinding poverty and hunger to deforest marginal lands in a kind of Darwinian scramble for survival (Knox 1989; Shoumatoff 1988). Such conceptualizations blame the victims – people dispossessed and marginalized by the forces of social and ecological change – and isolate the practice of shifting cultivation from the complex realities of land-use practices in specific places at specific times. Moreover, such accounts of shifting cultivators neglect historical considerations of the political economy of land-use practices and tenure patterns which, I contend, are central to an understanding of regional tropical deforestation. As Peluso (1992: 19) maintains, forest degradation and poverty are not isolated or self-perpetuating conditions. They are symptomatic outcomes of agrarian change indicating complex social conflicts over resource rights, distribution, access.

The population–resources component of this "blame the victim" ideology as defined by the World Bank and most of the UNCED participants at Rio's Earth Summit, is a variant of the neo-Malthusian argument. A neo-Malthusian polemic emerged in the 1950s, as population increased rapidly in the Third World; variants re-emerged as explanations of environmental degradation in the early 1970s, with the rise of an environmental movement in the United States and Western Europe (Meadows *et al.* 1972; Paddock and Paddock 1967); they have returned with a vengence in the 1990s (see World Bank 1992). In this view, increasing consumption of fossil fuels, the industrialization of agriculture, and associated pollution problems, are all effects of population growth (Buttel *et al.* 1990: 59). Ehrlich and Ehrlich (1990), in perhaps the most influential variant, relate population growth to global warming, acid rain, depletion of the ozone layer, loss of soil fertility, and groundwater depletion. Population growth remains a popular explanation for world hunger: for example, Brown (1989) links population growth with environmental degradation and food scarcity, arguing that rapid population growth has outstripped farmers' ability to produce. Brown (Brown and Young 1990: 77) argues, "Feeding people adequately in the nineties will depend on quickly slowing world population growth to bring it in line with the likely increase in food output." They conclude that the equilibrium between world food production and population has been upset by rapid population growth, particularly in the South.

Neo-Malthusian models frequently conceptualize population growth as a variable independent of time and place. In the eighteenth century, Thomas Malthus (1798/1929) postulated that because food and the "passion between the sexes" are necessary for human existence, population inevitably pressures the means of subsistence and, unless controlled, necessarily outstrips production. This universalizing, ahistorical explanation is relentless in its definition of the poor as "irrational." Not only does it find population growth resulting from inherent tendencies in people, it conveniently omits consideration of how other social processes, such as colonialism and development, shape the use of productive resources and, even more conveniently, tends to underplay the role of overconsumption in the North. In this conceptualization, the poor are stripped of their humanity and their history. Hence, population growth conveniently shifts the blame, under the cover of liberal concern, to poor subsistence cultivators of the South.

Blaikie and Brookfield (1987: 34) challenge this approach from a political ecology perspective. They assert that population growth does not invariably lead to environmental degradation but is one factor among others, varying in significance in regionally specific ways. They stress the importance of examining political economy within a regional context and the importance of geographic scale spanning the local to the global as a means of defining environmental change in terms of its full contextual complexity.

With the work of Humbert (1923, 1927) and Perrier de la Bathie (1917, 1927), Madagascar has become a well-known example of the destruction of tropical flora by fire, shifting cultivation, and overgrazing. Humbert (1927) assumed Madagascar was largely deforested due to hundreds of years of these practices. However, the linkage between population growth, shifting cultivation, and forest clearance has only recently been made in the case of Madagascar (Jackson 1971; Knox 1989; Rossi 1979; Shoumatoff 1988), and throughout the Third World more generally (see Russell 1988; UNEP 1980).

THE POLITICAL ECOLOGY OF TROPICAL DEFORESTATION

As Peet and Watts indicate in their introduction (Chapter 1 in this volume), the poststructural critique of development discourse reveals the ideology and power relations embedded in Western rationality's claim to universal truth. "Truth" was a metaphor for modern rationality as defined in Western discourses. Within the context of modernization and Malthusian theory, binary dualisms construct "truth" in terms of traditional/modern and irrational/rational as they relate to the practices of shifting cultivation and the productive and reproductive strategies of indigenous cultivators.

In the case study presented here, shifting cultivation is situated in relation to other land-use practices and their impacts on the tropical rainforests of eastern Madagascar. I focus on the colonial period, when shifting cultivation was legally banned in the interest of forest conservation, and population growth was slowed, yet forests disappeared rapidly. The chapter focuses also on how the colonial state defined the shifting cultivation–deforestation nexus, and how this definition changed in the post-colonial period. Moreover, I reveal how groups other than the state within Malagasy society interpreted and reacted to the changing definitions of the link between population, shifting cultivation, and deforestation.

Western discourses about the "rationality" of land-use practices permeated administrative definitions and policy on deforestation in Madagascar in both the colonial and post-colonial periods. A counter-discourse of rationality also emerged, grounded in peasant resistance to the state's ban on shifting cultivation. In their discussion of everyday resistance, social movements, and discourses of protest in the work of James Scott (1985, 1990), Peet and Watts argue for uncovering discourses of protest and resistance of subaltern peoples and linking these alternative discourses with critiques of the hegemonic ideas on development and the environment. Through this case study, I demonstrate how the emergence of an alternative discourse critical of the Western "rational" discourse of wise forest use and conservation is a part of peasant resistance to threats to livelihood. I link considerations of peasant resistance to discourses defining what is "rational" and "irrational" in the struggles over environmental and agrarian land-use practices. The case study suggests that the political ecology approach must now also include an examination of notions of the rational and irrational as they appear in discourses marking the social struggles over environmental and agrarian practices. This is particularly important in discourses defining and explaining tropical deforestation, whether global or local.

THE POLITICAL ECONOMY OF DEFORESTATION IN COLONIAL MADAGASCAR

War, resistance, and famine accompanied Madagascar's annexation as a French colony in 1895. Many people fled to the forests, surviving there for years as shifting cultivators. Irrigated, marsh, and rain-fed rice fields were abandoned, and from virtually the dawn of colonial rule, the island's people faced chronic rice shortages. Meanwhile, the state was eager to increase revenues and exports to France and elsewhere in Europe and Africa, as well as generate new markets for French goods in the colony.

Increasing numbers of rain-fed fields were abandoned – the casualties of labor shortfalls, low producer prices, cyclones, and drought – and food security in the eastern region was eroded. Due to its labor demands and attractive producer prices, coffee cultivation increased in popularity among European settlers and some Malagasy farmers. The resulting introduction of coffee cultivation also contributed to shortfalls in rice production. Razoharinoro-Randriamboavonjy (1971), Althabe (1982), and Rakotoarisoa and Richard (1987) have noted the tension between cash cropping and rain-fed rice regimes in terms of claims on land and labor time.

Chronic shortfalls in rice production may also be partially attributed to the region's physical geography. There are few, small, low-lying fertile valley bottoms and riverbanks on the eastern coast suitable for either irrigated rice or coffee cultivation. Where there are, the most fertile were used for coffee production. Population densities were increased by in-migration of the Antandroy and Antaifasy from the south, who moved north to find wage work to pay taxes. This led to the expansion of agricultural land for subsistence rice production through shifting cultivation. The first notable shortfalls in rice production for the market occurred in 1911 due to increased exports, often exacerbated through state requisition. In the tropical forest areas of the east coast, shortfalls in rice production were attributed to the state's ban on shifting cultivation as former surplus producing areas were transformed into net importers during the colonial period (Razoharinoro-Randriamboavonjy 1971).

Thus, deforestation in Madagascar is directly related to the introduction of coffee cash cropping. Still the island's chief export crop, coffee was planted on the east coast, the island region with the largest remaining forest cover. The soil erosion rates on coffee lands are nearly double those on subsistence plots, because broad expanses of bare soil under the coffee bushes are particularly vulnerable to violent storms during the rainy season (Temple 1972). However, coffee production was not declared an "irrational" land-use practice by the state. As the most fertile areas became devoted to export crop production, cultivators cleared forested slopes for subsistence. The imposition of taxes and the resulting in-migration increased the region's demand for both locally grown and imported rice. The province of Toamasina, where most of the island's rainforests are found, became a rice-importing region, and subsistence production was carried out through shifting cultivation (Figure 7.1).

From the outset, the colony's agricultural production was geared primarily to exporting: coffee, rice, and beef were of particular importance (Isnard 1971). The Central Highlands became the primary irrigated rice-growing region for both subsistence and export. Cloves, vanilla, and sugar cane were cultivated in the north; cattle, rice, and maize were major crops in the west; the arid south became a labor reserve. The state's emphasis on export production led to a pattern of uneven economic development and regional fragmentation (Hugon 1987; Isnard 1971) which created increasing production pressures and demands on environmental resources.

THE STATE'S PROHIBITION OF SHIFTING CULTIVATION

Estienne de Flacourt (1661: 23) was the first European to describe shifting cultivation (*tavy*) in Madagascar:

> They plant their rice in the hills and valleys, after having cut the woods which are largely of certain coarse canes which are called Voulou through-out the island and in the great Indies Bambu or Mambu. When dry, they are set afire and burn with a noise to make the earth tremble for a mile around . . . when the woods have been burned, all ground is covered with ashes, which are moistened by the rain. After some time they plant the rice in a curious manner. It is that all the women and girls of a village help each other in planting, marching side by side as a front, each having a pointed stick in hand with which they punch holes in the ground, dropping into each two grains of rice, covering the whole with the foot, all doing the same thing in unison, dancing and singing.

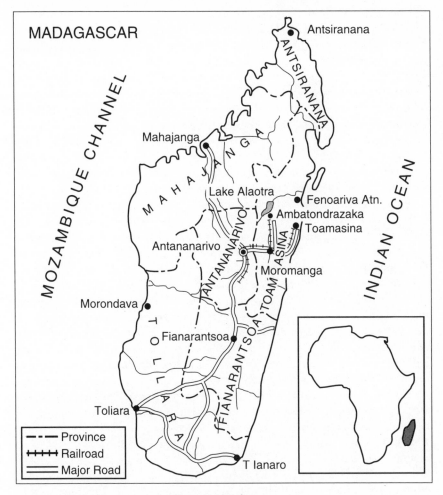

Figure 7.1 Madagascar

Early geographic accounts of shifting cultivation characterize it as unplanned, aimless, nomadic, unproductive, and uneconomical in the utilization of land and labor and destructive of the environment (Whittlesey 1937). The discourse of colonial conservation parallels this academic view. Writing about the forests of Madagascar in 1890, one observer on a missionary tour expressed "a hope that the present wholesale destruction of the forest by the natives may be soon effectually stopped by the Government, and that its valuable resources may be speedily utilized" (Baron 1890: 211).

On tour a year after Madagascar became a colony, Lieutenant Michel (1897) wrote that, in the eastern areas of the island, inhabitants had virtually destroyed the last two remaining forested areas due to their preference for forest lands for shifting rice cultivation. He recommended strong penalties to stop what he regarded as a barbarous and deplorable practice. In 1909, Governor General Victor Augagneur wrote in a circular to the administrative heads of provinces, districts, and military outposts: "My attention has been drawn to the enormous yearly damage that the natives (*indigènes*) do to the forests through the practice of *tavy*." Laws prohibiting the burning of the forests and lands in various regions of the island were enacted in 1881, 1900, and 1902. In 1913, clearing land by fire for *tavy* was universally prohibited. In a circular dated 29 April 1913 to the heads of the provinces and districts, Governor General Picquie (1913) wrote, "*Tavy* is condemned because of the enormous damage it does to the forests and will, before long, lead to the disappearance of the beautiful forests of the Colony." He promoted efforts to "lead the natives to progressively abandon their mountainous, rainfed fields for lowland marsh rice cultivation." He envisioned a transformation of agricultural production that would permanently attach farmers to the land, assuring regular and abundant harvests. On 10 July 1915 the administration announced it would recruit farmers from the Central Highlands to develop irrigated rice production on the east coast, in the hope that this would shift east coast rice farmers away from *tavy*.

The colonial state's stated objective was to save primary forest for "rational forest resource management," forcing the Malagasy into other forms of rice cultivation considered more intensive and sustainable. In fact, the colonial vision proved difficult to implement for ecological and social reasons. The ecosystems of the mountainous, eastern coast are quite different from those of the drier, more mountainous Central Highlands. In many areas, there were insufficient low-lying areas suited either to marsh rice or irrigated agriculture and, accordingly, the enforcement of the ban was uneven. For example, in Beforona, a mountainous region with little low-lying land, the colonial administration permitted residents to practice *tavy* on previously deforested mountain flanks. But in nearby Vatomandry, where ecological conditions made it easier to establish irrigated rice cultivation, *tavy* was completely prohibited. However, as more peasants moved from Vatomandry to Beforona exactly to practice *tavy*, the administration was forced to ban shifting cultivation completely there too.

Also it was easier, from the state's standpoint of timely and efficient tax collection, to group the Malagasy into settled villages near major transportation routes. *Tavy* was generally practiced by scattered, extended family groups, with members living in temporary shelters near their fields as harvest approached. In the eyes of colonial officials, *tavy* encouraged tax evasion or, at the very least, it made tax collection more difficult and time consuming; it was also a major headache when procuring workers for forced labor parties.

The colonial discourse concerning "rational" forest use glossed over the power relations rooted in state strategies of natural and human resource control and extraction. "Rational" use, for the state, meant utilizing forest resources for railroad ties and fuel, for hardwood exports, for logging concessions, as well as nature reserves and parks. Shifting cultivation for survival was irrational and destructive in both colonial and post-colonial discourses.

MALAGASY INTERPRETATIONS: RESISTING THE BAN

The nationalist newspaper *L'Opinion* interpreted the ban on shifting cultivation as an administrative strategy designed to force the Malagasy into wage work by depriving them of independent means of subsistence. The prohibition of *tavy* figured in the revolt of 1904 in Antaiska and in the massacres of 1947 near Anosibe and Moramanga (Vérin 1954). Peasant unrest followed the ban in the east. Peasants in Moramanga ignored the prohibition and spent time in prison or paid fines in lieu of abandoning shifting cultivation. Illegal burning of primary and secondary forest and prairie became a symbol of peasant protest against state authority. The Malagasy also circumvented the prohibition by involving colonial settlers. Taking wage work on colonial concessions, Malagasy practised shifting cultivation there, leaving colonial landowners to argue with the state as to whether or not the ban could be enforced on private property.

The Malagasy distinguished between subsistence work on the one hand, and export production and wage work on the other. Wage work was synonymous with enslavement. By comparison, of all subsistence farm work, rice cultivation was culturally and materially the most significant. It necessitated a wide-ranging array of social relations spanning field preparation, sowing, harvesting, and processing activities which confirmed the value of collective labor. Observers as widely separated in time as Flacourt (1661) and Feeley-Harnik (1991) said that work parties were festive and often playful. For the people of the east coast in particular, *tavy* represented an ideal way of life inherited from the ancestors (Beaujard 1985). The state's ban elevated *tavy* to a symbol of independence and liberty from colonial rule: Gérard Althabe's research (1982) among communities on the east coast confirms the transformation of *tavy* from subsistence activity to symbol of resistance to state authority and affirmation of Malagasy identity. While practising *tavy,* farmers did not wear clothes or use tools manufactured in Europe. For the Malagasy, reverence and remembrance of the ancestors were, and continue to be, a part of collective cosmology and identity (see Bloch 1971; Feeley-Harnik 1991). Prayers and sacrifices are offered to God (*Zanahary*) and the ancestors at planting time to affirm the existence of the past in the present through ritual practice (Kottak 1980; Ratovoson 1986). The location of the fields as part of ancestral lands can also be significant. There is a boundary between the cash crop and subsistence fields. *Tavy* thus became the spatial metaphor of Malagasy identity and resistance in the wake of the ban.

The colonial state's perspective on shifting cultivation emphasized the necessity of its prohibition as an economic goal to ensure rational forest management, which ultimately would increase state revenues and increase the supply of local rice grown by other forms of sedentary agriculture. This economic objective was articulated in terms of forest conservation and the necessity for the ban. The Malagasy interpretation of the ban differed completely from that of the colonial state. Peasants interpreted the ban as a form of labor control compelling them to work for wages and buy rice, thus losing their independence. Moreover, the ban annihilated the sacred space where the living engaged in dialogue with the ancestors. Mass revolts and resistance, as well as scattered, individual acts of non-compliance, spoke directly to this.

Resistance to the ban thus meant more than pitting the right to subsistence over forest conservation; it embraced issues of power, labor control, culture, and Malagasy identity. Thus multi-faceted forms of resistance to the ban on shifting cultivation cut across lines of class, gender, and ethnicity and emerge in both urban and rural settings. Resistance was individual and collective, overt and covert, passive and aggressive. Shifting cultivation became a signifier of resistance to colonial rule nourishing an emerging nationalism which exploded in the revolt of 1947. This analysis resonates with Scott's assertions (1985, 1990) concerning the complex nature of the cultural politics of resistance, and reveals how instances of peasant resistance can be combined with an examination of discourses of protest and alternative rationalities. Not surprisingly, the French failed to eradicate the practice; likewise, the post-colonial state is beset with the same difficulties.

Notions of rationality marked the colonial discourse of deforestation and shifting cultivation, thus obscuring the political struggle over the fate of the forest and over the economic and spatial autonomy of shifting cultivators. Shifting cultivation can be socially and culturally rational. The colonial definition of shifting cultivation as irrational ignores the livelihood needs and culture of the Malagasy. A social definition of shifting cultivation emphasizes a cultural construction of the environment. Contending definitions of shifting cultivation in relation to deforestation tend to stress one or the other sides of the culture/nature dualism. Neither attempts to conceive culture and nature in terms of mutual interdependence.

THE EXTRACTION OF FOREST PRODUCTS AND DESTRUCTIVE LOGGING PRACTICES

In 1921, the colonial state opened the island's forests to concessionary claims for exploitation (Service Forestier 1922). In its 1922 annual report, the colonial Forest Service objected to the state's action, claiming that owners of concessions "mined" forest resources for short-term gain. Exploitation meant the pillage and destruction of some of the most beautiful and accessible forests on the island as the search for precious woods such as ebony, rosewood, and palisander intensified. Fines were much lower than actual damages, and often owners clearcut and destroyed vast areas beyond the boundaries of their concessions. In one case, near the capital city of Antananarivo, 1,100 hectares were cut beyond the perimeter of a concession. The report noted the irony of clearcutting on massive concessions while the prohibition against *tavy* was in effect. Forest concession owners were both European and Malagasy, and ownership cut across gender and class lines. Malagasy women formed a sizable proportion of those requesting concessions as a means of subsistence, while former military men demanded them as payment for service to the state and *la patrie*. Forest products gathered for export included raffia, beeswax, honey, lichens, and camphor. The Forest Service was unable to regulate resource extraction due to shortages of labor and capital, as well as lack of political will. Infractions such as clearcuts and the burning of forests were often overlooked by forestry inspectors who stayed at the homes of concession owners while touring their districts.

Roughly 70 percent of the primary forest was destroyed in the thirty years between 1895 and 1925 (Hornac 1943–4). Gathering forest products, logging, burning, grazing, shifting cultivation, and export crop production – all these contributed to the destruction. The construction of railroads and their operation relied heavily on timber, intensifying the demand for wood. The Forest Service in Madagascar was established to ensure a stable and sufficient supply of wood for railroad operation. The creation of natural reserves was legislated in 1927, and forest reserves were created by law in the early 1930s, but these tangible responses to calls for forest conservation and preservation came too late to preserve more than small pieces of the forest ecosystem.

POPULATION GROWTH RATES AND DEMOGRAPHIC CRISIS, 1900–41

As primary forests disappeared between 1900 and 1941, the national population growth rate was at, or below, the replacement level due to malnutrition, famine, bubonic plague, tuberculosis, syphilis, and alcoholism. The Malagasy press identified labor conscription, famine, and the introduction of the plague into the Central Highlands by Europeans as key factors in what academics defined as "a demographic problem" or "depopulation" (Chevalier 1952: 32–3). In fact, settlers were plagued by labor shortfalls in urban and rural areas. In 1930, according to colonial estimates, population density was approximately 6 people per square kilometer, and 12 people per square kilometer in the most densely populated areas of the Central Highlands and the east coast. Between 1900 and 1958 the island's population grew from 2.5 million to 4.6 million (Chevalier 1952: 33; Jackson 1971: 71). From 1900 to 1940, estimates of forest clearance range from 3–7 million hectares (Boiteau 1982; Hornac 1943). According to Forest Service reports of 1930, I estimate that approximately 4 million hectares were cleared during this period due not to population growth *per se* but to logging, forest product extraction, export crop production, shifting cultivation, grazing, and burning. Between 1941 and 1982, another 4 million hectares of forest were transformed (Green and Sussman 1990). Deforestation rates during the colonial period, roughly match those of the post-colonial period, suggesting that population growth alone cannot explain these similar rates.

CONCLUSION

The forests of Madagascar are dwindling to the vanishing point, as an impoverished and rapidly swelling human population slashes and burns the forests simply to survive.

(Angier 1992)

Between 1900 and 1940, Madagascar's forests were transformed into fields for cash crops, subsistence plots for wage workers, and timber concessions. These land-use practices also changed land tenure and access. In turn, this affected forest cover. A legacy of denudation which marks the island's contemporary landscape has little to do with population growth or shifting cultivation. Indeed, in the case of Madagascar, population stagnation coincided with historically unprecedented deforestation rates during the colonial period. The discourse of the population–deforestation connection enters the debate only after the Second World War. The international environmental movement focused on the disappearance of the rainforests of Africa, Asia, and Latin America, linking this with the threat of global warming and the loss of biodiversity. Growing numbers of poor peasants using "destructive" practices like shifting cultivation were seen as threatening Madagascar's treasury of unique flora and fauna, which included lemurs and the rosy periwinkle – a plant essential for the treatment of leukemia. Yet the ban on shifting cultivation had not solved the problem in the first half of this century, and probably would not do so now. My argument is not that population growth is unimportant, but rather that it was of negligible importance as cause during a forty-year period in the early part of the century in which approximately 4 million hectares of forest were felled. Population growth is not a universal, root cause of deforestation; instead the impacts of population growth are mediated through a variety of social institutions on multiple, overlapping scales (Palloni 1992). Similarly, in contemporary times, Marcus Colchester's (1993) analysis of the political ecology of West Africa's forests indicates that the political nature of deforestation has been ignored and reveals that shifting cultivation and population growth are not main factors of deforestation in Gabon, the Congo, and the Central African Republic. He concludes that only a "radical transformation of the political economies of the region," not technical fixes, will address poverty and environmental degradation (1993: 166).

I contend that contemporary "post-"colonial discourses about deforestation mirror the colonial rhetoric: there is a division concerning the "irrational" ways peasants use forests and the "rational," "wise use" management symbolized by state control. "Rationality" in the discourse of forest management is part of a "regime of truth" promulgated by dominant powers (Peet and Watts, Chapter 1 in this volume). The notion of "truth" both reveals and obscures the political economy of deforestation in Madagascar. Western definitions of "reason" and "rationality" can construct – as well as destruct – landscapes and places. Through the invocation of Western "reason," landscapes are altered and conserved; in the name of "development" or "conservation" resources are extracted and controlled.

Invocations of Western reason and rationality actually reinforce domination, destruction, and deforestation while purporting to accomplish the opposite. Such forms of "reason" are masks for exploitation. They are contested and resisted in this case by alternative definitions of landscape use. These definitions link landscapes to indigenous physical and cultural survival. The political ecology approach unites considerations of cultural materialism with considerations of ecological change. It reveals how local forms of resistance to the dominating discourses of reason also involve struggles over the meanings of nature, culture, and landscape.

This chapter demonstrates how the meanings of shifting cultivation have changed among various groups in line with shifting political and economic interests. For the Malagasy peasants, shifting cultivation was a link to the ancestors, an affirmation of identity, symbol of and means of resistance to state authority. For the colonial state, shifting cultivation was a destructive practice transforming forests into degraded grassland and impeding state-led forest extraction, labor control, and tax collection. Recently, on the international level, as exemplified by multilateral agencies such as the World Bank, shifting cultivation has been defined as primitive, inefficient, and destructive in itself and then later redefined as destructive in terms of growing numbers of shifting cultivators. The link recently made between shifting cultivation and population growth is compelling evidence of yet another variation of neo-Malthusian themes concerning human–environment interactions. This sort of interpretation has become increasingly common in debates concerning global issues such as tropical deforestation. Through a focus on a crucial global environmental problem, I link discourse analysis – both colonial and post-colonial – with an examination of peasant resistance and response during the colonial period. This case study demonstrates not only the inadequacy of the neo-Malthusian approach to deforestation but also how alternate rationalities and resistances emerge to counter and contest the hegemonic discourse of domination and development that define rational land use.

As David Harvey (1974) once noted, the use of specific kinds of definitions and interpretations of the relationship between population and resources can have profound political implications. Abstracting shifting cultivation and population dynamics from specific regional contexts and omitting a consideration of external processes, such as colonial capitalism and imperialism, may serve an ideology of repression, feed the fires of prejudice and fear, and promote what are exactly the wrong kinds of solution. In addition, a postmodern awareness of the politically instrumental uses of "reason" and "rationality" in definitions and debates about deforestation reveals an important new dimension to struggles over resource access and control in both the colonial and post-colonial spheres.

NOTE

I would like to thank the editors, Dick Peet and Michael Watts, for their assistance in the preparation of this chapter. This research was funded by the Rockefeller Foundation and the Graduate School Research Fund of the University of Washington.

REFERENCES

Agarwal, B. 1988. "Neither sustenance nor sustainability," in B. Agarwal (ed.) *Structures of Patriarchy: State, Community and Household in Modernising Asia.* London: Zed Books, pp. 83–120.

Althabe, G. 1982. *Oppression et libération dans l'imaginaire: les communautés villageoises de la côte orientale de Madagascar.* Paris: François Maspero.

Angier, N. 1992. "Bizarre baby raises hopes for an endangered primate," *New York Times* 19 May.

Augagneur, V. 1909. "Circulaire au sujet des incendies de frets," Reglementation relative aux tavy, incendies et feux de brousse 1903–1930. Madagascar Série D 6(7) D27. Archives d'Outre Mer, Aix-en-Provence. Archives Nationales de France.

Baron, R. 1890. "A Malagasy forest," *Antananarivo Annual and Madagascar Magazine* 4, 2: 196–211.

Bassett, T.J. 1988. "The political ecology of peasant–herder conflicts in the Northern Ivory Coast," *Annals of the Association of American Geographers* 78: 433–52.

Beaujard, P. 1985. "Riz du ciel, riz de la terre: idéologie, système politique et rizicultures dans les 'royaumes' tanala de l'ikong du 18–19 siècles," *Etudes rurales* 99–100: 389–402.

Blaikie, P. and H. Brookfield. 1987. *Land Degradation and Society.* London and New York: Methuen.

Bloch, M. 1971. *Placing the Dead.* London and New York: Seminar Press.

Boiteau, P. 1982. *Contribution à l'histoire de la nation Malgache.* Paris and Antananarivo: Editions Sociales, reprint of the 1958 edition.

Brown, L.R. 1989. "Feeding six billion," *World Watch* September/October: 32–40.

Brown, L.R. and J.E. Young. 1990. "Feeding the world in the nineties," in L.R. Brown, A. Durning, C. Flavin, H. French and J. Jacobson (eds.) *State of the World 1990.* New York and London: Norton, pp. 59–78.

Bryant, R.L. 1992. "Political ecology: an emerging research agenda in Third-World studies," *Political Geography* 11, 1: 12–36.

Buttel, F.H., A.P. Hawkins and A.G. Power. 1990. "From limits to growth to global change." *Global Environmental Change* 1, 1 (Dec.): 57–66.

Carney, J. and M. Watts. 1991. "Disciplining women: rice, mechanization and the evolution of Mandinka gender relations in Senegambia," *Signs* 16, 4: 651–81.

Chevalier, L. 1952. *Madagascar: populations et ressources.* Institut national d'études démographiques. Travaux et Documents. Cahier no. 15. Paris: Presses Universitaires de France.

Colchester, M. 1993. "Slave and enclave: towards a political ecology of equatorial Africa," *The Ecologist* 23, 5: 166–73.

Conklin, H. 1961. "The study of shifting cultivation," *Current Anthropology* 2, 1 (Feb.): 27–61.

Dove, M.R. 1983. "Theories of swidden agriculture, and the political economy of ignorance," *Agroforestry Systems* 1: 85–99.

Ehrlich, P.R. and A.H. Ehrlich. 1990. *The Population Explosion.* New York: Simon & Schuster.

Feeley-Harnik, G. 1991. *A Green Estate: Restoring Independence in Madagascar.* Washington: Smithsonian Institution Press.

Flacourt, E. de. 1661. *Histoire de la Grande Isle Madagascar avec une Relacion de ce qui s'est passé en années 1655, 1656, & 1657, non encore vue par la première.* Paris: Chez Pierre Bien-fait.

Green, G.M. and R.W. Sussman. 1990. "Deforestation history of the eastern rain forests of Madagascar from satellite images," *Science* 248: 212–15.

Harvey, D. 1974. "Population, resources, and the ideology of science," *Economic Geography* 50, 3: 256–77.

Hornac, J. 1943–4. "Le déboisement et la politique forestière à Madagascar," Mémoire de Stage. Mémoire de l'École Coloniale ENFOM. Archives d'Outre Mer, Aix-en-Provence. Archives Nationales de France.

Hugon, P. 1987. "La crise économique à Madagascar," *Afrique contemporaine* 144: 3–22.

Humbert, H. 1923 *Les Composées de Madagascar.* Caen: Imprimerie E. Lanier.

—— 1927. *La Destruction d'une flore insulaire par le feu: principaux aspects de la végétation à Madagascar.* Mémoires de l'Académie Malgache. Monograph, vol. 5.

Hurst, P. 1990. *Rainforest Politics: Ecological Destruction in South-East Asia.* London: Zed Books.

Isnard, H. 1971. *Géographie de la décolonisation.* Paris: Presses Universitaires de France.

Jackson, R.T. 1971. "Agricultural development in the Malagasy Republic," *East African Geographical Review* 9: 69–78.

Knox, M.L. 1989. "No nation is an island," *Sierra* May/June: 80–4.

Kottak, C.P. 1980. *The Past in the Present.* Ann Arbor, MI: University of Michigan Press.

Malthus, T. 1798. *An Essay on the Principle of Population.* Reprinted 1929. London: Macmillan.

Meadows K.L., J. Randers and W.W. Behrens II. 1972. *The Limits to Growth.* New York: Universe Books.

Michel, Lieut. 1897. "Excursion dans la province de Andevorants, Colonie de Madagascar," *Notes, Reconnaissancese et Explorations* 2, 9: 467–71. Tananarive: Imprimerie Officielle.

Multinational Monitor. 1992. "Who's to blame?" (Editorial), July/August: 5.

Murphy, A.B. 1991. "Regions as social constructs: the gap between theory and practice," *Progress in Human Geography* 15, 1: 22–35.

Paddock, W. and P. Paddock. 1967. *Famine 1975!* New York: Little Brown.

Palloni, A. 1992. "The relation between population and deforestation: methods for drawing causal inferences from macro and micro studies," CDE Working Paper 92–14, Center for Demography and Ecology, University of Wisconsin-Madison.

Peluso, N.L. 1992. *Rich Forests, Poor People.* Berkeley: University of California Press.

Perrier de la Bathie, H. 1917. "Au sujet des tourbières de Marotampona," *Bulletin de l'Académie Malagache* 1: 137–8.

—— 1927. *Le Tsaratanana, l'Ankaratra et l'Andringitra.* Mémoires de l'Académie Malgache. Monograph, vol. 3.

Peters, W.J. and Neuenschwander, L.F. 1988. *Slash and Burn: Farming in the Third World Forest.* Moscow, ID: University of Idaho Press.

Pickles, J. and M.J. Watts. 1992. "Paradigms for inquiry?" in R.F. Abler, M.G. Marcus, and J.M. Olson (eds.) *Geography's Inner Worlds.* New Brunswick, NJ: Rutgers University Press, pp. 301–26.

Picquie, A. 1913. "Circulaire au sujet de l'interdiction d'employer le feu pour la préparation des tavy et des terrains de culture." Reglementation relative aux tavy, incendies et feux de brousse 1903–1930. Madagascar Série D 6(7) D27. Archives d'Outre Mer, Aix-en-Provence. Archives Nationales de France.

Pred, A. 1984. "Place as historically contingent process: structuration and the time-geography of becoming places," *Annals of the Association of American Geographers* 74: 279–97.

—— 1986. *Place, Practice and Structure.* Totowa: Barnes & Noble.

Pudup, M.B. 1988. "Arguments within regional geography," *Progress in Human Geography* 12: 369–90.

Rakotoarisoa, J. and Richard, J. 1987. *Le Café dans le sud-est de Madagascar.* Paris: Ministère de la Coopération.

Ratovoson, C. 1986. "Religions traditionelles, environnement, espace et développement à Madagascar," paper presented at the Symposium pour les exigences religieuses et impératifs de développement dans les sociétés malgaches, Académie Malgache, 15–19 December.

Razoharinoro-Randriamboavonjy, Mme. 1971. "Note sur Madagascar pendant la deuxième guerre mondiale: la question du riz," *Bulletin de Madagascar* 304: 820–68.

Rosenblatt, D. 1992. "Black gold in western Siberia: the oil industry and regional development," M.A. thesis, Dept. of Geography, University of Washington.

Rossi, G. 1979. "L'erosion à Madagascar: l'importance des facteurs humains," *Cahiers d'Outre-Mer* 128–32: 355–70.

Russell, W.M.S. 1988. "Population, swidden farming and the tropical environment," *Population and Environment* 10, 2: 77–94.

Sayer, A. 1989. "The 'new' regional geography and problems of narrative," *Environment and Planning D: Society and Space* 7, 3: 249–362.

Scott, J.C. 1985. *Weapons of the Weak: Everyday Forms of Peasant Resistance.* New Haven, CT: Yale University Press.

—— 1990. *Domination and the Arts of Resistance: Hidden Transcripts.* New Haven, CT: Yale University Press.

Service Forestier. 1922. *Rapport.* Madagascar 5(18) D5. Archives d'Outre Mer, Aix-en-Provence. Archives Nationales de France.

Shoumatoff, A. 1988. "Our far-flung correspondents: look at that," *New Yorker.* 7 March: 62–83.

Temple, P.H. 1972. "Measurements of runoff and soil erosion at an erosion plot scale with particular reference to Tanzania," *Geografiska Annaler*, Series A Vol. 54A No. 3–5.

Thrift, N. 1983. "On the determination of social action in space and time," *Society and Space* 1: 23–57.

United Nations Environmental Programme (UNEP). 1980. *Tropical Woodlands and Forest Ecosystems: A Review.* UNEP Report No. 1. Nairobi: UNEP.

Urry, J. 1987. "Society, space, and locality," *Society and Space* 5: 435–44.

Vérin, P. 1954. "La Destruction de la fôret dans la zone orientale de Madagascar," Mémoire de Stage (Mars–Octobre). Mémoires de l'Ecole Coloniale ENFOM. Archives d'Outre Mer, Aix-en-Provence. Archives Nationales de France.

Watts, M.J. 1991. "Mapping meaning, denoting difference, imagining identity: dialectical images and postmodern geographies," *Geografiska Annaler* 73 B(1): 7–16.

Whittlesey, D. 1937. "Shifting cultivation," *Economic Geography* 13, 1: 35–52.

World Bank. 1992. *World Development Report: Development and the Environment.* Oxford: World Bank.

8

CONVERTING THE WETLANDS, ENGENDERING THE ENVIRONMENT

The intersection of gender with agrarian change in Gambia

Judith A. Carney

The startling pace of environmental change in the Third World during recent decades has directed increasing attention in regional studies to political ecology. This research framework combines the broad concerns of ecology with political economy, especially the ways that institutions like the state, market, and property rights regulate land use practices (Blaikie and Brookfield 1987; Bryant 1991; Peet and Watts, Chapter 1 in this volume). The growing association of environmental change with female-based social movements and gender conflict within rural households, however, suggests the need for improved understanding of gender relations and the domestic sphere since class as well as non-class struggles over resources are frequently mediated in the idiom of gender (Carney and Watts 1991; Guyer 1984; Shiva 1989; Sontheimer 1991). This poststructuralist emphasis on gender and household relations offers political ecology a better conceptualization of the complex and historically changing relations that shape rural land-use decisions (Peet and Watts, Chapter 1 in this volume).

A central insight of poststructuralist research over the past fifteen years is the need to extend the definition of politics from the electoral politics of the state and/or between classes to one that includes the political arenas of the household and workplace (Guyer and Peters 1987; Hart 1991: 95). This emphasis brings attention to the crucial role of family authority relations and property relations in structuring the gender division of labor and access to rural resources (Carney and Watts 1991; Guyer 1984; Moore 1988; Sharma 1980). However, as development interventions, environmental transformations, and markets place new labor demands and value on rural resources, these socially constructed relations of household labor and property rights often explode with gender conflict (Guyer 1984; Jackson 1993). Struggles over labor and resources reveal deeper struggles over meanings in the ways that property rights are defined, negotiated, and contested within the political arenas of household, workplace, and state (Hart

1991: 113). By linking property rights and gender conflict to environmental change, this chapter brings the poststructuralist concern with power relations and discourse to political ecology (Peet and Watts, Chapter 1 in this volume).

The environmental transformation of the wetlands of Gambia, a small country in West Africa, provides the setting for this chapter, an examination of the gender-based resource struggles accompanying irrigation schemes. I use multiple case studies of two forms of wetland conversion – irrigated rice schemes and horticultural projects – to trace the disputes that have surfaced during the past fifty years in Mandinka households over women's land rights. The analysis reveals repeated gender conflicts over rural resources as male household heads concentrate landholdings in order to capture female labor for surplus production. Mandinka gender conflicts on the wetlands involve disputes over women's traditional land rights within the common property system, thereby illustrating the significance of struggles over meaning for contemporary struggles over labor and resources.

Building upon previous research in several Gambian wetland communities, the chapter is divided into five sections. The first presents the environmental context of the Gambian wetlands, the extent and significance of wetland farming, as well as women's labor in ensuring its productive use. The next section provides an historical overview of environmental and economic changes modifying women's access to Gambian wetlands. This follows an account of recent policy shifts addressing the country's environmental and economic crisis. Two case studies then detail the relationship between economic change and the forms of women's resistance to the process of land concentration. The chapter concludes by analyzing how wetland commodification has made women's access to resources increasingly tenuous despite income gains.

THE ENVIRONMENTAL CONTEXT OF THE GAMBIAN WETLANDS

Gambia, a narrow land strip 24–50 kilometers (14–30 miles) wide and nearly 500 kilometers (300 miles) long, encloses a low-lying river basin that grades gradually into a plateau where the altitude seldom exceeds 100 meters (325 feet) (Figure 8.1). The plateau forms about one-third of the country's land base and depends upon rainfall for farming (Carney 1986: 21). Precipitation during the months of June to October averages 800–1,100 millimeters (31–43 inches) and favors the cultivation of millet, sorghum, maize, and peanuts. As with neighboring Sahelian countries, the Gambian rainfall regime fluctuates considerably between years and within a season. Between the 1960s and 1980s, for example, annual rainfall declined by 15–20 percent and became increasingly distributed in a bi-modal seasonal pattern (Hutchinson 1983: 7). The recurrence of a two-week, mid-season dry spell during the month of August increased cropping vulnerability on the uplands and dependence on lowland farming (Carney 1986: 25–30).

166

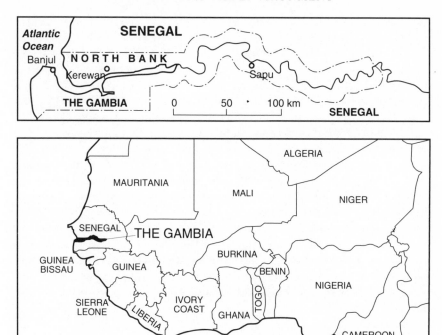

Figure 8.1 Gambia

These wetlands are critical for understanding human livelihood and survival in the unstable rainfall setting of the West African Sudano–Sahelian zone. Lowland environments permit a multiple land-use cropping strategy which utilizes other forms of water availability, thereby freeing agricultural production from strict dependence on rainfall. Constituting nearly 70 percent of the country's land mass, the Gambian wetlands make available two additional environments for agriculture: (1) the alluvial plain flooded by the river and its tributaries; (2) a variety of inland swamps which receive water from high water tables, artesian springs, or occasional tidal flooding (Carney 1986: 20–1) (Figure 8.2). Farming these critical environmental resources enables an extension of crop production into the dry season or even year-round. Gambian wetlands are traditionally planted to rice, although vegetables are frequently grown with residual moisture following the rice harvest (Carney 1986: 82; Dunsmore 1976: 208–11).

While Gambia abounds in lowland swamps, not all are suitable for farming. Riverine swamps coming under marine tidal influence are permanently saline within 70 kilometers (42 miles) of the coast, seasonally saline up to 250 kilometers (150 miles) and fresh year-round only in the last 150 kilometers (90 miles) of the Gambia River's course (Carney 1986: 33). The suitability of inland swamps for crop production, moreover, depends on the influence of differing

167

moisture regimes for groundwater reserves. Consequently, although Gambia contains over 100,000 hectares (247,000 acres) of lowland swamps, only about a third can be reliably planted (ALIC 1981: 19; CRED 1985: 127; GGFP 1984). Until the mid-1980s most of the available swampland was farmed to rice, with about 20,000–25,000 hectares (49,400–61,750 acres) planted along the river floodplain and another 6,000–8,000 hectares (14,800–19,760 acres) cultivated in inland swamps (FAO 1983: 17; Government of Gambia 1973–91).

Wetland cultivation is thus pivotal to the Gambian farming system, enabling crop diversification over a variety of micro-environments and a reduction in subsistence risk during dry climatic cycles. Wetland agriculture, however, requires considerable attention to forms of water availability as well as edaphic and topographic conditions. In Gambia this knowledge is embodied in women who have specialized in wetland cultivation since at least the early seventeenth century and have adapted hundreds of rice varieties to specific micro-environmental conditions (Carney 1991: 40; Gamble 1955: 27; Jobson 1904 [1623]: 9). This cumulative *in situ* knowledge of lowland farming underlies Gambia's regional importance as a secondary center of domestication of the indigenous West African rice, *Oryza glaberrima*, cultivated in the area for at least 3,000 years (Porteres 1970: 47).

GENDER, ENVIRONMENT, AND ECONOMY

Although lowland swamps and rice production are traditionally women's domain, prior to the mid-nineteenth century both men and women were involved in upland and lowland cropping systems. Men assisted in field clearing for rice cultivation while women weeded upland cereal plots (Carney and Watts 1991: 657; Weil 1982: 45–6). The abolition of the Atlantic slave trade and the turn to "legitimate commerce" in the nineteenth century led to Gambia's incorporation into the world economy through commodity production. By the 1830s peanut cultivation proliferated on the uplands (Carney 1986: 77–8). The imposition of British colonial rule by the end of the century brought taxation and fiscal policies, thereby accelerating reliance on peanuts as a cash crop. These political-economic changes resulted in an increasingly specialized use of agricultural space and a more gendered division of labor. The effects of colonial rule became most evident among the rice-growing Mandinka, Gambia's dominant ethnic group and principal wetland farmers.

By the end of the century, Mandinka men's growing emphasis on peanut cultivation resulted in a reduction in millet and sorghum production for household subsistence (Carney 1986: 92; Jeng 1978: 123–4; Weil 1973: 23). Women compensated for upland cereal shortfalls by augmenting rice production in lowland swamps. The gender division of labor became increasingly spatially segregated with the cash crop concentrated on the uplands under male control and women's farm work largely oriented to lowland rice, which emerged as the dietary staple (Carney 1986: 89–91; Weil 1973) (see Figure 8.2). The specialized

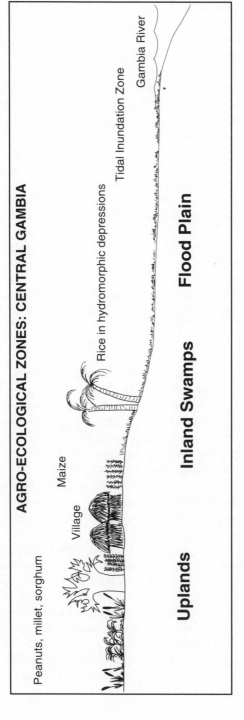

Figure 8.2 Agro-ecological zones, central Gambia
Source: Carney (1986)

use of agricultural land and concomitant disruptions in the gender division of labor accompanying nineteenth-century commodity production provide the setting for understanding twentieth-century gender conflicts among the Mandinka over commodification of the Gambian wetlands.

Policy interest in wetland environments began in the early decades of the twentieth century when colonial officials began documenting farming practices in diverse lowland settings (Carney 1986: 126–7). The objective was to improve household subsistence security and generate rice surpluses which would feed an expanding pool of migrant male laborers whose seasonal influx (*c.* 20,000) accounted for the pace of peanut expansion on the uplands (Carney 1986; Swindell 1977). Initial colonial efforts aimed at improving swamp accessibility, opening up new areas for cultivation, and increasing rice yields through improved seeds. By the 1960s swamp development projects had culminated in an expansion of rice planting to some 26,000 hectares (65,000 acres) (Carney 1986: 178). But limits had been reached on the degree to which women could carry the subsistence burden. Further gains in food availability rested on altering the gender division of labor by drawing men into rice growing. The colonial government's inability to persuade Mandinka men to take an active part in rice cultivation prompted Governor Blood to lament: "[Rice] is still almost entirely a woman's crop and I doubt whether more woman hours can be devoted to this form of cultivation" (NAG 1943). With absolute limits reached on available female labor for rice cultivation, swamp rice projects came to a halt (Carney 1986: 139; Carney and Watts 1991: 660).

In 1949 the colonial government initiated another approach to surplus rice generation by implementing a large-scale irrigation scheme on the site of the present-day Jahaly Pacharr project. The Colonial Development Corporation (CDC) scheme departed from the earlier swamp rice improvement project in one important way: land was removed from female growers through a 30-year lease program in an effort to force men into rice cultivation (Carney 1986: 126; Carney and Watts 1991: 666). The project failed due to a poorly designed irrigation system and lack of male and female interest in wage work; yet the CDC scheme is notable for foreshadowing the post-independence emphasis on irrigation as well as the gender-based conflicts that would surface in subsequent wetland development projects.

These conflicts center on resistance by male household heads to repeated attempts by the colonial and post-colonial state to force them into rice cultivation, a change in household labor patterns crucial for producing a marketable surplus of the dietary staple. Gambian rice projects have accordingly exploded with gender conflict as males attempt to deflect the labor burden in rice growing onto their wives and daughters while simultaneously making new claims to the surpluses produced by female labor. They have facilitated this objective by manipulating customary tenure "laws" to reduce women's individual land rights in developed wetlands, and thus female control over a portion of their labor benefits. This gender conflict over the past fifty years has resulted in the steady erosion of women's crop rights and mounting female militancy in technologically improved wetlands.

The stage for the ensuing conflicts was set in the 1940s when colonial development policies improved swampland access and rice productivity. Male household heads and village elites responded to swamp development by calling into question women's long-standing land rights on the wetlands. In one case that reached the colonial authorities, Mandinka men argued that "if women mark the land and divide it, it would become 'women's property' so that when a husband dies or divorces his wife, the wife will still retain the land, which is wrong. Women must not own land" (Rahman 1949: 1). Women's land rights were being contested on the grounds that wetland development would enable females to alienate developed swampland from the domestic landholding.

Such claims, however, ignored women's pre-existing rights to individual plots within the customary tenure system. Presenting female land rights as a threat to domestic property relations obfuscated the real issue at stake with colonial rice development: control over the surplus produced by female labor. Under the guise of safeguarding the domestic landholding from a female land grab, men severed women's individual land rights (*kamanyango*) in developed swamps. New land development was placed under another customary tenure category, *maruo*, or land whose product contributes to household subsistence. A similar conflict arose following the CDC project failure: women claiming the benefits to the plots they farmed (*kamanyango*), household heads declaring the project's improved area *maruo* (Dey 1980: 252–3). The significance of the *maruo* designation for resource struggles is that its application means females no longer receive income benefits from their farm work. The surplus, instead, comes under control of the male household head who controls decisions on its disposition.

A brief review of the meaning of the two Mandinka terms for property rights and control over surplus illuminates the issues in dispute. Land in rural Gambia is held in communal tenure but controlled at the village level by male lineages who trace their descent to the village's first settlers. As unmarried daughters, women receive land rights from their father's landholding lineage. When they marry and move to their husband's household, land rights are exercised through the husband's kinship group.

The struggles over land use on improved wetlands derives from the multiple meanings assigned to the Mandinka term *maruo*. At the most general level, *maruo* refers simultaneously to the household landholding as well as to the labor obligations of family members towards collective food production. As men argued in the swampland development projects, no part of the family landholding may be alienated from household control. But this claim conflates discussion of family property control with the intricacies of the communal tenure system which, by custom, accords all able family members a few individual plots (*kamanyango*) in return for laboring on the greater number of fields dedicated to household subsistence (*maruo*). The salient issue is that such individual plots grant junior male and female family members the right to keep the benefits from their labor and to sell the product if they so desire. Men traditionally meet their *maruo* work obligations on the uplands through cultivation of millet, sorghum, and maize as well as peanuts, which are often sold for cereal purchases. Male *kamanyango* production is usually devoted to the cultivation of peanuts for sale. Mandinka women, whose work chiefly occurs on the wetlands, fulfill both their *maruo* and *kamanyango* production through rice cultivation.

Kamanyango plots are therefore a critical issue in Gambia, where rural society is largely polygynous, young men compete with their fathers for wives, male and female budgets are frequently separate, and women are customarily responsible for purchases of clothing and supplemental foods crucial for the well-being of their children. While the steady erosion in female *kamanyango* rights with subsequent rice development projects has resulted in males assuming many of the responsibilities formerly met by their wives, women often acutely experience the loss of their economic independence. The shift in wetland resource control frequently exacerbates intra-household conflict between co-wives, their husbands and fathers-in-law as females are forced into negotiating for the allocation of household resources received from their labor in rice swamps.

THE ENVIRONMENTAL AND ECONOMIC CRISIS: POLICY SHIFTS

Since independence in 1965, Gambia has experienced rainfall declines and accelerated environmental degradation of its uplands, a massive influx of foreign aid for development assistance (1968–88), policy shifts favoring commodification of the wetlands, and an International Monetary Fund (IMF) structural adjustment program (1985–95). These changes have shaped post-independence accumulation strategies and gender conflicts among rural households.

Gambia entered independence with a degraded upland resource base and a vulnerable economy. The results of the longstanding monocrop export economy were evident throughout the traditional peanut basin, once mantled with forest cover but substantially deforested during the colonial period (Mann 1987: 85; Park 1983: 4). Reliance on peanuts to finance mounting rice imports grew more precarious in the years after independence: peanut export values fluctuated considerably but through the 1980s grew less rapidly than the value of food imports (Carney 1986: 254; FAO 1983: 4). Farmers responded to declining peanut revenues through an intensification of land use – namely, by reducing or eliminating fallow periods in peanut cultivation. The result was accelerated land degradation, particularly in the north bank region, oriented to Senegalese peanut markets with generally higher producer prices (see Figure 8.1). Today land degradation on the north bank is more advanced than in the rest of the nation (Gambia German Forestry Project 1988; Government of Gambia 1977.

All this brought renewed attention to the wetlands. The 1968–73 Sahelian drought coincided with an escalation of capital flows from multilateral banks and financial institutions to the Third World (Shiva 1989: 220; Thrift 1986: 16).

The changing pattern of global capital accumulation impacted the Gambian wet-lands in the form of river basin development and irrigated farming. International development assistance brought far-reaching changes to the critical wetland food production zone. Nearly 4,500 hectares (11,115 acres) of riverine swamps were converted to irrigation schemes and another 1,000 hectares (2,470 acres) of inland swamps to horticultural projects (Carney 1992: 77–8). Although affecting less than 10 percent of total swampland, these conversions in land use have had profound consequences for food production, female labor patterns, and access to environmental resources. But irrigated rice schemes and the introduction of technology to implement year-round cultivation have not reversed the country's reliance on imported rice. By the 1990s, only 40 percent of the land under this new strategy of development remained in production with just 10 percent under double-cropping (Carney 1986: 278). As domestic production lags, milled rice imports steadily climb, currently accounting for more than half the country's needs (Government of Gambia 1973–91). Population growth rates exceeding 3 percent per annum suggest demographic pressure on agricultural land; yet the failure to achieve food security is not the result of a Malthusian squeeze. Rather, it is the outcome of the changing use and access to resources which concentrates land within the communal tenure system and denies women benefits from improved rice production.

By the 1980s, women's economic marginalization from irrigated rice develop-ment resulted in non-governmental organizations (NGOs) targeting them for horticultural projects developed on inland swamps. The policy emphasis on horticulture intensified with the debt crisis of the 1980s and the implementation of the IMF-mandated structural adjustment program in 1985 to improve foreign exchange earnings and debt repayment. Economic restructuring has reaffirmed Gambia's comparative advantage in peanuts while favoring the conversion of hydromorphic swamps to horticulture (Government of Gambia 1987; Harvey 1990: 3; Landell Mills Associates 1989; McPherson and Posner 1991: 6; UNC-TAD 1986).

The respective policy emphases of the past twenty years have commodified the wetlands and accelerated changes under way in customary use and access to environmental resources. As the irrigation schemes provide new avenues for income generation within rural communities, women's access to improved land for income benefits is increasingly contested. The next two sections present an overview of the two post-drought wetland policy shifts, illustrating how customary laws are reinterpreted to reduce female access to productive resources, and the forms of women's resistance to a deteriorating situation.

"DROUGHT-PROOFING" THE ECONOMY: IRRIGATED RICE DEVELOPMENT

In 1966, the Gambian government, with bilateral assistance from Taiwan, initiated a wetland development strategy aimed at converting tidal floodplains into irrigated rice projects. The rationale for this development was to promote import substitution by encouraging domestic rice production. Rice imports had reached 9,000 tons per annum, and foreign exchange reserves had seriously eroded with declining world commodity prices for peanuts. The 1968–73 Sahelian drought revived late-colonial interest in irrigation and mobilized foreign aid for investment in river-basin development and irrigated agriculture (CILSS 1979; CRED 1985: 17; Derman 1984; Franke and Chasin 1980: 148–51; UNDP 1977). Hailed as a way of buffering the agricultural system from recurrences of a similar disaster, irrigation projects created also a steady demand for imported technical assistance, machinery, spare parts, and other foreign inputs. This "drought-proofing" strategy embodied in Gambian irrigation schemes targeted rice, prioritized by the post-independence government (CRED 1985: 22; Government of Gambia 1966). From the 1970s to the mid-1980s, following the Sahelian drought, the World Bank, the mainland Chinese government and the International Fund for Agricultural Development (IFAD) continued the Taiwanese development strategy by implementing double-cropped irrigated rice schemes on more than 4,000 hectares of women's tidal swamps (Figure 8.3).

Despite the contrasting ideological perspectives of the donors involved, all development strategies adhered to a remarkably similar course by introducing a Green Revolution package for increased production to male household heads (Dey 1981: 109). Developed at a cost of US $10,000–25,000 per hectare, and inserted into a pre-existing gendered form of agricultural production and land use, these schemes failed to deliver their technological promise while increasing dependence on imported inputs and spare parts (Carney 1986: 275; CRED 1985: 273). Donor production targets required double-cropping and thus a shift in agricultural production to year-round farming. Although male heads of households were taught this new form of production, cropping calendars could be followed only if women provided their agronomic expertise by joining their husbands in irrigated rice farming. By placing men in charge of technologically improved rice production, the donors hoped to encourage male participation; instead, they unwittingly legitimized male control over the greater surpluses gained from double-cropping.

As in colonial swamp development projects, gender-based household conflicts erupted over which family members were to assume the labor burden in irrigated rice cultivation. Male household heads claimed female labor under the customary category, *maruo*, but irrigation projects now meant that the claim was invoked for year-round labor. *Maruo* labor claims for household subsistence had historically evolved within the confines of a single agricultural season. There was no precedent for women to perform *maruo* labor obligations during *two* cropping periods when production would yield men a marketable surplus (Carney 1988: 306). Irrigation projects were commodifying rice production, but income gains depended on female labor availability.

Women contested the changing lexicon of plot tenure and the enclosure of

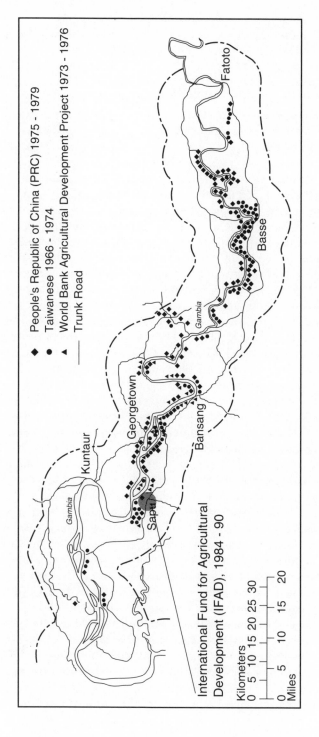

Figure 8.3 Irrigated rice projects in Gambia
Source: Carney (1986)

People's Republic of China (PRC) 1975 - 1979
Taiwanese 1966 - 1974
World Bank Agricultural Development Project 1973 - 1976
Trunk Road

International Fund for Agricultural
Development (IFAD), 1984 - 90

Kilometers
0 5 10 15 20 25 30
Miles
0 5 10 15 20

Kuntaur
Georgetown
Bansang
Sapu
Basse
Fatoto
Gambia

traditional *kamanyango* and *maruo* swamp into irrigation schemes. For them "development" meant the delivery of female labor for intensified rice farming without concomitant income gains. The reinterpretation of customary tenure by male household heads and village elites aimed at ensuring continued female access to rice land, but only as workers on plots whose benefits would flow to men as disposable surpluses. The donors' uninformed view of the Gambian household-based production system was to prove the projects' nemesis.

Female rice farmers responded in three principal ways to loss of *kamanyango* plots and efforts to augment their labor burden: (1) by relocating *kamanyango* production to unimproved swamplands where they could generate small surpluses for sale; (2) when alternative swampland for rice farming was not available, by agreeing to perform *maruo* labor obligations on irrigated rice plots during the dry season cropping cycle in exchange for using the same plot without irrigation during the rainy season for *kamanyango* production; or (3) by laboring year-round on irrigated schemes but demanding remuneration in rice for labor performed during one of the cropping seasons (Carney 1994). The first two responses involved an attempt by women to re-establish *kamanyango* rights while the third focused on substituting the value of those rights with claims to part of the surplus.

In the earliest phase of Gambian irrigation schemes, donor agencies developed swampland on a small scale (*c.* 30 hectares). Decisions over land use remained with community households. This facilitated women's efforts to reassert *kaman-yango* land rights, even though anticipated productivity rates suffered during the rainy season when women either planted individual fields outside the scheme or within the project, without irrigation. The third response mentioned above characterized the 1,500 hectare Jahaly Pacharr irrigation scheme, implemented by IFAD in 1984, which removed land from village control. Deploying a legal mechanism reminiscent of the earlier colonial CDC project, the IFAD project negotiated a thirty-year lease. Most of the available swampland of forty contiguous villages became absorbed within the large-scale project, thereby proletarianizing numerous women from floodplain rice cultivation. The lease proved instrumental to the efforts of government officials to exert new labor discipline among project households and to ensure year-round irrigated rice production. Land-use decisions now were placed under project management jurisdiction, and participation in the project made more attractive by raising the producer rice price. Continued household access to project land was linked to prompt repayment of inputs and mechanization charges advanced on credit; these rates, in turn, were based on anticipated productivities which could only be achieved by carefully adhering to the irrigation labor calendar (IFAD 1988). Households unable to comply with these terms faced eviction.

The project's mandate to double-crop as a condition for participation placed intense pressure on household labor, which the subsequent land designation as *maruo* could not easily resolve. Previous irrigation schemes had frequently accommodated women's *kamanyango* claims at the cost of year-round pumped production. Confronted by a legal mechanism that threatened eviction for households falling short of production targets, women now faced enormous pressure within the household to augment their labor burden. Because the IFAD project had incorporated most of the region's available swampland, pre-existing *kamanyango* land access came to an end. Gender-based conflicts exploded throughout the project area as women resisted the erosion of their right to derive benefits from a greatly augmented work burden. While ethnicity, class, and differences among types of irrigated land within the IFAD project shaped the ensuing patterns of conflict resolution, Mandinka women responded by demanding either *kamanyango* plots within the project or 10 percent of the rice produced as payment for their work (Carney 1994). Nevertheless, about a quarter of Mandinka households failed to honor female demand for individual plots or remuneration in kind for year-round *maruo* labor. The result was women's outright refusal to work on the family's irrigated plots.

These dispossessed women consequently pursued two complementary economic strategies for income generation: the formation of work groups to carry out the project's labor-demanding tasks of transplanting, weeding, and harvesting; and a shift in *kamanyango* production to upland cultivation. By organizing work groups for hire, women managed to bid up their daily wage rate within the project and take advantage of peanut land made available as men intensified their work in the now more remunerative rice scheme (Webb 1989: 66). But their efforts to obtain upland *kamanyango* plots proved not always successful as they came into direct competition with the claims of junior males for individual land rights. Women consequently placed considerable effort in capturing the support of NGOs to develop village vegetable gardens for income generation (Carney 1986: 311).

In summary, rice development unfolded initially on riverine floodplains. As these areas became technologically improved and commodified, male household heads reinterpreted women's pre-existing crop rights and benefits to gain access to their labor for the intensified work burden. Irrigated rice development simultaneously undermined women's customary access to rice land for income generation while enabling male household heads to capture surplus value. In the effort to discipline the Mandinka peasantry for domestic rice import-substitution, the Gambian state had ruptured the relationship between women's knowledge systems and agronomic expertise that had regulated wetland cultivation for centuries. The gender conflicts underlying project rice farming contributed to repeated delays in cropping schedules and to lower yields.

Project management efforts to evict farmers for non-payment of advanced credits suffered a setback after 1985 when the Gambian government underwent successive IMF-imposed structural adjustments. The producer rice price dropped in favor of peanuts, and fertilizer prices quadrupled over pre-reform values. Like its predecessors, the much-heralded IFAD project now operates principally during the dry season. IMF reforms have resulted in males shifting labor back to peanuts during the wet season and low rice yields during the dry season from insufficient fertilizer application (Carney 1994, fieldwork). Despite the underutilization of project land, few households have faced eviction as the costs of rice production fail to lure new recruits to the scheme. Efforts to remove unproductive households from project land also proved difficult in the heightened political sensitivity throughout rural Gambia of the Jahaly Pacharr project, which is viewed as an example of an uncaring central government attempting to "take" land from peasants (Carney 1991, fieldwork).

COMPARATIVE ADVANTAGE AND HORTICULTURAL DEVELOPMENT

Shortly after the 1968–73 Sahelian drought the Gambian government promoted economic ventures in inland swamps that grew over the years into a major focus of donor assistance and income generation within the country. During the 1970s, the government encouraged onion-growing schemes among village women's groups as a means to increase household incomes in the peri-urban corridor and north bank district, geographically proximate to the capital (Ceesay *et al.* 1982) (see Figure 8.1). During the next decade women's vegetable gardens emerged as a major focus of donor support within the country (see Schroeder and Suryanata, this volume Chapter 9). By the 1990s over 340 small (0.5–2 ha) and medium-scale (5–15 ha) vegetable gardens were developed by NGOs and multilateral donors (DeCosse and Camara 1990; Giffen 1987; Nath 1985; Smith *et al.* 1985; Sumberg and Okali 1987). The entry of private growers into the burgeoning horticultural sector, along with incipient women vegetable growers groups (not funded), accounts for an expansion of market gardening that currently exceeds 1,000 hectares (Carney 1992: 79).

The boom in market gardening on Gambian wetlands results from the confluence of several policy directions over the past fifteen years. Following independence, Gambia began developing its pristine beaches for international tourism; by the 1990s over 100,000 Europeans were taking a six-hour flight to vacation along the Gambian coast between November and April each year (N'Jang 1990). The initial onion projects successfully linked local production to the tourist sector and awakened donor agencies to the possibilities of expanding vegetable production to meet the dry season tourist demand. These developments meanwhile were unfolding against a growing clamor within the international donor community for women in development (WID) projects. The emergent WID focus in Gambia was pioneered by NGOs who countered male control over irrigation schemes by implementing horticultural projects on unimproved inland swamps previously sown to rice. The donors aimed to bolster female income earning opportunities by improving seasonally wet swamps with wells for dry season vegetable cultivation.

Policy support for diversifying wetland agriculture into horticulture received additional impetus from the IMF-mandated structural adjustment program. Geographic proximity to Europe encouraged policy makers to exploit Gambia's comparative advantage as a winter fruit and vegetable supplier, as did favorable tariffs and the removal of export taxes on fresh produce (Government of Gambia 1987; Jack 1990; UNCTAD 1986). Seeds of non-traditional horticultural crops such as lettuce, tomatoes, green peppers, carrots, eggplants, beans, cabbages, and tropical fruits were distributed, and marketing strategies focused on hotels, the expatriate community, neighboring Senegal, and Europe.

By the 1990s horticultural production had expanded to rain-fed areas in the peri-urban corridor located near the international airport, with boreholes dug to reach underground aquifers. With few exceptions, the projects are operated by the state, senior government officials, and resident Lebanese and Indian landowners, and are oriented to European export markets. In the five years following IMF economic reforms, annual fruit and vegetable exports to Europe alone grew to 3,000 tons, a value exceeding US $1 million (Jack 1990). The same period witnessed growing involvement by multilateral donors (EEC, Islamic Development Bank, UNDP, and the World Bank) in women's horticultural production and marketing along the coastal corridor (Barrett and Browne 1991: 244; Carney 1992: 78; Ceesay *et al.* 1982; Government of Gambia 1987; World Bank 1990). Despite this most recent form of donor support, Gambian women's horticultural projects remain concentrated in rural areas, on inland swamps of small areal extent (0.5–2 ha), and oriented to local and regional markets.

Although the policy emphasis on converting inland swamps to horticulture dates from the 1970s, Gambian women have long been involved in vegetable production (Schroeder 1993). They were observed marketing vegetables during the dry season as far back as the mid-fifteenth century, while eighteenth-century travelers noticed vegetable cultivation in inland swamps following the rice harvest (Adanson 1759; Crone 1937: 48; Park 1983). Dry season horticultural production received encouragement during the colonial period, but its expansion was limited by the elementary technology employed for irrigation, *shadufs*, in which river water is lifted with a pole and bucket by hand (Carney 1986: 144). Although colonial horticultural programs targeted men, their failure left vegetable growing in women's hands. Females remained the country's principal producers, using residual moisture from inland rice swamps early in the dry season to cultivate traditional crops such as bitter tomatoes, okra, sorrel, and hibiscus for subsistence.

Donor support for well construction from the 1980s has enabled an extension of the vegetable-growing period in inland swamps. Deep dug, concrete-lined wells have revolutionized Gambian horticultural production by tapping water tables for dry season cultivation. Vegetable gardening no longer remains a seasonal activity, as it was prior to donor involvement. Women's village gardens receiving NGO assistance grow vegetables during the entire dry season and in some cases, all year round.

The provision of reliable water supplies through well-digging is central to NGO efforts to implement a rural development strategy aimed at improving women's incomes. By promoting village gardens for women's groups interested in commercialized vegetable cultivation, NGOs have launched a development strategy that targets women who were ignored in the previous wetland policy approach. NGO support for well construction has proved crucial to women's negotiations with male landowners and village elites for access to land for a communal vegetable garden. At a cost of US $3,000–4,000 per hectare, these wells provide communities with a valuable infrastructure to ensure a permanent water source for dry season agricultural production (Nath 1985: 6; Schroeder 1989: 13; Sumberg and Okali 1987).

Arrangements to secure female access to improved village gardens, however, vary between communities and depend on the availability of land locally, as well as the swamp's land-use history. Consequently, in rural communities with NGO-supported gardens, women are granted either year-round usufruct for cash cropping vegetables or *kamanyango* dry season rights, with the plot reverting to subsistence cereal production in the rainy season. Once access to land is accomplished, NGOs provide assistance for constructing the concrete-lined wells and barbed wire fences (for protection from livestock damage). When completed, female growers are credited the seeds and tools for vegetable farming.

Vegetable gardening is a labor-intensive process. During the dry season it requires two daily waterings – averaging about two hours per session – weeding, and pest control, as well as transporting the bulky and highly perishable produce to weekly markets. But in a country where rural per capita income averages US $130, efforts are often well-rewarded (World Bank 1981). Schroeder (1992: 4) records women vegetable growers attaining gross incomes between US $67–265 during the dry season, with more than half reporting incomes exceeding their husband's earnings from peanuts. These income differentials are the new source of contemporary gender conflict in north-bank vegetable gardens.

An examination of three areas where market gardens figure prominently in women's economic options reveals the effects that improved water availability and increased income opportunities play in fueling contemporary gender conflicts. These include: (1) the area around Kerewan on the north bank, the site of the pilot onion schemes, which borders neighboring Senegal; (2) across the Gambia River, the peri-urban corridor close to the capital, Banjul, where tourist hotels, government offices, and expatriate community are concentrated; and (3) 260 kilometers up-river from the capital on the river's south bank, the IFAD-funded Jahaly Pacharr project, centered at Sapu (see Figure 8.1). The first two areas, the original foci for horticultural development, enjoy numerous marketing opportunities, while women growers in the IFAD project rely principally on weekly markets (*lumos*) for vegetable sales. While income returns from vegetable marketing in the peri-urban corridor approximate the low range cited for north-bank growers, incomes in the IFAD project are only US $15–35 for dry season production (Carney 1991, fieldwork).

Each of these areas offers men different income opportunities. Jobs in government, the tourist sector, petty commerce, and transportation are concentrated in the capital. In the IFAD project and north-bank areas, men derive incomes chiefly from agriculture – peanuts and irrigated rice in the former, peanuts in the latter. As noted above, within the IFAD project men's control over irrigation schemes and peanut farmland has severely limited women's access to village land for vegetable gardens. The explosion over the past five years of banana cultivation by men within the IFAD project, directly along banks of riverine tributaries, and in inland swamps outside the project area has reduced wetland availability even further. Village women in the IFAD scheme accordingly have severely restricted access to potential garden land; when they do manage to negotiate land access, it is usually only for dry season vegetable cultivation.

Female growers in the capital face an altogether different situation. The concentration of tourist hotels, an expatriate community, and international horticultural trade provide numerous marketing opportunities. But potential income benefits are limited by the proliferation of vegetable projects and an excessive number of female participants in each scheme. Local markets are saturated with women selling vegetables, resulting in meager returns and a continuous search for new outlets. An important outcome of the explosion of vegetable gardens within the peri-urban corridor is the growing involvement of women's groups in contract farming production relations with large growers. While the latter arrangement provides an outlet for excess production, prices are driven down by large-scale growers and traders, who set conditions and control distribution networks (Carney 1992: 80–2).

Vegetable gardening nonetheless remains attractive to women whose alternative income-earning prospects are limited. While structural adjustment programs have led to a 10 percent reduction in employment within the government sector catapulting men into increasing involvement in horticultural production, women have generally maintained usufruct to village land for gardening because donor representatives, located in the capital, are poised to defend them. The case of the north bank, detailed by Schroeder and Suryanata (Chapter 9 in this volume), differs, however, and provides yet another illustration of the complex intersection of gender with environmental transformation and economic change.

Proximity to the land border with Senegal and declining peanut production associated with upland environmental degradation underlie the gender conflicts. As with peanuts, most vegetable production flows across the border to Senegal, where horticultural import–export distribution networks and the internal demand for vegetables are more developed, and prices higher (Mackintosh 1989: 15). NGO improvement of inland swamps with wells has resulted in new avenues for income generation that sharply conflict with the WID objectives of NGOs.

NGO-funded vegetable projects in north-bank communities have transformed the inland swamps and the social relations regulating pre-existing cropping and labor patterns. Well construction, in effect, has widened the seasonal window that formerly regulated vegetable cultivation. Crops are no longer confined to the autumnal planting period following the rice harvest; vegetables can be planted throughout the dry season and frequently year-round, since profits from cross-border sales currently compensate for displaced rice production (Schroeder and Watts 1991: 62).

As north-bank horticultural projects have augmented women's earnings, female rights of disposal over their income and access to vegetable land have come under increasing threat. Schroeder and Suryanata (Chapter 9 in this volume) reports men deferring to women the burden for costs formerly met by males and their capture of part of women's earnings through unpaid loans. Additionally, male landholders in numerous communities contest women's access rights to vegetable land through the planting of economically valuable trees (e.g. mango and orange) within the vegetable gardens. After five to ten years the canopy closes, blocking the sunlight needed for vegetable growth. Tree planting therefore facilitates the conversion of land use from vegetable gardens to orchards, enabling male landlords to reclaim the improved plots for their own economic strategy based on tree crops within a decade (Schroeder 1992: 9).

By making verbal agreements with NGOs for women's vegetable gardens landlords acquiesce to female demand for *kamanyango* land rights. But these rights to a vegetable garden are honored for a limited number of years – that required to capture women's labor for watering adjacent fruit trees during the initial growth period. The use of economically valuable trees to recapture garden plots as male *kamanyango* over the long run, however, is not lost on women. Schroeder (Chapter 9 in this volume) documents the gender confrontations that have occurred with orchard planting which include women cutting back mango and orange trees as they begin shading out vegetables, deliberate setting of fires to fatally damage fruit trees, and sending delegations to local officials for legal action.

The inland swamps of the north bank, formerly used by women for subsistence rice production, are being increasingly commercialized to vegetables. But the process is unfolding within a region of limited economic opportunity and severe environmental degradation. While NGOs attempted to address the gender equity issue ignored in the first wetland development phase, this second development approach indicates that women's gains over the long run are indeed precarious.

CONCLUSION

The structural dislocation of a monocrop export economy and attendant food shortages brought government attention to the Gambian wetlands during the late colonial period. The pattern of swamp development implemented during colonial rule foreshadowed a large-scale emphasis on the wetlands that materialized with the influx of foreign capital coincident with the Sahelian drought. During the past twenty-five years, wetland development through irrigation projects has transformed Gambian agriculture from a seasonal to year-round activity, enabling agricultural diversification, surplus cereal production, and new avenues of income generation among rural households.

The promise of irrigated agriculture, however, depends upon the ability of peasant households to restructure family labor to the dictates of irrigated farming – a labor regime that requires a greater work burden during the entire calendar year. As claims to family labor evolved in the context of a limited wet season, institutional mechanisms within the household-based production system were deformed to mobilize family labor for year-round agriculture. Use of the term *maruo* for technologically improved swamps proved central to obtaining a female labor reserve for the intensified work burden in irrigated farming. The term strengthens prevailing patriarchal power relations while facilitating the concentration of benefits produced by female labor within the household.

Women contest the semantics of *maruo* precisely because it provides a mechanism for the loss of their customary rights. They are acutely aware that the rules of access to and control over environmental resources are not a codification of immemorial tradition, but rather the outcome of struggle and negotiation with husbands, male community leaders, state and donor officials (Berry 1986: 5; Okoth-Ogendo 1989: 14). This awareness has sharpened over the past twenty-five years as irrigation projects imbued wetlands with new economic value. Gambian women are not engaging men in mere semantic discussion as they struggle for *kamanyango* rights – their actions reveal growing recognition that commodification of the wetlands is steadily eroding female economic and social status within the household and village community.

The two case studies of irrigated agriculture illustrate the multiple ways in which women contest and renegotiate their access to resources. Struggles in rice schemes have centered on village women reaffirming claims to a portion of the surplus by requesting seasonal plot use as *kamanyango*; remuneration in the form of paddy rice for year-round labor availability; or, when labor benefits are denied, outright refusal to work on the household's irrigated fields and entry into local wage markets. As the scale of wetland development increased, women joined forces with those from other village households experiencing similar difficulties. They appealed to local governmental officials for intercession on their behalf and met that failure by creating village work groups for hire to improve the daily wage rate in irrigated rice. As NGOs expanded their activities in female vegetable projects, village women actively sought their assistance. While these efforts fall short of a broad-based social movement, they have irrevocably changed the way women regard their socioeconomic position in contemporary rural Mandinka households.

In vegetable projects female fortunes appear much improved. But as Schroeder and Suryanata detail in this volume (Chapter 9), women growers find their incomes from garden cultivation being claimed in new ways by their husbands, who, in some cases, refuse to pay back the loans given to their wives or abrogate their contributions toward household expenses. Moreover, the increasing emphasis on orchards for income generation indicates that women's *kamanyango* gardening rights may only be exercised for a limited number of years – equivalent to the time required for hand-watering of trees until the plot's land use converts to mature orchards. Despite income gains and growing militancy, women's earnings in vegetable gardening appear precarious over the long run.

These case studies indicate that a process of land concentration is occurring in Gambian wetlands improved with irrigation. Concentration is not the result of absolute land scarcity and overpopulation but rather a response to household labor shortages and new income opportunities with irrigated agriculture. The designation *maruo* for irrigated land reveals how land is enclosed to create an artificial scarcity for accessing female labor. This unusual form of enclosure permits women access to irrigated land while denying them benefits from their work. Land concentration consequently involves the conversion of wetlands from land with multiple female rights to the surplus product to land with a single claim over the surplus produced by multiple female laborers.

In contrast to the pattern described by Humphries (1990: 38–9) for eighteenth-century England, land enclosure in Gambia has resulted in very different outcomes for women. While land enclosures in England pushed rural men into waged work and left their wives and children to defend traditional rights to rural resources, it is women in contemporary rural Gambia who are increasingly proletarianized, as men gain control of both customary rights over female labor and the income from irrigated lands. The forms of economic change promoted by national and international organizations are intensifying household labor demands in wetland cultivation, thereby spearheading a form of enclosure that uses the *maruo* designation to weaken women's customary rights to rural resources so that male heads of households can capture their labor for individualized accumulation.

In outlining the social and historical processes of changing land-use strategies on the Gambian wetlands, this chapter reveals that more than the environment is being transformed. So too are the social relations that mediate access to, and use of, land within rural households. As commodification transforms the use of wetland environments, the social relations that produce these environments are also restructured (Berry 1989: 41). The process of wetland conversion in Gambia clearly illustrates these changes in women's reduced control over lowland resources. The contemporary pattern of accumulation unfolding in the Gambian wetlands centrally depends on controlling access to technologically improved swamps by dissolving women's customary land rights and by imposing new work routines that undervalue and intensify their labor contribution on irrigated land. Women, however, are resisting their newly assigned role as cheap labor reserves. Changing self-perception among women in rural Gambia has resulted in protests, household- and village-level negotiations that are modifying resource strategies and agrarian practices on the irrigated wetlands.

NOTE

This research was funded in part by the African Studies Center of the University of California, Los Angeles. In addition, I would like to acknowledge the thoughtful comments of the editors.

REFERENCES

Adanson, M. 1759. *A Voyage to Senegal, the Isle of Goree and the River Gambia.* London: Nourse.

Arid Lands Information Center (ALIC). 1981. *Environmental Profile of The Gambia.* Tucson: Office of Arid Lands Studies.

Barrett, H. and A. Browne. 1991. "Environment and economic sustainability: women's horticultural production in The Gambia," *Geography* 776: 241–8.

Berry, S. 1986. "Concentration without privatisation: agrarian consequences of rural land control in Africa," in paper presented at the Conference on Agricultural Policy and African Food Security: Issues, Prospects and Constraints, Toward the Year 2000. Champaign-Urbana, IL: University of Illinois, Center for African Studies.

—— 1989. "Social institutions and access to resources," *Africa* 59, 1: 41–55.

Blaikie, P. and H. Brookfield. 1987. *Land Degradation and Society.* New York: Methuen.

Bryant, R. 1992. "Political ecology: an emerging research agenda in Third World studies," *Political Geography* 11, 1: 12–36.

Carney, J. 1986. "The social history of Gambian rice production: an analysis of food security strategies," Ph.D. dissertation, University of California, Berkeley.

—— 1988. "Struggles over crop rights within contract farming households in a Gambian irrigated rice project," *Journal of Peasant Studies* 15: 334–49.

—— 1991. "Indigenous soil and water management in Senegambian rice farming systems," *Agriculture and Human Values* 8: 37–58.

—— 1992. "Peasant women and economic transformation in The Gambia," *Development and Change* 23: 67–90.

—— 1994. "Contracting a food staple in The Gambia," in P. Little and M. Watts (eds.) *Peasants under Contract: Contract Farming and Agrarian Transformation in Sub-Saharan Africa.* Madison: University of Wisconsin Press, pp. 167–87.

Carney, J. and M. Watts. 1991. "Disciplining women? Rice, mechanization, and the evolution of Mandinka gender relations in Senegambia," *Signs* 16: 651–81.

Ceesay, M., O. Jammeh and I. Mitchell. 1982. *Study of Vegetable and Fruit Marketing in The Gambia.* Banjul, Gambia: Ministry of Economic Planning and Industrial Development and the World Bank.

Center for Research on Economic Development (CRED). 1985. *Rural Development in the Gambian River Basin.* Ann Arbor: CRED.

CILSS (Permanent Interstate Committee for Drought Control in the Sahel). 1979. *Development of Irrigated Agriculture in Gambia: General Overview and Prospects. Proposals for a Second Program 1980–1985.* Paris: Club du Sahel.

—— 1990. *Study on Improvement of Irrigated Farming in The Gambia.* Paris: Club du Sahel.

Crone, G.R. 1937. *The Voyage of Cadamosto.* London: The Hakluyt Society.

DeCosse, P. and E. Camara. 1990. *A Profile of the Horticultural Production Sector in The Gambia.* Banjul, Gambia: Department of Planning and Ministry of Agriculture.

Derman, W. 1984. "USAID in the Sahel," in J. Barker (ed.) *The Politics of Agriculture in Tropical Africa.* Beverly Hills, CA: Sage.

Dey, J. 1980. "Women and rice in The Gambia: the impact of irrigated rice development projects in the farming system," Ph.D. dissertation, University of Reading, UK.

—— 1981. "Gambian women: unequal partners in rice development projects?" *Journal of Development Studies* 17: 109–22.

Dunsmore, J.R. 1976. *The Agricultural Development of The Gambia: An Agricultural, Environmental and Socio-economic Analysis.* Land Resource Study No. 22. Surrey: Ministry of Overseas Development, Land Resources Division.

Food and Agriculture Organization (FAO). 1983. *Rice Mission Report to The Gambia.* Rome: FAO.

Franke, R. and B. Chasin. 1980. *Seeds of Famine*. New Jersey: Allanheld.

Gambia German Forestry Project (GGFP). 1988–91. Unpublished project data.

Gamble, D. 1955. *Economic Conditions in Two Mandinka Villages: Kerewan and Keneba*. London: Colonial Office.

Giffen, J. 1987. *An Evaluation of Women's Vegetable Gardens*. Banjul: Oxfam.

Government of The Gambia (GOG). 1966. *Five-year Plan for Economic and Social Development*. Banjul: Ministry of Economic Planning and Industrial Development.

—— 1973–91. Sample surveys of agricultural production. Banjul: Public Planning and Monitoring Unit.

—— 1977. *The Gambia: Land and Vegetation Degradation Survey: The Need for Land Reclamation by Comprehensive Ecological Methods*. Banjul: Ministry of Agriculture.

—— 1987. *Donors' Conference on the Agricultural Sector in The Gambia*. Banjul: Ministry of Agriculture.

Guyer, J. 1984. "Naturalism in models of African production," *Man* 19, 3: 371–88.

Guyer, J. and P. Peters. 1987. "Introduction" to Special Issue: Conceptualizing the Household: Issues of Theory and Policy in Africa, *Development and Change* 18, 2: 197–214.

Hart, G. 1991. "Engendering everyday resistance: gender, patronage and production politics in rural Malaysia," *Journal of Peasant Studies* 19, 1: 93–121.

Harvey, C. 1990. *Improvements in Farmer Welfare in The Gambia: Groundnut Price Subsidies and Alternatives*. Institute of Development Studies (IDS) Discussion Paper #277. Sussex: IDS.

Humphries, J. 1990. "Enclosures, common rights, and women: the proletarianization of families in the late eighteenth and early nineteenth centuries," *Journal of Economic History* 2, 1: 17–42.

Hutchinson, P. 1983. *The Climate of The Gambia*. Banjul: Ministry of Water Resources and the Environment.

International Fund for Agricultural Development (IFAD). 1988. *Small-scale Water Control Program*. Rome: IFAD.

Jack, I. 1990. *Export Constraints and Potentialities for Gambian Horticultural Produce*. Report prepared for the National Horticultural Policy Workshop. Banjul: Ministry of Agriculture.

Jackson, C. 1993. "Women/nature or gender/history? A critique of ecofeminist 'development'," *Journal of Peasant Studies* 20, 3: 389–419.

Jeng, A.A.O. 1978. "An economic history of the Gambian groundnut industry 1830–1924: the evolution of an export economy," Ph.D. dissertation, University of Birmingham, UK.

Jobson, R. 1904 [1623]. *The Golden Trade*. Devonshire: Speight & Walpole.

Landell Mills Associates (LMA) 1989. *A Market Survey for Gambian Horticultural Crops in the UK, Sweden, the Netherlands and the Federal Republic of Germany*. London: Commonwealth Secretariat.

Mackintosh, M. 1989. *Gender, Class and Rural Transition*. Atlantic Highlands, NJ: Zed Books.

McPherson, M. and J. Posner. 1991. "Structural adjustment in sub-Saharan Africa: lessons from The Gambia," paper presented at the 11th annual symposium of the Association for Farming Systems Research-Extension, Michigan State University.

Mann, R. 1987. "Development and the Sahel disaster: the case of The Gambia," *The Ecologist* 17: 84–90.

Moore, H. 1988. *Feminism and Anthropology*. London: Polity.

Nath, K. 1985. *Women and Vegetable Gardens in The Gambia: Action Aid and Rural Development*. Working Paper No. 109, African Studies Center. Boston: Boston University.

186

National Archives of The Gambia (NAG). 1943. Department of Agriculture files, 52: 47/50.

N'Jang, A. 1990. *Characteristics of Tourism in The Gambia*. Banjul: Ministry of Information and Tourism.

Okoth-Ogendo, H. 1989. "Some issues of theory in the study of tenure relations in African agriculture," *Africa* 59, 1: 56–72.

Park, M. 1983 [1799]. *Travels into the Interior of Africa*. London: Eland.

Porteres, R. 1970. "Primary cradles of agriculture in the African continent," in J. Fage and R. Oliver (eds.) *Papers in African Prehistory*. Cambridge: Cambridge University Press, pp. 43–58.

Rahman, A.K. 1949. Unpublished notes on land tenure in Genieri, courtesy of David Gamble.

Schroeder, R. 1989. *Seasonality and Gender Conflict in Irrigated Agriculture: Mandinka Rice and Vegetable Production in The Gambia*. Co-evolutionary Terrains: History, Ecology and Practice in Malagasy and Gambian Rice Systems, Working Paper No. 4. New York: Rockefeller Foundation.

—— 1992. *Shady Practice: Gendered Tenure in The Gambia's Garden/Orchards*. Yundum: Department of Agriculture Horticultural Unit and Oxfam America.

Schroeder, R. and M. Watts. 1991. "Struggling over strategies, fighting over food: adjusting to food commercialization among Mandinka peasants," *Research in Rural Sociology and Development* 5: 45–72.

Sharma, U. 1980. *Women, Work and Property in North-West India*. London: Tavistock.

Shiva, V. 1989. *Staying Alive*. London: Zed Books.

Smith, F., I. Jack and R. Singh. 1985. *The Survey of Rural Women's Vegetable Growing and Marketing Programme*. Banjul: Action Aid.

Sontheimer, S. 1991. *Women and the Environment: A Reader*. London: Earthscan.

Sumberg, J. and C. Okali. 1987. *Workshop on NGO-sponsored Vegetable Gardening Projects in The Gambia*. Yundum: Department of Agriculture Horticultural Unit and Oxfam America.

Swindell K. 1977. "Migrant groundnut farmers in The Gambia: the persistence of a nineteenth-century labor system," *International Migration Review* 11, 4: 452–72.

Thrift, N. 1986. "The geography of international economic disorder," in R.J. Johnston and P.J. Taylor (eds.) *A World in Crisis?* New York: Basil Blackwell, pp. 12–67.

United Nations Commission on Trade and Development (UNCTAD). 1986. *Post-harvest Handling and Quality Control for Export Development of Fresh Horticultural Produce (Gambia)*. Geneva: UNCTAD/GATT.

United Nations Development Program (UNDP). 1977. *Development of the Gambia River Basin: Multi-disciplinary Mission and Multi-donor Mission. Programme of Action*. New York: UNDP.

Webb, P. 1989. *Intrahousehold Decisionmaking and Resource Control: The Effects of Rice Commercialization in West Africa*. Washington, DC: International Food Policy Research Institute (IFPRI).

Weil, P. 1973. "Wet rice, women, and adaptation in The Gambia," *Rural Africana* 19: 20–9.

—— 1982. "Agrarian production, intensification and underdevelopment: Mandinka women of The Gambia in time perspective," in *Proceedings of the Title XII Conference on Women in Development*. Newark: University of Delaware.

World Bank. 1981. *The Gambia: Basic Needs in The Gambia*. Washington, DC: World Bank.

—— 1990. *Women in Development Project. Staff Appraisal Report*. Washington, DC: World Bank.

9

GENDER AND CLASS POWER IN AGROFORESTRY SYSTEMS

Case studies from Indonesia and West Africa

Richard A. Schroeder and Krisnawati Suryanata

[A]groforestry initiatives . . . have been sheltered in the discursive shade of trees as symbols of green goodness.

(Rocheleau and Ross 1995: 408)

Agroforestry systems are widely touted for their prodigious capacities. From a production standpoint, intercropping trees with underlying crops can fix nitrogen and improve nutrient cycling, enhance chemical and physical soil properties, add green manure, conserve moisture, and make generally efficient use of a range of limited yield factors. Similarly, from the standpoint of environmental stabilization, agroforestry systems may reduce erosion, provide alternate habitat for wildlife, and shelter a diverse range of plants; they are also sites where the critical knowledge systems of indigenous peoples are reproduced. In the context of 1990s environmentalism, an agroforestry approach that simultaneously boosts commodity production and contributes to stabilizing the underlying resource base is constructed as an unambiguous and unalloyed 'good' (Rocheleau and Ross 1995; cf. Schroeder 1995). Institutional actors in forestry and environmental agencies, as well as the major multilateral donor agencies such as the World Bank, have accordingly joined forces to promote and preserve agroforestry in many parts of the world.

We recognize that, in addition to favorable production and environmental capacities, agroforestry approaches also sometimes open up critical options for otherwise disenfranchised groups. Rocheleau (1987) demonstrates quite clearly how women mobilize agroforestry strategies to make the best use of the minimal landholdings allotted to them (cf. Leach 1994). Other authors have argued eloquently for the rights of indigenous peoples to perpetuate their livelihoods in agroforestry systems (Clay 1988). And Dove (1990) suggests that the diversity and complexity of so-called 'home garden' agroforestry systems, which incorporate a wide range of cultivars with high use-value but low exchange-value, provide peasant groups with the means effectively to resist the extractive propensities of the state.

This chapter challenges the assumption that environmentalist policies and development practices related to agroforestry are universally beneficial to local interests. Instead we seek to redirect attention to agroforestry as a site of contentious political struggle. Farmers often view trees and forests as "tenure liabilities," particularly when the state has criminalized their removal: "As long as trees [are] not-agriculture, not legally available for harvest and sale, and forests [are] unimproved lands of untouchable resources" (Rocheleau and Ross 1995; cf. Peluso 1992), they remain impediments to livelihood and effective resource control, and are resisted. In sharp contrast, proponents of agroforestry stress that trees are assets which not only enhance the value and quality of land resources, but vary the scope and seasonality of income streams and thus the viability of the economic units engaged in agroforestry production. The problem with this idealized view of agroforestry is that it minimizes the internal workings of property and labor claims, despite ample evidence that these are pivotal to successful management (Fortmann and Bruce 1988; Raintree 1987). By their very nature as spatially enclosed systems, agroforestries often encapsulate the social conflicts that permeate societies. This is especially the case in successional systems such as the British colonial invention, the *taungya* system, where one species, and hence one set of property claims, supersedes all others as the system matures (Bryant 1994; Goswami 1988; King 1988; Peluso 1992). Where agroforestry approaches are commercialized, they tend to extend and rigidify (Millon 1957; Raintree 1987) the tenurial rights of tree growers *vis-à-vis* competing resource users, such as cultivators of underlying crops, forest product collectors, and pastoralists. With such social and technical dynamics embedded in combinations of tree and understorey crops, the design and implementation of agroforestry systems, and especially the actions of tree holders, must be carefully analyzed.

At the minimum, there is a need to move beyond technocratic and managerial classification systems (Farrell 1987; Nair 1989, 1990) and distinguish between agroforestries on political-economic grounds. Systems such as those described by Dove, Clay, Rocheleau, and others as embodying culturally diverse knowledge systems and practices are fundamentally different in scope and purpose than contemporary strategies pressed into being by economic, forest management, and (more recently) environmental developers bent on merging environmental and commodity production objectives. There is, in other words, a striking contrast between systems that actually accentuate and preserve a diversity of species, uses, and claims, and those that practically narrow the range of options within each of these parameters.

This chapter looks at two contemporary agroforestry initiatives in Gambia and upland Java which illustrate problems of ignoring the social and political dimensions of agroforestry. Both systems involve the production of tree commodities. Both have been hailed as bold steps toward environmental stabilization: in Gambia, toward reversing the cumulative effects of drought and deforestation; in Indonesia, toward stabilizing slopes in order to reduce the silting up of reservoirs.

189

In both cases, an environmental discourse has served to mask the exclusionary objectives of fruit tree holders – male mango growers in Gambia, and a new class of 'apple lords' in Java – which are ultimately directed at entrepreneurial gain and control over key production resources. Our argument is that, while these agro-forestries often contribute in some measure to ecological goals, they nonetheless can also be seen as deliberate strategies of dispossession and private accumulation. The commoditization of tree cropping has driven a wedge between holders of tree and land/crop rights, and this polarization has in turn produced a range of agro-ecological and social contradictions. Such dynamics grow directly out of a more general "commercialization-cum-stabilization" ethos (Schroeder 1995) – the "market triumphalism" identified by Peet and Watts (Chapter 1 in this volume) – which erodes moral economies and replaces them with a morally indifferent (not to say bankrupt) stance which elevates profit taking above all other objectives, *including* ecological stability.

GENDERED AGROFORESTRY IN GAMBIAN GARDEN/ORCHARDS

Rights over resources such as land or crops are inseparable from, indeed are isomorphic with, rights over people . . .

(Watts 1992: 161)

Since the mid-1980s, agroforestry efforts in Gambia have primarily been focused on adding trees to hundreds of low-lying women's gardens originally established under the guise of "women in development" initiatives. A veritable boom in market gardening by women's groups grew out of a conjuncture of poor climatic conditions, foreign investment in women's programs, and numerous unconscionable national budget reductions mandated by a World Bank structural adjustment program. Average annual rainfall along the river basin has declined approximately 25–30 percent over a twenty-year period. During that time, the respective fortunes of the male and female agricultural sectors have reversed: hundreds of thousands of dollars have been invested in the women's garden sector by donors interested in promoting better nutrition and an increase in female incomes, while prices for male peanut producers (gardeners' husbands) have stagnated on the world market (Carney, Chapter 8 in this volume; Schroeder 1993). Despite the fact that women's gardens have become the basis for house-hold reproduction in many areas, they have since come under threat from male landholders interested in planting fruit orchards in the same locations.

Customary land law among the Mandinka residents of Gambia's North Bank Division, where research for this chapter was conducted in 1991, preserves a basic distinction between matrilineal and patrilineal land. Women's landholding rights are almost exclusively limited to swampland, where plots originally cleared by women are heritable property passing from mother to daughter. Patrilineal land, by contrast, consists both of upland areas, where men control virtually all arable land and grow groundnuts, millet, and maize, and some swampland, where rice is grown by female family members for joint household consumption. Such land is nominally controlled by men who are relatively senior in the lineage structure, although practical day-to-day production decisions are often taken by junior kin who are either delegated responsibility for cultivation or are granted use rights to plots prior to acceding to full landholding status as they grow older. Women's gardens, ranging in size from a fraction of a hectare to nearly 5 hectares, are almost all constructed on lineage land. Rights of access are granted on a usufruct basis to groups, although individual women operate separate plots within the communally fenced perimeters. The gardens are thus vulnerable to being reclaimed by landholders interested in planting tree crops. According to Mandinka custom, trees belong to those who plant them. Under circumstances such as the gardens in question, where the tree planter is also the landholder, the tree crop takes precedence over other forms of cultivation. (Tree crops may take precedence even in systems where the tree planter is *not* the landholder, as in the Javanese case outlined below.)

On the face of it, this situation appears clear cut: two groups of commodity producers vie for control of the same land and labor resources, as well as the development largesse generated through their respective production systems. Neither group has total power over the garden/orchard spaces (Schroeder, forthcoming): gardeners are dependent upon usufruct rights to land controlled by senior male members of landholding lineages, and would-be orchard owners are dependent upon the labor of women's groups, not just for irrigation, but for maintenance of fences and wells, clearing brush from garden/orchard plots, and protection from livestock incursions. The potential for conflict between gardeners and landholders is thus manifest in every production decision taken within the fence perimeters which bound the system (Schroeder, forthcoming). Each relocation of the fence line, each tree planted, each year's planting sequence and plot layout can be read as a strategic and spatial embodiment of power.

Conjugal conflict and intensified land use

Work in the horticulture sector has generated incomes for women gardeners that are roughly equivalent to the rural per capita income in Gambia (Schroeder 1993), and female household members have consequently taken on major new financial responsibilities. Of the women in the sample, 57 percent had purchased at least one bag of rice in 1991 to supplement home-grown food supplies; 95 percent buy all their own clothes, 84 percent buy all their children's clothes, and 80 percent had purchased Islamic feast day clothes for at least one member of their family – all responsibilities borne either solely or primarily by men prior to the garden boom. While all cash earned from vegetable sales is nominally controlled by women, growers' husbands have, nonetheless, devised a complex system of tactics for alienating female earnings, or otherwise directing them toward ends of their own choosing (Schroeder 1994). These include a range of loan-seeking strategies, each carrying its own measure of commitment to

191

repayment, and its own underlying threat of reprisal if the loan is not forth-coming. Gardeners' husbands also increasingly default on customary financial obligations they feel their wives can assume due to improved financial circum-stances (Schroeder 1994). The key point here is that the social pressure for women to share garden incomes with other family members mounted steadily throughout the early stages of the garden boom, and vegetable growers responded by both expanding and intensifying production.

Attempts to resolve *intra-household* tensions often displace the conflict to the spatial arena of the garden perimeters. The technical innovations accompanying the garden boom included replacement of poor quality stick and thorn fences and hand-dug, unlined wells serving individual plots with communal wire and concrete structures that do not have to be replaced on an annual basis. These enhancements reduced prohibitive recurrent expenses, removed some of the threat of encroachment by grazing livestock, and improved access to ground-water. While these improvements stabilized the vegetable production system in several key respects, the narrow selection of crops cultivated and relatively poor market returns meant that gardeners were unable to adequately meet their husbands' demands for greater financial support. Moreover, even as marginal increases were achieved, a strongly 'pulsed' income stream left women vulnerable to their husbands' loan requests. Growers consequently reverted to more complicated intercropping strategies that prolonged the market season and spread income over several months. Planting fruit trees and production of new crops such as cabbage, bitter tomatoes, and sweet peppers opened up sizable new markets and improved the seasonality of the income returns from gardens. The potential of these intercropping strategies could only be met with an expansion of garden territory, however. Requests to enclose new areas for gardening purposes and the *de facto* conversion of garden space into a more complex agroforestry system caused male landholders to re-evaluate the garden boom and its long-term effects. From the landholders' perspective, fruit production in the gardens threatened to confer a sense of *permanence* and *legitimacy* upon women's usufruct rights. Like the Javanese case below, the interests of tree holders and landholders began to diverge, with tree holders – in this instance, women gardeners – apparently holding the upper hand.

Shady practice

When an expatriate volunteer was posted in the area in 1983, local gardeners seized upon the opportunity to lobby for material support to expand two existing garden sites. Ensuing efforts to implement plans to rebuild and enlarge the community's two primary fenced perimeters were thwarted, however, when the landholder on one of the sites objected to the fact that his landholding prerogatives were being violated by the provisions of the proposed project. Increasing tensions eventually resulted in the detention of three garden leaders and a spontaneous protest demonstration on the part of several hundred gardeners, which resulted in the issuing of a temporary injunction against gardening on the site. In the court's ruling, nearly all substantive claims by the vegetable growers were upheld. The sole exception involved allegations made by the landholder that the women had planted dozens of fruit trees within the perimeter without authorization. His insistence that they be removed won the court's backing, and women were ordered to remove all trees at his request. Within a day or two of the decision, the land-holder visited the garden and ordered several dozen trees removed. Then, in an action that foreshadowed much of what was to come in the north bank's garden districts, he immediately replanted several dozen of his own trees within the perimeter. By locating seedlings directly on top of garden beds already allocated to vegetable growers, his expectation was that water delivered by growers to the vegetable crop would support his trees until the ensuing rainy season (a sort of indirect subsidy).

This controversy marked a watershed in the political ecology of gardening on the north bank. Not only were several hundred women involved in the demon-stration at the police station, but the case also received attention from politicians at the highest levels of government. Every step taken by the landlord and every aspect of the women's claims to use rights were carefully scrutinized and debated throughout the area. This led other landholders to reappraise their own stance with respect to their management of low-lying land resources. Most telling, it set a precedent for landholders in the attempted use of female labor to establish private fruit tree orchards.

Within a few years of this incident, both gardeners' and landholders' attitudes toward agroforestry practices had changed. From the gardeners' perspective, the relative economic benefits of tree planting and vegetable growing shifted decisively in favor of gardens. As the leader of one of the oldest garden groups in the area put it: "We are afraid of trees now. . . . You can have one [vegetables or fruit] or you can have the other, but you can't have both." Thus, in order to minimize shade effects, growers began cutting back or chopping down trees – in many cases, trees which they themselves had planted – in order to open up the shade canopy and expose their vegetable crops to sunlight. At the same time, landholders saw a new opportunity developing for themselves. Whereas they had initially resisted tree planting on the grounds that it reduced their future land-use options, the "capturing" of a female labor force to water trees, manure plots, and guard against livestock incursions within the fenced perimeters led landholders to wholeheartedly embrace fruit growing.

In 1983, a new garden site was established immediately adjacent to an older site where gardeners had already begun to feel the effects of shade canopy closure. Given the land pressure at the time, many women from the older site took second plots in the new site. Under what was then still a somewhat novel arrangement, the garden was converted into a garden/orchard, with a dense stand of trees in a grid pattern over the entire area. The understanding was that ownership of the trees would be divided between the landholder and gardeners on an alternating basis; every other tree, in effect, belonged to the landholder.

Within five or six years, however, the prospect of shade canopy closure appeared in the new garden. Gardeners had already determined that vegetables brought them a greater return than any harvest they could expect from their trees. Consequently, many of the maturing trees were either drastically trimmed or simply removed, including, apparently, many of the trees belonging to the land-holder. In response, the landholder banned tree trimming in his garden, only to find his young trees still being destroyed as women burned crop residues to clear plots for each new planting season. While some of this destruction was doubt-less accidental, the landholder claimed that growers deliberately hung dry grass in tree branches so that fires set to clear plots would fatally damage trees. A survey of tree density on the site revealed that fully half of the original orchard no longer exists, so it is clear that vegetable growers were at least partially successful in defending their use rights.

By 1991, the situation regarding garden/orchard tenure was somewhat uncertain. Survey data from a dozen gardens show clear trends toward tighter control of garden spaces by orchard entrepreneurs, and a major emphasis within orchards on mango trees – the species most likely to cause shade problems for gardeners sharing the space. Landholders opening new gardens in the late 1980s tended to do so only under the strict conditions that women agree in advance to water the landholder's tree seedlings and vacate their temporary use rights when the trees matured. Of the twelve sites surveyed, only three remained solely under gardeners' control. All others had either already been, or were about to be, planted over with tree crops. Some 60 percent of the prime low-lying land in the vicinity of the communities surveyed was thus at risk of being lost to shade within the decade. At the same time, at the end of the 1991 rainy season, gardeners chopped and burned their plots clear almost at will in nearly all of the surveyed sites. This would suggest that, tougher rhetoric and recent clamp-down notwithstanding, the struggle to claim control over garden land in the area is ongoing.

In sum, this brief comparison of the north bank's garden/orchards establishes that trees can be used as a means for claiming both material and symbolic control over garden lands. Tree planting on garden beds, moreover, is a mechanism for landholders to alienate surplus female labor and subsidies embodied in concrete-lined wells and permanent wire fences. In this respect, the Gambian case differs from the apple-based agroforestry system in Java described below, where land-holders often lack the capital to build the infrastructure necessary to convert their lands to orchards. At the same time, shade effects from tree planting threaten to undermine the productivity of gardeners, who now play key roles in providing for the subsistence needs of their families.

On balance, the agroforestry system practised by women gardeners seems of greater value than the successional systems landholders have imposed. Viewed from a production standpoint, garden-based agroforestry practised by women appears to generate a greater absolute income than a monocrop mango system, as well as a more seasonally varied income stream, one better suited to meeting

194

the myriad financial challenges rural families face throughout the year. From an environmental standpoint, since the orchards in the successional schemes are small, they have little impact on climate change and deforestation problems they were ostensibly intended to address. On the micro-scale, the women's systems are clearly more diverse than the men's. Soil quality is typically better, by dint of the incorporation of countless headpan-loads of compound sweepings and manure. Moreover, the evidence shows that, given the chance, gardeners routinely incorporate fruit trees into their crop mix, and that they effectively mánage the ecological competition between vegetables and trees implied by intercropping, *if* they actually control decisions over the selection of species, the location of trees, and rights of trimming or removal, which is to say, the substance of the labor process and property rights. Such social relations are precisely what is overlooked in theories of agroforestry that construct all forms of tree planting in the same terms, namely as beneficial interventions with unambiguous stabilizing effects on local environments.

AGROFORESTRY AND CLASS RELATIONS IN A JAVANESE VILLAGE

Conventional wisdom suggests that upland Java faces an imminent ecological crisis under increasing population pressure. Poor, subsistence households seek to increase their immediate income by using cropping patterns that accelerate soil erosion from their rain-fed farms (USAID and Government of Java 1983). Rainfall intensities are extremely high in Java, contributing to severe soil erosion (Carson 1989). One survey in the mid-1980s estimated that 2 million hectares, or one-third of Java's cultivated uplands were severely degraded, and that the problem was increasing at a rate of 75,000 hectares annually (Tarrant *et al.* 1987).

Since the early 1980s, however, dramatic economic and land-use changes have occurred in many upland villages in Java. As urban incomes have risen, improving the market for fresh fruit, upland farmers have expanded cultivation of commercial fruit trees. A Jakarta-based newspaper reported that throughout the 1980s, domestic demand for fruit increased at the rate of 6.5 percent per annum (*Pelita*, 1 Sept. 1991). Development planners concerned with stabilizing the environment of upland Java viewed this with optimism, as tree planting and agroforestry have always been associated with lower soil erosion rates.

Agroforestry has indeed been an essential component for upland development programs in Indonesia (Mackie 1988). Nonetheless, adoption of tree cropping in response to these programs was modest at best. Conversion of upland farming systems depended heavily on government subsidies (Huszar and Cochrane 1990; McCauley 1988), and farmers often reverted to old practices soon after a project ended. Soil erosion rates from Java's uplands remained high, much to the confusion of planners who failed to understand how peasant-based agroforestry programs could meet with so little success in a country famed in environmental circles for its home gardens.

Where the more narrowly constructed environmental initiatives failed to arrest erosion, however, a commercial 'fruit boom' had dramatic stabilizing effects. The following case study examines the development of apple-based agroforestry in Gubugklakah, a high mountain village in the upper watershed of the Brantas River in East Java. In much of this region, economic depression during the 1930s, followed by war in the 1940s, and subsequent disease outbreaks and soil fertility exhaustion (Hefner 1990), have caused widespread poverty and land degradation. Since the introduction of apples in the late 1970s, however, many farmers have adopted sophisticated soil conservation measures to support fruit production.

Unprotected sloping soils in this region erode at the rate of 2 cm per year, exposing and destroying roots within the lifetime of apple trees (Carson 1989). Construction of bench terraces is thus a prerequisite to apple farming, and small-holders and large growers alike have built terraces in the anticipation of growing apples. By the time apple seedlings are planted, the completion of backsloping terraces and closed ditches between terraces has accounted for roughly 1,000 person days of labor investment per hectare. During heavy rainfall, virtually all mud carried by water runoff collects in the ditches of each terrace bench. After the rain, farmers return the mud to the terraces, thus minimizing the loss of topsoil and fertilizers.

Apple-based farming has markedly changed the agronomic and conservation scene. Approximately three-quarters of the land in Gubugklakah has been converted into terraced apple orchards or apple-based agroforestry. Of the remaining lands, about half have already been terraced. Overall, close to 90 percent of lands in the village have been 'stabilized' in this manner within the last two decades. Government officers both at district and provincial levels, struggling in their efforts to reduce soil erosion from Java's upper watersheds, have applauded this development, and Gubugklakah has often been cited as a model of successful upland management practices (Carson 1989; KEPAS 1988).

Changing social relations of apple-based production

There is no landlord in Gubugklakah, but we have plenty of apple-lords. This is a good arrangement because nobody loses all means to make a living. A small farmer can still grow vegetables even when the trees on his land are leased-out.

(Former Village Head, 1991)

Temperate fruit fill a particular, albeit small, niche in the urban market of Indonesia, and apples are the most important temperate fruit crop in Indonesia. In 1980, the Indonesian government banned the imports of many categories of food, including most fresh fruits. As a result, domestically produced temperate fruit such as apples enjoyed a buoyant market. In the few areas suitable for growing apples, such as Gubugklakah, an economic boom followed. One of the challenges in growing temperate fruit in the tropics is finding ways to prevent bud dormancy in the absence of variation in temperature and daylength. Intensive labor and chemical input is necessary before apple trees can bear fruit. Workers must defoliate and modify plant architecture to stimulate buds to flush. Cultivation of apple trees in the tropics also relies on the frequent application of heavy doses of pesticides and fungicides. With heavy inputs of labor and fertilizers, apple trees in Java can be harvested twice each year.

Apples are intercropped with underlying vegetable crops, including leeks, scallions, garlic, cabbages, and potatoes. Unlike Gambia, customary law in Java does not distinguish land tenure rights along gender lines. In 1991, 94 percent of the lands in Gubugklakah were owner operated, with an average holding of 0.53 hectare. A few large farms of more than 2 hectares belonged to the richest 6 percent, and they covered only about a quarter of lands in the village, which is fairly typical of the region (cf. Hefner 1990). While the seemingly egalitarian distribution pattern indicates that the most recent economic boom has not resulted in land accumulation by richer peasants, this finding belies the ongoing struggle, not over land, but over the utilization of space beneath the apple trees. Just as in Gambia, boom conditions produced tensions and competition between apple growers and vegetable gardeners.

Close to 80 percent of all landowning households have planted apple trees in their vegetable gardens. Tree planting did not cause intra-household tensions as in Gambia. Instead, conflicts developed along class lines, as apple trees were favored by capital-rich farmers. The high commercial value of apples has reinforced the separation of tree tenure from land tenure. Apple trees constitute a valuable asset with higher marketability than land itself, and are often exchanged independently of land. In times of emergency, rights over trees, especially mature trees at fruit-bearing stage, can quickly be liquidated to raise cash. Among less productive trees under three years old, 91 percent are owner operated, as compared to 69 percent among the more productive ones that are four years and older.

Tree transfers under such circumstances have contributed to a process of rapid economic differentiation without apparent land accumulation (Suryanata 1994). Despite the fact that the pattern of land distribution has remained relatively undisturbed, a new class of 'apple lords' has emerged as the village's dominant power. The richest 15 percent control only 50 percent of the land in the village, but 80 percent of the apple harvest. Similarly, although only 21 percent of the village's households were landless, 68 percent did not have any access to apple harvest. Despite the fact that the largest landholding was only 5 hectares, the largest apple farmer operated close to 15,000 trees growing on 20 hectares of land.

Mechanisms for the transfer of tree assets vary. Tree seedlings themselves are sometimes sold and transplanted, but the transfer of rights to trees *and* the space they occupy is more common. Although the land tenancy rate in this village was only 6 percent of all individual landholdings, close to 20 percent were operated under some form of *tree* tenancy, and that figure appears to be growing. By transferring only the tree tenure, a landowner retains the rights to other uses of the land. A structural tension is nonetheless created between the two land management systems.

Two specific forms of tree transfer have emerged. The institution of tree sharecropping (*maro apel*) began about a decade ago in Gubugklakah, and is a modified form of a credit arrangement, once common among vegetable growers. Sharecroppers provide the capital, and in most cases, the labor and skills necessary for the cultivation of apple trees. Landowners provide the land but retain the rights to grow annual crops underneath the trees until it is prohibitively difficult to do so. The terms of tree sharecropping specify how profit from apple production is to be divided, and rules on other access to the land where the trees are standing. In contrast to vegetable sharecropping, the longevity of apple trees and their permanent tenure preclude terminating the contract at a season's notice, unless landowners compensate their tenants for the trees, a practical impossibility in most cases given their high value.

Tree leasing (*sewa apel*) is a post-boom phenomenon. As capital-rich apple growers began to acquire management skills and reduce production risks, they increasingly favored fixed-rent leasing. Persistent credit needs of smaller-scale owner-operators have accordingly created a rental market for apple trees. The typical arrangement involves capital-rich growers leasing apple trees from landowning, capital-poor peasants. Invariably, the reason for leasing out trees is a pressing need for cash, which may arise from crises or basic demands of household reproduction, such as the illness or death of a family member, children's education, and house building expenses. It may also arise from the desire to possess luxury goods such as motor vehicles which have become more common as the new prosperity has contributed toward changes in consumption patterns (cf. Lewis 1992). Most often, the need to lease out apple trees arises from the inability to maintain young trees that have absorbed investment capital, but not yet produced any return. Renting out the trees is the only option if a farmer does not want to lose the investment made thus far. If a farmer owns several fields, tree leasing of one plot may be a way to raise capital to finance the operation costs for another field. The rent is typically negotiated and payable in advance, albeit within the context of a renter's market. In most cases, the liquidity crisis puts the lessor in a disadvantaged position, resulting in a very low rent relative to the potential yield.

The duration of tree lease contract ranges from one harvest to as long as fifteen years (thirty harvests under a double crop regime). If a lessor needs extra cash before the contract expires, the lessor can choose to extend the contract in return for an agreed sum of money, or a share of the net profit of an agreed number of harvests. The lessor's bargaining position then, however, is far weaker than when the contract was first established. The lessee is in a position to negotiate a lower rent, impose more restrictions on growing field crops, or advance a permanent tenure claim to the trees. With the reduced amount of resources available to a lessor household after it enters into the contract, the likelihood of needing further credit extensions before the lease term expires is fairly high. Of the twenty-nine cases of tree leasing in the study, more than half have renegotiated their contracts before the original terms expired, resulting in increased benefits for tree lessees. As one lessee in Gubugklakah put it in 1991:

> In 1984 I rented 900 apple trees from my neighbor for twenty harvests. Five harvests into the lease, he wanted to borrow more money. In return he would stop growing vegetables on this land. I agreed to suspend the lease for one season, and share the net profit of the sixth harvest. Because of this adjustment, when the lease expires I gain the right to sharecrop the trees even though I did not plant them.

Agroforestry and labor control

After a long string of failures in stabilizing the environment in Java's sloping uplands, improved market incentives for tree products have presumably enhanced the adoption rate of tree planting. At the outset, apple-based agroforestry in Gubugklakah seemed to offer a sustainable and equitable solution to the problems of poverty and soil erosion that characterized the village twenty years ago. Indeed, the case appeared to counter arguments that link agricultural commoditization with environmental degradation (Blaikie 1985; Grossman 1981), insofar as apple cultivation provided incentives for land improvement and rehabilitation, while simultaneously bringing economic prosperity.

In sharp contrast to this vision, however, the new land-use system is neither environmentally sound nor equitable. Instead of developing into a system with a high biological diversity that requires low inputs, apple-based farming systems are increasingly simplified, and require extensive use of chemicals. While this system does play a role in reducing soil erosion, the reduction does not come from the vertically intermingled plant cover as in traditional home gardens, but from the heavy labor input for constructing and maintaining terraces.

As apple trees mature, spatial conflict and competition between apple trees and vegetables increases. Village surveys showed that in *owner*-operated fields, expanding canopies and intensive maintenance of apple trees do not rule out intercropping with vegetable crops. By contrast, in fields under tenancy contracts the ecological competition between vegetables and trees becomes more pronounced. Apple lords blame the traffic of disinterested landowner/vegetable growers for causing blemishes in apples that lower their market value. Meanwhile, the frequent trampling by apple workers uninterested in the undergrowth often damages vegetable crops. In such struggles, tree lessees invariably come out as winners. Their advantages are exercised either through formal terms in the contract extensions or through the reckless practices of apple workers that impose an environment hostile to the vegetable crop. As a result, just as in Gambia, many fields have effectively turned into monoculture apple orchards which deprive landowners of access to their own land.

The system's equity soon deteriorates as input costs are driven up by the increasing demand for a controlled environment. The spatial conflict peculiar to the configuration of apple-based agroforestry also serves as a means of labor control for the "apple lords." Labor need is highest during the first ten weeks of each season when the trees are defoliated, fertilized, and pruned. Competition for hiring wage laborers escalates during peak operations. Apple lords growing more than 1,000 trees secure laborers in dependent wage-labor relations akin to patron–client relationships. Patrons offer benefits that include loan provisions with low or no interest, access to fodder from patrons' fields, or year-round guarantees of employment. Under such terms, about 24 percent of the lessor/landowners also work as paid laborers for their tree lessees. These arrangements provide landowners with the opportunity to personally ensure that apple maintenance does not cause trampling damage to the vegetable crops. The land-owner's residual rights are thus appropriated by the apple patron and *returned* to the landowner as part of a labor contract. Thus, while the new labor relation may partially mitigate the effect of lost control over trees, it does so only under terms which increase the dependency of landowners on their creditors/tree lessees, deepening the imbalance of power between them.

A combination of tenure multiplicity and intercrop dynamics unique to agroforestry have actually facilitated economic polarization in this village. Tree leasing in particular slowly dispossesses capital-poor landowners from any land-based production, as access to growing field crops is increasingly suppressed by the lessees. In addition, apple cultivation often pushes vegetable growers into dependent wage-labor relationships. Despite their formal landowning status, they have formed a new class of 'propertied labor' (cf. Watts 1994) as the original multi-purpose agroforestry system has given way to monoculture apple orchards, controlled by the richest few.

CONCLUSION

It is easy to invoke the environmental crisis and the poor people's energy crisis to open up new avenues for reductionist science and commodity production.

(Shiva 1988)

We argued in this chapter that agroforestry approaches are not always the unalloyed good they are sometimes made out to be. In practice, 'stabilization' efforts involving tree crops are often highly ambiguous. Our two case studies examined agroforestry practices premised on the commoditization of tree crops and the assumption that market incentives enhance the rate of tree planting (Murray 1984). Both cases, however, show the contradictions of efforts to stabilize the environment through the market as commoditization leads to shifting patterns of resource access and control. In each place, this process takes on different characteristics, producing different forms of social friction and resistance depending on local social structure and institutions. In Gambia, gender conflict between husbands and wives has grown out of multiple tenure claims to patrilineal land which intensified with the commoditization of fruit trees. By contrast, the tree boom in upland Java was the cause of inter-*class* tenure conflict as commercialization polarized the village's peasantry. Both case studies illuminate the need to recognize basic political ecological considerations, such as identifying clearly on whose behalf stabilization efforts are undertaken, specifying who is in the position to define stability and determine when in fact it is achieved.

In the case of Gambia's garden boom, in each of the hundreds of garden perimeters springing up over the past two decades, the ecological and economic significance of wells, fences, soil improvements, and tree stands must be assessed in light of competing local, national, and international interests. Wells, fences, and soil improvements provide the necessary conditions for vegetable production and thus serve the needs of both vegetable growers and their families heavily dependent on vegetable incomes. But such improvements also tie female labor to a specific spatial domain, thereby stabilizing conditions which allow landholders to establish orchards. The addition of the tree crop, in turn, negates the value of the infrastructure for gardeners, effectively *de*stabilizing their productive base, and actually compounding problems within a broader political economic context by attracting the intervention of outside donors interested in claiming the land improvements as their own (Schroeder 1993). Similarly, Javanese farmers on the western slope of Mount Bromo have built elaborate terrace systems to stabilize their land resources and accommodate commercial apple-based farming. The presence of high-value apple trees, however, is conducive for the development of tree-leasing contracts and a gradual dispossession of land resources, and thereby helps capital-rich apple lords to establish and accumulate apple orchards. As a result, while the threat of soil erosion to downstream interests may have been reduced, the value of this 'stable' environment to the landowners themselves has been shrinking.

Viewed from a slightly broader perspective, the loan-seeking behavior of men on Gambia's north bank has forced their vegetable-growing wives to intensify horticultural production through expansion of fence enclosures and tree planting. Landholders – a select group of men who hold senior positions in family lineages – have finessed the issue of enclosure in a way that allows them to control women's labor and capture subsidies intended for the construction of garden infrastructure. Non-governmental donor agencies use landholders' leverage over vegetable growers to meet their own objectives of land stabilization via tree planting (Lawry 1988; Mann 1989; Norton-Staal 1991; Thoma 1989; Worldview International Foundation 1990). And the state and multilateral donors build on NGO successes to meet national goals in environmental stabilization, agricultural diversification, and full-scale economic readjustment (Agroprogress International 1990; Government of Gambia n.d., 1990; Thiesen *et al.* 1989; Thoma 1989; USAID 1991). This implies, quite simply, that

201

developers at all levels pin their hopes, indeed stake their very legitimacy in some cases, on the continued mobilization of unpaid female labor. Once again, Java offers a striking parallel. After decades of failure in promoting tree cropping by upland smallholders, district and provincial governments point to the recent growth of fruit-based agroforestry as an indicator of success in meeting the goals of environmental stabilization and economic development. Commercial agroforestry has become a model for upland development, and donor-assisted programs have funded new research and development efforts directed at fruit trees suitable for upland farming. At the national level, the government is interested in exploiting the growing international markets for tropical fruit and thereby increasing its non-traditional exports. As in Gambia, these various interests are premised on the development of a new class of 'fruit lords' who can mobilize the labor of capital-poor landowners to their own ends.

We contend on the basis of this evidence that there is a contradiction at the heart of commercial agroforestries undermining their effectiveness as strategies of resource stabilization. The strengths of agroforestry systems do not lie exclusively in the ways they enhance productivity or reverse degradation; they also rest in the opportunities afforded for sheltering a multiplicity of claims and uses. From a political ecological point of view, agroforestry systems are strongest when people can manage their resources independently, beyond the scope of powerful interests that often converge when commercial incentives increase the rigidity and exclusivity of claims.

REFERENCES

Agroprogress International. 1990. *Project preparation consultancy for an integrated development programme [European Community] for the North Bank Division.* Bonn: Agroprogress International.

Blaikie, P. 1985. *The Political Economy of Soil Erosion in Developing Countries.* London: Longman.

Bryant, R. 1994. "The rise and fall of *taungya* forestry: social forestry in defence of the Empire," *The Ecologist* 24, 1: 21–6.

Carson, B. 1989. "Soil conservation strategies for upland areas in Indonesia." East–West Center Environment and Policy Institute, Occasional Paper 9. Honolulu.

Clay, J. 1988. *Indigenous Peoples and Tropical Forests: Models of Land Use and Management from Latin America.* Cambridge, MA: Cultural Survival Inc.

Dove, M. 1990. "Socio-political aspects of home gardens in Java," *Journal of Southeast Asian Studies* 21, 1: 155–63.

Farrell, J. 1987. "Agroforestry systems," in M. Altieri (ed.) *Agroecology: The Scientific Basis of Alternative Agriculture.* Boulder, CO: Westview Press.

Fortmann, L. and J. Bruce (eds.). 1988. *Whose Trees? Proprietary Dimensions of Forestry.* Boulder, CO: Westview Press.

Goswami, P. 1988. "Agro-forestry: practices and prospects as a combined land-use system," in L. Fortmann and J. Bruce (eds.) *Whose Trees? Proprietary Dimensions of Forestry.* Boulder, CO: Westview Press.

Government of Gambia. 1990. *National Natural Resource Policy.* Banjul.

—— n.d. *Executive Summary. Programme for Sustained Development. Sectoral Strategies.* Banjul: Gambia Round Table Conference.

Grossman, L. 1981. "The cultural ecology of economic development," *Annals of the Association of American Geographers* 71, 2: 220–36.

Hefner, R. 1990. *The Political Economy of Mountain Java.* Berkeley: University of California Press.

Huszar, P. and H. Cochrane. 1990. "Subsidization of upland conservation in West Java: the Citanduy II Project," *Bulletin of Indonesian Economic Studies* 26, 2: 121–32.

KEPAS. 1988. *Penelitian agroekosistem lahan kering Jawa Timur.* Bogor, Indonesia: KEPAS.

King, K. 1988. "Agri-silviculture (*taungya* system): the law and the system," in L. Fortmann and J. Bruce (eds.) *Whose Trees? Proprietary Dimensions of Forestry.* Boulder, CO; Westview Press.

Lawry, S. 1988. *Report on Land Tenure Center Mission to The Gambia.* Madison: University of Wisconsin Land Tenure Center.

Leach, M. 1994. *Rainforest Relations: Gender and Resource Use among the Mende of Gola, Sierra Leone.* Washington, DC: Smithsonian Institution Press.

Lewis, M. 1992. *Wagering the Land: Ritual, Capital and Environmental Degradation in the Cordillera of Northern Luzon, 1900–1986.* Berkeley: University of California Press.

McCauley, D. 1988. *Citanduy Project completion report, annex V: Policy analysis.* USAID: Jakarta.

Mackie, C. 1988. *Tree Cropping in Upland Farming Systems: An Agroecological Approach.* USAID/Indonesia, Upland Agriculture and Conservation Project.

Mann, R. 1989. "Africa: the urgent need for tree-planting," Methodist Church Overseas Division, unpublished manuscript.

Millon, R. 1957. "Trade, tree cultivation and the development of private property in land," *American Ethnologist* 57: 698–712.

Murray, G. 1984. "The wood tree as a peasant cash crop: an anthropological strategy for the domestication of energy," in R. Charles and A. Foster (eds.) *Haiti Today and Tomorrow.* New York: University Press of America, pp. 141–60.

Nair, P. 1989. *Agroforestry Systems in the Tropics.* Boston: Kluwer Academic Publications and ICRAF.

—— 1990. "The prospects for agroforestry in the tropics." World Bank Technical Paper No. 131. Washington, DC: World Bank.

Norton-Staal, S. 1991. *Women and Their Role in the Agriculture and Natural Resource Sector in The Gambia.* Banjul: USAID.

Peluso, N. 1992. *Rich Forests, Poor People: Resource Control and Resistance in Java.* Berkeley: University of California Press.

Raintree, J. (ed.). 1987. *Land, Trees and Tenure.* Nairobi, Kenya and Madison, WI: ICRAF and University of Wisconsin Land Tenure Center.

Rocheleau, D. 1987. "Women, trees and tenure: implications for agroforestry research and development," in J. Raintree (ed.) *Land, Trees and Tenure.* Nairobi, Kenya and Madison, WI: ICRAF and University of Wisconsin Land Tenure Center, pp. 79–121.

Rocheleau, D. and L. Ross. 1995. "Trees as tools, trees as text: struggles over resources in Zambrana Chacuey, Dominican Republic," *Antipode* 27, 4: 407–28.

Schroeder, R. 1993. "Shady practice: gender and the political ecology of resource stabilization in Gambian garden/orchards," *Economic Geography* 69, 4: 349–65.

—— 1994. "'Gone to their second husbands': marital metaphors and conjugal contracts in The Gambia's female garden sector," paper presented at the African Studies Association Annual Meeting, Toronto.

—— 1995. "Contradictions along the commodity road to environmental stabilization: foresting Gambian gardens," *Antipode* 27, 4: 325–42.

—— (forthcoming) "'Gone to their second husbands': marital metaphors and conjugal contracts in The Gambia's female garden sector," *Canadian Journal of African Studies.*

Shiva, V. 1988. *Staying Alive: Women, Ecology and Development.* London: Zed Books.

Suryanata, K. 1994. "Fruit trees under contract: tenure and land use change in upland Java," *World Development* 22, 10: 1567–78.

Tarrant, J., E. Barbier, R. Greenberg, M. Higgins, S. Lintner, C. Mackie, L. Murphy and H. van Veldhuizen. 1987. *Natural Resources and Environmental Management in Indonesia: An Overview.* USAID: Jakarta.

Thiesen, A., S. Jallow, J. Nittler and D. Philippon. 1989. "African food systems initiative," project document. Gambia: U.S. Peace Corps.

Thoma, W. 1989. *Possibilities of Introducing Community Forestry in The Gambia, Pt. 1.* Gambia–German Forestry Project, Deutsche Gesellschaft for Technische Zusammenarbeit (GTZ). Feldkirchen, Germany: Deutsche Forstservice.

USAID. 1991. *Agricultural and Natural Resource Program, Program Assistance Initial Proposal.* Banjul, Gambia.

USAID and Government of Indonesia. 1983. *Composite Report of the Watershed Assessment Team.*

Watts, M. 1992. "Idioms of land and labor: producing politics and rice in Senegambia," in T. Bassett and D. Crummey (eds.) *Land in African Agrarian Systems.* Madison: University of Wisconsin Press, pp. 157–93.

—— 1994. "Life under contract: contract farming, agrarian restructuring, and flexible accumulation," in P. Little and M. Watts (eds.) *Living under Contract: Contract Farming and Agrarian Transformation in Sub-Saharan Africa.* Madison: University of Wisconsin Press, pp. 21–77.

Worldview International Foundation. 1990. *WIF Newsletter* 3, 1: 4.

10

FROM CHIPKO TO UTTARANCHAL

Development, environment, and social
protest in the Garhwal Himalayas, India

Haripriya Rangan

Demands for the creation of a new state of Uttaranchal comprising eight
Himalayan districts, presently part of the state of Uttar Pradesh in India, have
recently escalated in stridency and violence. Grafitti on the walls of public
buildings in the region's towns and cities proclaim, "We ask today [for the creation
of Uttaranchal] in a friendly spirit; tomorrow we'll demand it with guns." Youths
have raided police outposts for rifles and ammunition; government buildings have
been vandalized and torched. There is, in the words of a national newspaper,
a movement "slowly but surely taking on the dimensions of a bloody stir"
(*Sunday* 1994: 44–7). For the men and women from Garhwal and Kumaon
Himalayas who picket the central government offices in New Delhi, statehood
is the means of extricating their region from its current backwardness. Lack of
development coupled with inefficient administration, they assert, is the cause
of high unemployment and increasing marginalization of hill regions in the state
of Uttar Pradesh.

Popular protests over local development issues are neither new nor rare in
post-independence India (Hauser 1993; Mitra 1992). But the Uttaranchal
movement occurs in a region known as home of the Chipko movement. Hailed
by academics and environmental activists throughout the world for its grassroots
environmental mobilization in the Indian Himalayas, Chipko was seen as a
powerful critique of a modernizing Indian state whose economic development
policies were considered ecologically unsustainable and callously indifferent to
the traditional needs of peasant subsistence (Bahuguna 1982, 1987; Berreman
1989; Bhatt 1988, 1991; Dogra 1980, 1983; Gadgil and Guha 1992; Guha
1989; Shiva 1989; Shiva and Bandhyopadhyay 1986, 1987, 1989; Weber 1988).
Some said that Chipko symbolized a new ecological consciousness illuminating
the path towards a "green earth and a true civilization" (Weber 1988). Yet nearly
twenty years later, Chipko has been displaced by Uttaranchal, a widespread
regional movement that, rather than arguing for environmental protection
or alternatives to development, demands the creation of a separate state for
promoting economic growth in the Himalayan regions of Uttar Pradesh.

This chapter illustrates how the Chipko and Uttaranchal movements are linked by a common tissue of concerns over regional economic development. The Chipko movement emerged in the Garhwal Himalayas during the 1970s in response to economic policies and regulations that introduced stricter controls over access to resource extraction and constrained opportunities for local economic development in the region. Chipko's initial attempts to alter regulations and demand economic concessions were overwhelmed by its subsequent fame as a grassroots ecological movement. Chipko's ecological successes resulted in new environmental regulations that compounded the lack of economic opportunities and development in the region. Today, demands for creating an Uttaranchal state are, once again, directed against regulations and policies that severely constrain local and regional economic development. I show that the growing violence in the Uttaranchal movement expresses the frustration of some participants who realize that their protests have neither gained wider political support, nor succeeded in forcing the Indian government to negotiate the issue of statehood.

In this chapter I also challenge some contemporary views that see new social movements in the Third World as grassroots agents seeking alternatives *to* development (Escobar 1995). My chapter argues that social protests in post-independence India are, contrary to these views, centrally concerned with access to development, and forcing the state to assume greater responsibility in addressing problems of uneven regional development and social equity (Brass 1994; Hauser 1993). Even as the welfare state and the Nehruvian model of state-led economic development retreat in the face of neo-liberal reforms in India, the state remains integral for mediating new reconfigurations of markets and civil society (Peet and Watts, Chapter 1 in this volume). Growing economic disparities between regions and rapid class differentiation within them constrain access to jobs for the unemployed and reveal a lack of local economic opportunities. Erosion of food security stimulates political action against the state and for institution-building in civil society. The idea of development is not so much subverted by these protests as integrated into an expanded "moral economy" (Thompson 1993; Wells 1994) of communities fighting to overcome political and economic marginality. Demands for regional autonomy in Uttaranchal, calls for ethnic and caste-based equality, popular environmental protests in India and arguably in other parts of the world, all have a common concern in pressuring states to intervene on their behalf to ensure equitable access to the potential benefits of economic development. New social movements in the Third World, I argue, are not *against* the idea of development, they are *part* of it. Thus even the most radical ideas celebrating the flowering of new social movements and calling for sustainable use of resources need to recognize that the broad notion of development – as a means of gaining access to social equality, economic well-being, and political recognition of marginalized communities – remains central to any project of a liberation ecology.

DEVELOPMENT AND ITS CRITICS

According to some contemporary critics, development is a totalizing and hegemonic discourse that perpetuates social and economic inequalities between rich and poor regions and social groups (Escobar 1992, 1995; Mies and Shiva 1993; Shiva 1989; Trainer 1989; Yapa 1993). Development, as one such critic says:

> has to be seen as an invention and strategy produced by the "First World" about the "underdevelopment" of the "Third World," and not only as an instrument of economic control over the physical reality of much of Asia, Africa, Latin America. Development has been the primary mechanism through which these parts of the world have been produced and have produced themselves, thus marginalizing or precluding other ways of seeing or doing.
>
> (Escobar 1992)

This argument regarding development is problematic. Deconstructing "development" in an academic exercise that reduces the term to its narrowest possible definition and reveals it as an instrument of domination, does make the term assume the form of an ominous, omnipotent, unchanging power; an elemental, ahistorical force beyond human control, shaping the fates of humans everywhere in the world. But the very fact that the term "development" has a complex genealogy encompassing a diversity of meanings (Williams 1983) indicates that it is as dynamic as life's processes – coming into being as an idea, changing over time, diversifying in meaning, becoming a contested terrain, diffusing through translation and re-emerging in different forms in different regions. Development is a dynamic process that involves states, markets, and civil societies to varying degrees in actively reshaping social relations and institutions. Imposing a reductionist definition of instrumental control on the term not only denies its historicity but also ignores the diverse ways in which ideas of development, despite their origins in Western thought, have been translated, appropriated, refashioned, and reconfigured by local circumstances (Watts 1993).

It has never been possible to ignore the realities of difference – regional, cultural, linguistic, religious, ethnic, class, and caste – in India. There the idea of development (referred to in the vernacular as *vikas*, meaning both the process of moving towards the dawn of a new social era, and the social era itself) has been used as a secular, democratic means for opening the political arena to the claims of various groups in civil society. Development is charged with the promise of change towards greater social equality and prosperity for all citizens. It has been taken up by disadvantaged groups as a means of gaining political recognition and access to economic empowerment. Social transformation is a glacial process in India, but the vocabulary of development, including the universal franchise, has allowed poorer classes and lower-caste groups to question, with growing assertiveness (Hauser 1993; Naipaul 1990), structures of social and economic inequality such as the caste system, "one of history's most well-entrenched and ornately elaborate ideological systems of legitimizing inequality and exploitation" (Bardhan 1985: 82). This slow process of social change has its historical roots in the way ideas of development and democracy were refigured and reworked in India as part of the post-independence struggles.

Development emerged as the leitmotif of political discourse in post-independence India as groups on the Left, Right, and Centre confronted each other with their interpretations of the "developmental imperative" for the country (Chakravarty 1987; Mitra 1992). There was little debate or argument over the precise definition of the term. Nationalist elites agreed that development – broadly meaning the reorganization of state, market, and social institutions – was necessary for overcoming persistent economic stagnation, high levels of poverty, and structural weaknesses in the Indian economy. Nationalist leaders, influenced in part by the Soviet experience and the legacy of Keynes, saw the state as the central actor in achieving this purpose. State intervention in the economy was to follow a broadly "socialist," rather than a free-market approach, using central planning as a means of "avoiding the unnecessary rigours of an industrial transition in so far as it affected the masses resident in India's villages" (Chakravarty 1987: 3). The Indian Planning Commission, established by the national government to systematically study the country's economy, identified sectoral priorities in its five-year plans for working out a coherent set of investment policies for economic development.

Rational state-led planning was, however, framed by parliamentary democracy in post-independence India. The situation faced by the independent Indian government was unlike the experiences of most other liberal democratic societies in the Western world, where the state had, in some sense, served as the "executive committee of the bourgeoisie" during the formative phase of capitalist development (Braudel 1977), and only later extended universal franchise to include all citizens (Bowles and Gintis 1986). On the one hand, Indian development inevitably involved a process of social and economic differentiation, by which some sections of society would benefit more than others (Herring 1989; Weiner 1989). On the other, the constitutional guarantees of universal franchise and social equality meant that the Indian state also needed to intervene to correct these imbalances and maintain its democratic credentials for ensuring continued political support of the populace. The nation's leaders recognized that this awkward predicament – of guaranteeing social equality while contributing to economic inequality – allowed the state to exercise some degree of flexibility. Development as nation-building ensured government a central role in the economy and, as a consequence, a key role in political accommodation between dominant classes and other groups within civil society. The Indian state emerged as the mediator – appealing to democratic traditions and invoking the collective goal of national development – whenever conflicts in civil society grew violent or gave rise to secessionist demands. The Indian state also created, in part by design, in part unintentionally, conditions of greater social fluidity that allowed less-privileged groups to maneuver for greater political representation and access to the benefits of economic development (Bardhan 1985; Mitra 1992).

Development came to be accepted by most of India's population as a legitimate activity promoted by the state. Yet its complex meanings, values, and benefits are constantly negotiated and contested in the public realm. Since social and economic change is seen as part of the state's agenda, political parties seeking power have found it necessary to employ the language of development to mobilize support among India's vast rural constituencies. Development has become, therefore, the means by which political allegiance of rural elites is gained. Rural elites function as intermediaries in the development process, their status depending in large part on their ability to occupy the interstices between and within the state, market, and civil society, and also in demonstrating an ability to direct the flow of developmental resources from the state to their localities (Bayly 1976; Hauser 1993; Mitra 1992). In rural localities, development assumes a generic expression referring to government-sponsored infrastructural activities, such as the construction of roads and bridges, provision of public utilities, social welfare programs, and investment subsidies for promoting economic growth. If local elites are seen to fail in bringing development to rural communities, their status and authority may be challenged, thus allowing other contending groups to force their way into the political arena. Because the "discourse" of development typically carries a broader symbolism of social justice and economic well-being, it confuses (or perhaps condenses within it) the conceptually rigid boundaries between state, markets, and civil society. It simultaneously creates a space for institutional participation and provides the language for radical critique. Social protests and movements in post-independence India have typically not argued against development but have always been part of its process.

DEVELOPMENT PLANNING IN INDIA

The first decade and a half of development planning (1950–65) reflected the economic and political theoretical debates of the time. Even though critics of India's development policies argue that the Gandhian approach was a viable alternative to Nehru's overwhelming zeal for modernization (Gadgil and Guha 1992), it is necessary to keep in mind that Nehru's Fabian socialism and enthusiasm for catching up with the "West" were tempered by the real political lessons learned from Gandhi as well as from active participation in the nationalist movement. Gandhi's ideas were indeed drawn from close experience and understanding of the problems of rural India, but there were few scholars at the time who could have confidently illustrated that his economic thought was based on a compelling and substantive theoretical foundation. Even fewer Indian scholars and nationalists, at the time of independence, would have argued for the low levels of production and consumption implicitly assumed in Gandhian economic thought. In contrast, India's modernizers focused on a more realistic assessment

209

of the structural weaknesses persisting in the national economy, and on "the fetters of the past" such as the caste system that needed to be decisively broken (Ambedkar 1945; Chakravarty 1987). Indian planners stressed the need for a strong industrial sector that would establish the foundation for a healthy and self-sustaining economy. Agriculture was regarded as a "bargain sector" with enormous potential for growth following the necessary institutional changes, infrastructural investments, and political mobilization by state and central governments (Chakravarty 1987: 21–3; Sen 1960, 1984). The *Second Five-Year Plan* (1956–60) spelt out the need for state administrations to engage in land reforms to underwrite increases in agricultural output and rural incomes. Farming co-operatives, community development, national extension programs, and public investment in irrigation were all seen as essential catalysts for unleashing the productive potential in agriculture (Government of India 1956). Gandhian ideas of promoting alternative employment in rural areas were co-opted by the plans through resource allocations to home-based artisanal and handloom industry (Chakravarty 1987: 21).

The course laid out by development planners was, however, radically reshaped by contingent events during the 1960s. The beginning of the *Third Five-Year Plan* (1961–6) coincided with the Sino-Indian and Indo-Pakistan wars (1962–5) that were fought over boundary demarcations and territories in the northwestern and eastern Himalayan regions (Woodman 1969). In addition, prolonged drought caused by two successive monsoon failures in 1965 and 1966, combined with ongoing price inflation resulted in near-famine conditions. These events led to a fiscal and development crisis for the Indian government. As defense spending increased, and the immediate catastrophe of famine was partly averted by large-scale import of foodgrains from the United States under PL 480, Indian planners realized that their development policies had inadequately assessed the growing imbalance between demand and supply of food. Population growth, limited potential for spatial expansion of cultivation, and collapse of regional crop specialization led planners to reformulate national economic policies towards ensuring agricultural productivity (Chakravarty 1987: 23). Planners were also concerned with stemming the net inflow of foreign aid over the following decade in order that the economy could be more self-reliant (Government of India 1961).

Based on these reassessments, the *Fourth Plan* (1969–74, developed after a series of annual plans between 1966 and 1969) redefined national economic priorities. A two-way flow of inputs between the agricultural and industrial sectors was seen as necessary for increasing the production of basic necessities and the growth of national income (Government of India 1970). The plan's emphasis shifted from the capital goods sector – which was already producing to excess capacity – to increasing the production of food, fuel, and other basic necessities (Chakravarty 1987: 33–5). The plan thus established the formal groundwork for promoting agricultural self-sufficiency through the Green Revolution strategy (Byres 1972; Crow *et al.* 1988; Harriss 1982; Mellor 1968;

Raj 1973; Srinivasan 1979; Subramaniam 1979). In short, agriculture was to receive greater capital investment in the form of credit, improved seeds, and chemical fertilizers. Fiscal restraints on public spending moved the focus away from large-scale irrigation projects and towards promoting small-scale, ground-water irrigation among farmers. Self-reliance in raw material production also extended to the forestry sector and involved extensive plantations of fast-growing tree species for meeting the rising internal demand for fuelwood, paper, pulpwood, and other forest resources (Government of India 1976).

Development planning emerged, therefore, from the necessity to deal with the social and economic problems faced by the nation, but was reshaped by contingent events such as wars, droughts, energy crises, and the unexpected outcomes of its own policies. India's planners set out their policies as broad guidelines within which social and economic transformation was to occur, reforming their strategies every five years in response to problems that were entirely new, unforeseen, or persisted stubbornly, despite reassessment and reformulation. Inter-regional disparities, or rather uneven regional development, was one problem persisting through the national planning process. Five-year plans, when translated into specific sectoral or regional policies, were substantially reshaped by political pressures exercised by different classes and electoral constituencies. Sectoral policies aimed at promoting industrial and agricultural growth led to widely divergent outcomes in different states and regions. Regional policies aimed at developing economically backward areas had varied impacts on labor, employment, and productivity of different sectors within, and between, states. Populist programs developed in response to protests voiced by rural elites in particular regions led to new national policies that either backfired or resulted in increased fiscal expenditure.

Development planning in India involved the ceaseless process of allocating and reallocating resources differentially among sectors, regions, and constituencies in an attempt to foster both economic growth and greater social-economic and political equality. Development, in short, was a dynamic force reconfiguring and reworking innumerable theaters, actors, and their roles, and also forcing new trajectories as regions, markets, and institutions within civil society contested and renegotiated its meanings.

UNEVEN DEVELOPMENT AND ITS OUTCOMES IN THE GARHWAL HIMALAYAS

At the time of independence in 1947, the major economic activities in the Garhwal Himalayas (see Figure 10.1) centered around subsistence agriculture, forestry, seasonal employment in pilgrimage services, and, to a smaller extent, in trans-Himalayan trade (Pant 1935; Rawat 1983, 1989; Walton 1910; Williams 1874).

Forests, comprising nearly 60 percent of the region (Stebbing 1932), were largely under the control of the Uttar Pradesh State. The State Forest Department

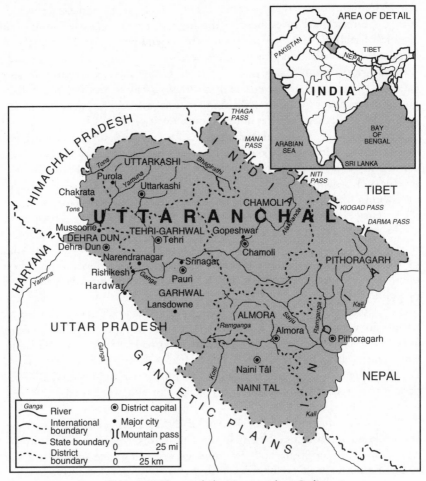

Figure 10.1 Uttaranchal region, northern India

controlled approximately two-thirds of this forested area, with the rest distributed under the authority of the State Revenue Department, village institutions, and a few private holdings (Rangan 1995; UP Forest Department 1989). During the first and second national five-year plans (1951–5, 1956–60) there was little economic change in the Himalayan districts. The eastern districts in the Garhwal region received a few Community Development projects aimed at providing basic infrastructure and encouraging the formation of village-level co-operatives for artisanal production (Khan and Tripathy 1976). Half-hearted attempts at land reform by the UP state government resulted in statutory ceilings on land ownership, and land redistribution allocated plots to lower-caste and landless households on government lands previously cleared or denuded. These statutes led private forest owners to rapidly strip their forests of all valuable resources and

212

commodities before their lands were acquired by the UP state government (Bora 1987; Dhoundiyal *et al.* 1993; Joshi *et al.* 1983; Khanka 1988; Swarup 1991).

The border wars between India, China, and Pakistan (1962–5) had a more direct effect on the Himalayan region within Uttar Pradesh. Following the war, the border with Tibet became a national security concern. Army bases and depots were rapidly established and connected by new roads along India's Himalayan frontiers. The region's economy was profoundly affected; nearly 10 percent of UP state lands and forests in the region were transferred to the Indian government for defense purposes (UP Forest Department 1989), thus withdrawing access for communities that depended on wage earnings in forestry or petty commodity extraction of forest resources. Trans-Himalayan trade, an activity that provided many border communities in the region with a substantial proportion of their income (Atkinson 1882; Pant 1935; Rawat 1983, 1989; von Fürer-Haimendorf 1981; Walton 1910; Webber 1902), came to an abrupt halt with the closing of the Indo-Tibetan border. Few economic alternatives remained. Subsistence agriculture was neither adequate nor remunerative for the majority of households in the Garhwal Himalayas. More than 75 percent of landholdings were less than 1 hectare in size (Government of Uttar Pradesh 1984), mostly rain-fed plots on steep, terraced mountain slopes where productivity largely depended on applications of cattle manure (Pangtey and Joshi 1987; Singh and Berry 1990). The scope for expanding cultivation of commodity crops was also limited, because it largely depended on the extent to which profits from trans-Himalayan trade and other activities such as small-scale timber extraction were reinvested by households in regionally specialized cash crops such as ginger, turmeric, chillies, and opium (Atkinson 1882; Rangan 1993; Rawat 1983, 1989; Walton 1910). Most households without access to capital or alternative employment within their localities were reduced to subsistence cultivation, augmenting their incomes with remittances by males who migrated in search of work (Bora 1987; Dobhal 1987; Joshi *et al.* 1983; Khanka 1988).

Green Revolution policies were mainly geared towards increasing agricultural productivity in the Indo-Gangetic plains and other grain-producing regions in India, and had little impact on the Garhwal Himalayas. Mountainous terrain, lack of infrastructure, and fragmented landholdings distributed across different ecological and altitudinal zones were hindrances to the introduction of Green Revolution techniques. Touted as scale-neutral by scientists and planners, the successful adoption of Green Revolution techniques crucially depended on access to both capital and irrigation. Enormous risks were involved in using capital-intensive inputs if steady and well-timed supply of irrigation was not assured. It was difficult for households in Garhwal – already reduced to dependence on subsistence cultivation and remittances for survival – to raise the necessary capital or collateral for credit, to obtain chemical fertilizers, invest in pumped irrigation (a high-cost investment on mountainous terrain), and purchase high-yielding seeds (Rangan 1993).

Economic marginalization due to the combined effects of the border wars, closure of trans-Himalayan trade, and lack of investible surplus in the region was accelerated by the implementation of forestry policies proposed in the *Fourth Plan* (1969–73). The plan directed State Forest Departments across the country to assume de facto authority over all forested and open lands owned by states (including forests and wastelands controlled by Revenue Departments) and expand plantations of fast-growing tree species that would provide the raw material needs for industry (Government of India 1976). Under the new system, the Forest Department limited rights of access and concessions for extraction from forests to fuelwood and fodder for household consumption, thereby precluding all forms of small-scale commodity extraction of forest resources. Costs of competing in Forest Department auctions paralleled the growing market demand for timber and other forest products escalating at an average rate of 8 percent per annum between 1950 and 1969 (UP Forest Department 1989). Small-scale extractors were thus marginalized from forestry activities because they now lacked physical access to Revenue forests, and did not have access to credit for competing in commercial extraction of resources from Reserved Forests. Commercial extraction was dominated by merchants and traders from outside the region who had the ability to mobilize finances through extensive credit networks, and retain control over wholesale and retail markets for forest commodities. Attempts by small-scale extractors to continue in forestry activities through organization of forest-labour co-operatives failed as timber traders rationalized their labor costs by recruiting migrant workers from western Nepal and other, even more impoverished regions (field interviews 1990–1).

This process of economic marginalization was compounded by a succession of natural disasters in the region. In 1971 and 1972, heavy monsoons caused floods, landslides and erosion, and extensively damaged terraced cultivation. Financial assistance from the UP state and central governments barely trickled through the interstices of the various institutions and agencies. Village leaders assailed the state government for its negligence, demanded immediate compensation for flood victims, and development assistance for the Himalayan districts. But their demands went largely unmet by a financially constrained and sluggish state administration (Dogra 1980). Resentment against the state government escalated further in 1973, when the UP State Forest Department rejected a petition made by an artisanal co-operative in eastern Garhwal (the *Dasholi Gram Swarajya Mandal*, Chamoli District) to increase allotments of ash trees for promoting local manufacture of agricultural implements. Village leaders discovered that their request had been rejected in favor of honoring a contract with a sporting goods firm that had purchased rights from the Forest Department to extract 400 ash trees from nearby Reserved Forests (Bahuguna 1981; Dogra 1980).

The Forest Department's intransigence was regarded as a confirmation of governmental apathy towards the welfare and development of communities in the Garhwal Himalayas. Village leaders and small-scale extractors from nearby areas protested at divisional offices and auction sites of the Forest Department, threatening to obstruct all extractive operations if their demands were not addressed. During the felling seasons from 1973 through 1975, village leaders exhorted their communities to prevent forest contractors from extracting timber. A number of stand-offs proved successful – men and women, regardless of age or caste, gathered in Reserved Forests adjoining their villages, preventing fellings by hugging trees and urging migrant laborers to return to their homes (Berreman 1989; Guha 1989; Jain 1984; Shiva and Bandyopadhyay 1987; Weber 1988). The Chipko movement was born.

VICTORIES AND VICISSITUDES OF THE CHIPKO MOVEMENT

The protests in the eastern districts of the Garhwal Himalayas earned the name *chipko* (which means to adhere, or stick to) from the way in which people hugged trees to prevent them from being felled by migrant work-gangs hired by forest contractors. Chipko's main focus centered on regaining access to small-scale forest extraction, and on pressuring the state government to provide developmental assistance to beleaguered communities in the region. The protestors were a heterogeneous constituency (Upadhyay 1990), with multiple political affiliations and even conflicting goals. Some demanded abolition of large-scale extraction by forest contractors, others were for promotion of locally organized forest-labor co-operatives, expanding rights of access and giving more concessions to local communities; yet others demanded a total ban on export of raw materials from the region (Aryal 1994; Dogra 1980; field interviews 1990–1). Village leaders and student activists affiliated with the Communist Party of India, for example, demanded higher wages for forest laborers and a ban on exports of forest resources (Aryal 1994; field interviews 1990–1); while those inspired by Gandhian ideals demanded timber subsidies and supply of other forest commodities at concessional rates for promoting local artisanal industries (Agarwal *et al.* 1982: 42–3; Bhatt 1988; Government of India 1985). Conversely, small-scale forest contractors argued that the Forest Department needed to privilege local entrepreneurs and forest-labor co-operatives by regulating external competition (Bahuguna 1981, 1988).

Negotiations with the Forest Department continued without much success in resolving any of these demands. Forest officers defended their position by claiming they were bound by law to comply with the production targets set by national- and state-level plans. Demands for greater access to small-scale extraction were turned down by invoking policy documents which stated that forests were national resources and could not be left open to reckless exploitation by local communities (Government of India 1952, 1976). Local forest contractors were denied exclusive access to commercial extraction on grounds that such policies would result in production monopolies, contribute to inefficiency, and increase the price of raw materials for industry (field interviews 1990–1).

Faced with an impasse in negotiations, Chipko's leaders and local activist groups resolved to bypass the UP state administration and appeal directly to the

Indian government. In 1975, Sunderlal Bahuguna, a local forest contractor, and one of several spokesmen of the Chipko movement, urged the central government to intervene before ecological problems threatened national security in the Himalayas. Employing the language of the state, interwoven with the vocabulary of popular protest (Mitra 1992), Bahuguna emphasized the importance of forests for strengthening national security and border defences, and for solving ecological problems faced by communities in the Himalayan regions. Chipko, he asserted, was a groundswell of popular outrage against the relentless commercial forces deforesting the region (Bahuguna 1981, 1982, 1987). Himalayan peasants, he stated, depended on forests to meet their simple needs, but their subsistence was being undermined by forest contractors who denuded the slopes for private profit without concern for the nation's security or for the sacred mountains. Bahuguna's narrative cast Chipko's followers as victims of natural disasters, the state, and the market. Floods, poverty, out-migration, and women's sufferings, their daily struggle to collect fuel and fodder for their households, were described as inevitable consequences of timber extraction by forest contractors.

Bahuguna's appeals proved effective. Politicians praised Chipko as the moral conscience of the nation, and promised immediate action to check the problems of ecological degradation and deforestation in the Himalayas (Government of India 1985). Bahuguna's ability to evoke a sympathetic response from the central government gained wide media coverage for the movement, which, in turn, brought support from scholars and activists in other parts of the country. Bahuguna was seen as the natural leader of the Chipko movement, the voice of a grassroots struggle that sought to protect the simple, peasant ways of life and restore the harmony between humans and nature in the Himalayas (Bahuguna 1987; Shiva and Bandyopadhyay 1986, 1987).

Chipko's ascent to fame, therefore, hinged on the central role played by rural elites like Bahuguna who could speak from interstitial spaces created by the state institutions, markets, and civil society, and seize the opportunities emerging from political and economic change. They were particularly successful when they articulated their demands in the state's vocabulary of national integrity, development, and democracy, and combined this language with symbolic acts of popular protest. Their protests gained wider audiences through simple, populist narratives that pitted peasants against the state and markets, but glossed over the heterogeneity of classes, interests, and constituencies within the movements. This skilled interweaving of state and populist rhetoric made Chipko the unquestioned icon of grassroots environmentalism in India and international environmental circles. Environmental scholars retold the Chipko story as India's civilizational response to ecological crisis in the Himalayas (Shiva and Bandyopadhyay 1987). International NGOs and scholars praised Chipko as an inspiration for environmental activists around the world, claiming that the ideals it represented were far more important that the aims it initially had set out to achieve (Weber 1988: 128–9). Chipko assumed legendary status when, following appeals to the Indian government, and after drawing widespread attention from scholars and the media, several pieces of legislation and constitutional amendments aimed at forest protection were introduced between 1975 and 1980.

THE AFTERMATH AND LEGACY OF CHIPKO

Chipko's growing reputation as an exemplar of grassroots environmentalism in the Third World diverted attention from the political and economic necessities undergirding the Indian government's willing capitulation. As the story of Chipko spread around the world and was retold time and again, it became detached from its specific demands regarding access to forest resources and local economic development. Details became difficult to remember. It seemed no longer important to recognize that Bahuguna's criticism of private forest contractors fortuitously coincided with a period in the 1970s when the central government launched a program of public sector expansion and nationalization unprecedented in Indian history (Bardhan 1985, 1991). Nationalization began after an internal struggle within the Congress Party in 1969 which led to a split, and resulted in Indira Gandhi (Nehru's daughter) assuming the leadership of the newly formed Congress-I (Indira). Mrs Gandhi won the elections in 1971 by appealing to two dominant constituencies – administrative classes and rural elites – for political support (Rudolph and Rudolph 1987). The Congress-I specified nationalization in populist terms, allying itself with the masses by arguing that government – as opposed to the narrow self-interests of private enterprise – would be more socially responsible and serve the public interest (Bardhan 1985, 1991; Chakravarty 1987). Bahuguna's appeals on the behalf of Chipko were received favourably by Mrs Gandhi because the criticism of forest contractors added popular endorsement for her agenda of nationalization and public sector expansion. She urged state governments to listen to Chipko's criticism of forest contractors and respond to the recommendations of the National Commission on Agriculture. The Commission advocated extensive afforestation measures on public and private lands, and for the creation of public sector firms to replace private businesses engaged in forest extraction (Government of India 1976; Lal 1989).

The UP state legislature responded to Mrs Gandhi's voluble support of the Chipko movement with the Forest Corporation Act of 1975. The Act authorized the creation of the UP State Forest Corporation, which was to function independently from the UP State Forest Department. The Corporation was charged with the responsibility of improving production efficiency and stabilizing market prices for timber and other raw materials. The UP State Forest Corporation was also to provide opportunities for local employment by using a network of forest-labor co-operatives for carrying out its activities (Government of Uttar Pradesh 1975).

This piece of legislation was soon followed by another. The UP Tree Protection Act of 1976 prevented the felling of all protected tree species (mainly valuable species identified by the Forest Department) on private lands. The Tree

Protection Act over-ruled earlier regulations requiring landowners to pay a nominal tax to district-level revenue authorities for selling trees harvested on their property (Government of Uttar Pradesh 1976). The Indian government passed a constitutional amendment, also in 1976, which deemed that state governments would require its prior consent for embarking on any project that involved large-scale conversion of forests to other land uses (Upadhyay 1990a). The legislation responded to the criticisms of environmental scholars and activists who argued that nearly 4.3 million hectares of forest areas in different parts of the country had been deforested by state governments within twenty-five years (1950–75) under the pretext of promoting industrial development and hydroelectric projects (Agarwal *et al.* 1982: 41). This indiscriminate destruction of forests by state governments, they claimed, needed to be regulated by the government of India. Four years later in 1980, the Indian government created the Ministry of Environment for addressing problems of deforestation, environmental risk, and ecological degradation in the country. This action was undertaken alongside the passage of the Indian Forest Conservation Act of 1980. The Act stipulated that state governments required permission from the Ministry of Environment for converting designated forest areas of more than 1 hectare to non-forest land uses. It also imposed a fifteen-year ban on felling green timber at altitudes above 1,000 metres in the Himalayan regions (Agarwal and Narain 1991; Lal 1989: 33–4; Upadhyay 1990a).

Ironically, international repute and legislative successes bore down heavily on communities in the Garhwal Himalayas. Since the green-felling ban followed soon after the passage of the Forest Corporation Act, most communities living at altitudes above 1,000 metres had little opportunity to find employment in forestry. Forest-labor co-operatives at higher elevations disbanded soon after the felling ban was imposed, while other co-operatives in the lower altitudes were riven by political rivalries and old disagreements within the Chipko movement that resurfaced among members. The UP Forest Corporation opted for labor contracts with "mates" (migrant labor agents) controlling groups of workers recruited from outside regions. Mates were invariably from western Nepal, with a smaller proportion coming from the neighboring state of Himachal Pradesh. The Forest Corporation defended its choice of migrant workers over locally organized labor by arguing that Nepalese workers were preferred for their skills, reliability, and industriousness. Some Forest Corporation managers were more candid, freely admitting that contract-laborers from western Nepal were used because they were willing to work for lower wages and were disciplined by their mates (field interviews 1990–1).

In addition to further reducing regional employment opportunities in forestry, the new legislation also weakened the UP State Forest Department. Revenues have steadily eroded along with its discretionary powers to settle local disputes over forest access, and to allocate use of classified forest areas for small-scale infrastructure projects proposed by village- or block-level institutions. The department's revenues have declined due to the fixed royalties (adjusted periodically for inflation), rather than market prices, that it receives from the State Forest Corporation. While the UP Forest Department's revenues have increased only at about 13 percent each year since 1980, its administrative costs and routine expenditure on forest conservancy and management have risen at an annual rate of about 18 percent since 1980 (UP Forest Department 1989). The green-felling ban affects a large proportion of its forests in Garhwal, and consequently, the Forest Department cannot legally carry out the necessary lopping, thinning, and clearing for maintaining valuable forest stock.

The felling ban has provided convenient space for a "timber mafia" to emerge in Garhwal, most of whom are reported to be local forest contractors engaged in illegal felling and theft from Reserved Forests (*Amar Ujala* 1991f). Forced to operate within a highly constrained budget, the Forest Department has been unable to increase its patrolling force, or even provide adequate equipment to forest rangers and guards to confront heavily armed mafia henchmen who smuggle timber from Reserved Forests. The green-felling ban has also made the Forest Corporation's operations in the higher altitudes more expensive by restricting extraction to dead or damaged trees. Labor and transport costs for felling damaged trees randomly distributed across high altitude areas exceed prices in wholesale markets. The greatest irony is that despite monopolizing market supply of timber, the Forest Corporation is plagued, like other public sector firms, by the rising costs of supporting an expanded bureaucracy, its losses exceeding profits due to inefficient management (*Amar Ujala* 1991b; field interviews 1990–1; *Hindustan* 1990). The social and economic problems of the region are compounded by the fact that the timber mafia controls a diversified portfolio of illegal activities such as the production and sale of locally brewed liquor in the hill districts. Even though several women's groups in Garhwal have protested against liquor-brewing (*Amar Ujala* 1991c, d, e; Jain 1984), the irritation for local communities is that this activity too, is carried out by migrant labor from western Nepal whose incomes from working for the mafia are rising rapidly in relation to local residents (field interviews 1990–1).

New forest legislation – which in some ways is the heart of Chipko's reforms – has in essence imposed a moratorium on most development activities in the region. Given that nearly two-thirds of the total land area in Garhwal is classified as forest (Government of Uttar Pradesh 1984), any development activity such as road-building, village electrification, or minor irrigation works requiring partial use or conversion of forest areas cannot proceed without permission from the national Ministry of Environment (Government of India 1986). Numerous examples of development projects held up or shelved due to bureaucratic delays litter the region's landscape: a small road connecting a few villages remains unbuilt because 1 kilometer of its length passes through a Reserved Forest; a minor irrigation channel passing through a Reserved Forest lies abandoned because the Ministry has delayed permission for over nine years; rural electrification projects have not been implemented in the inner ranges because permission was denied for clearing some forest tracts to set up transmission grids (Agarwal and Narain 1991; field interviews 1990–1). More recently, the Forest Conservation Act of 1988 deems it illegal for any agency, other than state or central government institutions, to even engage in afforestation projects without permission from the Ministry of Environment (Government of India 1987). All of this provokes great anger towards the new forest laws and the Ministry of Environment. Even though the Ministry defends its regulatory process by noting that cases requiring less than 9 hectares of land are sanctioned fairly quickly, village leaders nevertheless vehemently object. They argue that large-scale irrigation and hydroelectric projects have powerful coalitions and lobbies in state governments with the power to relentlessly pursue the Ministry and get their projects cleared. Small-scale development projects, conversely, do not have political clout and are indefinitely postponed or shelved altogether. By the late 1980s, regional political groups such as the *Uttarakhand Kránti Dal* (Uttarakhand Revolutionary Front) began publicly declaring their willingness to violate forest laws and clearcut areas on behalf of any community or village wishing to initiate development projects (Agarwal and Narain 1991; field interviews 1991).

THE RISE OF UTTARANCHAL

Throughout the 1980s, Chipko's celebrity status abroad was paralleled by growing resentment over the lack of development in the region, expressed periodically through outbursts of protests demanding the creation of Uttaranchal, a separate state comprising the eight Himalayan districts of Uttar Pradesh. Such demands were not new to the region. Mr P.C. Joshi, a member of the Communist Party of India and a political leader from Kumaon had, in 1952, sent a memorandum to Prime Minister Nehru requesting the creation of a separate state of Uttarakhand. Nehru himself was against the division of Uttar Pradesh but forwarded the memorandum to the States Reorganization Commission, which later rejected the demand (*Sunday* 1994). Subsequent Chief Ministers of UP State such as H.N. Bahuguna and N.D. Tiwari, who belonged to communities in the region, but were members of the Congress-I, set the issue aside during their tenure. For most national political parties assuming power in Uttar Pradesh, the attraction of ruling one of India's most populous states, which sends the largest numbers of elected representatives to the parliament, has outweighed concern for administrative efficiency and effective regional development within Uttar Pradesh. Ruling political parties at both national and state levels have routinely played to local sentiments, expressing solidarity and support for Uttaranchal during parliamentary or legislative elections, but ignoring the issue soon after they come into power. Over the past five years and after two elections, no ruling political party in the UP state legislature has gone beyond passing a resolution to look into the creation of Uttaranchal. The right-wing, Hindu fundamentalist party, the BJP (Bharatiya Janata Party) supported demands for Uttaranchal during the 1991 state elections, but went no further than changing the name of the Hill Development Department to the Uttaranchal Development

220

Department (*Sunday* 1994). In 1993, the new coalition government formed by the SJP-BSP (political parties mainly representing the interests of kulak farmers belonging to backward castes in northern India), accepted the recommendation for creating Uttaranchal, but have since responded with state violence when the protests erupted again.

One of the few concessionary gestures made towards regional autonomy from UP State was in establishing a Hill Development Department in the mid-1980s. The state agency was charged with the responsibility of planning and development assistance for various sectors of the regional economy. But since the agency operates under the umbrella of economic and environmental regulations imposed by previous legislation, its role has primarily focused on horticulture and tourism promotion. Village leaders and critics of the Hill Development Department argue that these projects continue to display poor understanding of the region's needs and local economic problems: the lack of adequate infra-structure and procurement network on the part of the Hill Department has rendered its price support for orchard and vegetable produce meaningless; households lacking access to markets, and those without finances for engaging private transportation of highly perishable produce outside the region, are saddled with harvests that must be sold at throwaway prices to local traders and merchants. Tourism development has proceeded slowly because of poor road networks, lack of electricity, and regulatory constraints imposed on conversion of forest lands to non-forest use. The economic incentives and subsidies offered for tourism promotion, as many village leaders point out, have been cornered by wealthy entrepreneurs from within and outside Garhwal to speculate in urban real estate or to use the low-interest financing to build large vacation homes, euphemistically labeled "hotels." Between 1990 and 1991, real estate values in Dehra Dun, the largest city at the foot of the Himalayas, and Mussoorie, a popular mountain resort for the wealthy during the summer, rose nearly 100 percent for residential land, and about 80 percent for commercial land (field interviews 1990–1).

The creation of Uttaranchal State is thus seen by village leaders and political activists as the only way by which some measure of local control can be exercised over the promotion of economic development in the region. But as formal dissent routinely erupts and subsides without gaining additional support for these demands from outside the region, youth organizations and activists accuse village leaders and political parties of incompetence and lack of commitment to their cause, and have turned increasingly towards militancy and violence (*Sunday* 1994).

CONCLUSION

Chipko in its early phase and latterly Uttaranchal are linked by common concerns over greater access to local resources and infrastructure for promoting economic development in the region. Yet the movements differ in important ways. The Uttaranchal movement has greater unanimity among diverse class and caste groups regarding the need for a separate state, and more widespread support within the region. In sharp contrast to Chipko, which received widespread political recognition and support from both the central government and outside environmentalists, Uttaranchal activists have been unsuccessful in gaining sympathy for their cause from these constituencies. Environmentalists in India and elsewhere, rapt and slavish in their adoration and assiduous pursuit of romance with Chipko's ecological reincarnation, have been oblivious to the processes of marginalization continuing in the region. They have typically been deaf to local criticism of Chipko's leader, Sunderlal Bahuguna, who is seen as reactionary and opposed to economic development (*Amar Ujala* 1991a; field interviews 1990–1). Paeans of praise for Chipko have drowned out militant local calls for tree-felling rather than tree-hugging, and strident appeals for revoking the green-felling ban in the Himalayas (Agarwal and Narain 1991; field interviews 1990–1; *Himachal Times* 1990a, b, c; *Indian Express* 1990). Their clamorous allegiance to Chipko's mythological existence prevents the protests and arguments of village leaders and activists from being heard outside the region. It seems now that guns, rather than verbal protests, can bring attention. In the momentary silences following the rattle of arms, voices can be heard exhorting the protestors to struggle for an expanded moral economy that *includes* development, democracy, and social justice (Dhoundiyal *et al.* 1993; field interviews 1994).

As the Uttaranchal movement sends out its message of secular development and social justice, scholarly criticism of "maldevelopment" (Shiva 1991), development as a mechanism of underdevelopment (Gadgil and Guha 1992), or calls for alternatives *to* development (Escobar 1995), bear the curious aura of a reactionary populism. It seems ironic that contemporary scholarly debates should clamor for a "post-development" era, just when voices from *the margins* – so celebrated in discourses of *difference* and alternative culture – are demanding their rights to greater access to a more generous idea of development.

REFERENCES

Agarwal, Anil, S. Narain *et al.* 1982. *The State of India's Environment.* New Delhi: Centre for Science and the Environment.

—— and Sunita Narain. 1991. "Chipko people driven to '*Jungle Kato*' [Cut the Forests] stir," *Economic Times* (New Delhi) 31 March.

Amar Ujala. 1991a. "Bahuguna's statements criticized," 8 Feb. (All article titles trans. from Hindi.)

—— 1991b. "Inefficient management of Forest Corporation causes wastage of timber in hill regions of UP," 18 Feb.

—— 1991c. "Women demonstrate against brewing of illicit liquor in Tehri Garhwal," 5 March.

—— 1991d. "Women strive yet again to ban liquor from hill regions," 20 April.

—— 1991e. "Demonstrations against sale of illicit liquor in Uttarkashi," 12 June.

—— 1991f. "Timber Mafia thumbs its nose at forest laws," 5 July.

Ambedkar, B.R. 1945. *Annihilation of Caste: With a Reply to Mahatma Gandhi*. Bombay: Bharat Bhushan Press.

Aryal, Manisha. 1994. "Axing Chipko," *Himal* 7, 1: 8–23.

Atkinson, E.T. 1882. *The Himalayan Gazetteer*, 3 vols, reprinted 1989. New Delhi: Cosmo Publications.

Bahuguna, Sunderlal. 1981. *Chipko: A Novel Movement for Establishment of a Cordial Relationship between Man and Nature*, 2 vols. Silyara, Tehri Garhwal: Chipko Information Centre.

—— 1982. "Let the Himalayan forests live," *Science Today* March: 41–6.

—— 1987. "The Chipko: a people's movement," in M.K. Raha (ed.) *The Himalayan Heritage*. New Delhi: Gian Publishing House, pp. 238–48.

Bardhan, Pranab. 1985. *The Political Economy of Development in India*. New Delhi: Oxford University Press.

—— 1991. "State and dynamic comparative advantages," *Economic Times* (New Delhi) 19–20 March.

Bayly, C.A. 1976. *The Local Roots of Indian Politics: Allahabad 1880–1920*. Oxford: Clarendon Press.

Berreman, Gerald. 1989. "Chipko: a movement to save the Himalayan environment and people," in C.M. Borden (ed.) *Contemporary Indian Tradition: Voices on Culture, Nature, and the Challenge of Change*. Washington, DC: Smithsonian, pp. 239–66.

Bhatt, Chandi Prasad. 1988. "The Chipko movement: strategies, achievements and impacts," in M.K. Raha (ed.) *The Himalayan Heritage*. New Delhi: Gian Publishing House, pp. 249–65.

—— 1991. "Chipko movement: the hug that saves," *Survey of the Environment 1991*. Madras: The Hindu.

Bora, R.S. 1987. "Extent and causes of migration from the hill regions of Uttar Pradesh," in Vidyut Joshi (ed.) *Migrant Labour and Related Issues*. New Delhi: Oxford and IBH.

Bowles, Samuel and Herbert Gintis. 1986. *Democracy and Capitalism: Property, Community, and the Contradictions of Modern Social Thought*. New York: Basic Books.

Brass, T. 1994. "The politics of gender, nature, and nation in the discourse of new farmer movements," *Journal of Peasant Studies* 21, 3: 27–71.

Braudel, Fernand. 1977. *Afterthoughts on Material Civilization and Capitalism*. Baltimore: Johns Hopkins University Press.

Byres, T.J. 1972. "The dialectic of India's Green Revolution," *South Asian Review* 5, 2: 99–116.

Chakravarty, Sukhamoy. 1987. *Development Planning: The Indian Experience*. New Delhi: Oxford University Press.

Crow, Ben, Mary Thorpe *et al.* 1988. *Survival and Change in the Third World*. London: Polity Press.

Dhoundiyal, N.C., V.R. Dhoundiyal, and S.K. Sharma. 1993. *The Separate Hill State*. Almora, U.P.: Shree Almora Book Depot.

Dobhal, G.L. 1987. *Development of the Hill Areas: A Case Study of Pauri Garhwal District*. New Delhi: Concept Publishers.

Dogra, Bharat. 1980. *Forests and People: The Efforts in Western Himalayas to Re-establish a Long-lost Relationship*. Rishikesh: Himalaya Darshan Prakashan Samiti.

—— 1983. *Forests and People: A Report on the Himalayas*. New Delhi.

Escobar, Arturo. 1992. "Imagining a post-development era? Critical thought, development and social movements," *Social Text* 10, 2/3: 20–56.

—— 1995. *Encountering Development: The Making and Unmaking of the Third World*. Princeton, NJ: University of Princeton Press.

Gadgil, Madhav and Ramachandra Guha. 1992. *This Fissured Land: An Ecological History of India.* Berkeley: University of California Press.

Government of India. 1952. *National Forest Policy.* New Delhi: Ministry of Agriculture.

—— 1956. *Second Five-Year Plan.* New Delhi: National Planning Commission.

—— 1961. *Third Five-Year Plan.* New Delhi: National Planning Commission.

—— 1970. *Fourth Five-Year Plan.* New Delhi: National Planning Commission.

—— 1976. *Report of the National Commission on Agriculture, Volume IX: Forestry.* New Delhi: Ministry of Agriculture and Irrigation.

—— 1985. *National Forest Policy.* New Delhi: Ministry of Environment.

—— 1986. *The Environment (Protection) Act, 1986* (Act No. 29 of 1986). New Delhi: Ministry of Law and Justice.

—— 1987. "Forest Rules Notification No. GSR14." New Delhi: Ministry of Environment.

Government of Uttar Pradesh. 1975. *The Uttar Pradesh Forest Corporation Act, 1974.* (UP Act No. 4 of 1975.) Lucknow: UP State Legislature.

—— 1976. *The Uttar Pradesh Tree Protection Act, 1976.* Lucknow: UP State Legislature.

—— 1984. *Statistical Diary of UP State.* Lucknow: Economics and Statistical Department.

Guha, Ramachandra. 1989. *The Unquiet Woods: Ecological Change and Peasant Resistance in the Himalaya.* New Delhi: Oxford University Press.

Harriss, J. (ed.). 1982. *Rural Development: Theories of Peasant Economy and Agrarian Change.* London: Hutchinson.

Hauser, Walter. 1993. "Violence, agrarian radicalism and electoral politics: reflections on the Indian People's Front," *Journal of Peasant Studies* 21, 1: 85–126.

Herring, Ronald. 1989. "Dilemmas of agrarian communism: peasant differentiation, sectoral and village politics in India," *Third World Quarterly* 11, 1: 89–115.

Himachal Times. 1990a. "Uttaranchal demanded: economic and political passions aroused," 11 April.

—— 1990b. "Amendment of Forest Act for developing hill areas," 2 April.

—— 1990c. "Consolidation of hill land holdings soon," 1 June.

Hindustan. 1990. "Crores of rupees wasted by the Forest Corporation" (trans. from Hindi) 31 August.

Indian Express (New Delhi). 1990. "UP to amend Tree Protection Act," 28 March.

Jain, S. 1984. "Women and people's ecological movement: a case study of women's role in the Chipko movement in Uttar Pradesh," *Economic and Political Weekly* 19, 41: 1788–94.

Joshi, S.C. *et al.* 1983. *Kumaon Himalaya: A Geographic Perspective on Resource Development.* Nainital: Gyanodaya Prakashan.

Khan, Waheeduddin Khan and R.N. Tripathy. 1976. *Plan for Integrated Rural Development in Pauri Garhwal.* Hyderabad: National Institute of Community Development.

Khanka, S.S. 1988. *Labour Force, Employment and Unemployment in a Backward Economy: A Study of Kumaon Region in Uttar Pradesh.* Bombay: Himalayan Publishers.

Lal, J.B. 1989. *India's Forests: Myth and Reality.* Dehra Dun: Natraj Publishers.

Mellor, J.W. 1968. "The evolution of rural development policy," in J.W. Mellor, T.F. Weaver, U. Lele, and S.R. Simon (eds.) *Developing Rural India: Plan and Practice.* New York: Cornell University Press.

Mies, Maria and Vandana Shiva. 1993. *Ecofeminism.* London: Zed Books.

Mitra, Subrata K. 1992. *Power, Protest and Participation: Local Elites and the Politics of Development in India.* London: Routledge.

Naipul, V.S. 1990. *India: A Million Mutinies Now.* London: Heinemann.

Pangtey, Y.P.S. and S.C. Joshi (eds.). 1987. *Western Himalaya: Environment Problems and Development,* Vols. I and II. Nainital: Gyanodaya Publishers.

Pant, S.D. 1935. *The Social Economy of the Himalayans.* London: Allen & Unwin.

Raj, K.N. 1973. "Mechanization of agriculture in India and Sri Lanka (Ceylon)," *Mechanization and Employment in Agriculture.* ILO: Geneva.

Rangan, Haripriya. 1993. "Of myths and movements: forestry and regional development in the Garhwal Himalayas," Ph.D. Dissertation. University of California, Los Angeles.

—— 1995. "Contested boundaries: forest classifications, state policies, and deforestation in the Garhwal Himalayas," *Antipode* 27, 4: 343–62.

Rawat, Ajay Singh. 1983. *Garhwal Himalayas: A Historical Survey: The Political and Administrative History of Garhwal 1815–1947.* Delhi: Eastern Book Linkers.

—— 1989. *History of Garhwal, 1358–1947: An Erstwhile Kingdom in the Himalayas.* New Delhi: Indus Publishing.

Rudolph, Lloyd and Susanne Hoeber Rudolph. 1987. *In Pursuit of Lakshmi: The Political Economy of the Indian State.* Chicago: Chicago University Press.

Sen, Amartya. 1960. *Choice of Techniques.* Oxford: Blackwell.

—— 1984. *Resources, Values and Development.* Cambridge, MA: Harvard University Press.

Shiva, Vandana. 1989. *Staying Alive: Women, Ecology and Development.* London: Zed Books.

—— 1991. *Violence of the Green Revolution: Third World Agriculture, Ecology and Politics.* London: Zed Books.

Shiva, Vandana and J. Bandyopadhyay. 1986. *Chipko: India's Civilisational Response to the Forest Crisis.* New Delhi: Intach.

—— 1987: "Chipko: rekindling India's forest culture," *The Ecologist* 17, 1: 26–34.

—— 1989. "The political economy of ecology movements," *IFDA Dossier* 71, May–June: 37–60.

Singh, S.P. and Anil Berry. 1990. *Forestry Land Evaluation: An Indian Case Study.* Dehra Dun: Surya Publications.

Srinivasan, T.N. 1979. "Trends in agriculture in India, 1949–50 to 1977–78," *Economic and Political Weekly* 14, 30–2.

Stebbing, E.P. 1932. *The Forests of India,* 3 vols. London: Bodley Head.

Subramaniam, C. 1979. *The New Strategy in Indian Agriculture.* New Delhi: Vikas Publishers.

Sunday. 1994. "The sound and the fury: is the Uttarakhand agitation taking a violent turn?" 6–12 November.

Swarup, R. 1991. *Agricultural Economy of Himalayan Region: With Special Reference to Kumaon.* Nainital: G.B. Pant Institute of Himalayan Environment and Development.

Thompson, E.P. 1993. *Customs in Common: Studies in Traditional Popular Culture.* New York: The New Press.

Trainer, F.E. 1989. *Developed to Death: Rethinking World Development.* London: Green Press.

UP Forest Department. 1989. *Forest Statistics: Uttar Pradesh.* Lucknow: Forest Administration and Development Circle.

Upadhyay, C.B. (ed.). 1990a. *Forest Laws, with Commentaries on Indian Forest Act and Rules, State Acts, Rules, Regulations, etc.,* 7th edn. Allahabad: Hind Publishing House.

Upadhyay, H.C. 1990b. *Harijans of the Himalaya: With Special Reference to the Harijans of Kumaun Hills.* Nainital: Gyanodaya Prakashan.

von Fürer-Haimendorf, Christopher. 1981. *Asian Highland Societies in Anthropological Perspective.* New Delhi: Sterling Publishers.

Walton, H.G. 1910. *British Garhwal: A Gazetteer, being Volume XXXVI of the District Gazetteers of the United Provinces of Agra and Oudh,* reprinted 1989. Dehra Dun: Natraj Publishers.

Watts, Michael J. 1993. "Development I: power, knowledge, and discursive practice," *Progress in Human Geography* 17, 2: 257–72.

Webber, Thomas. 1902. *Forests of Upper India and Their Inhabitants.* London: Edward Arnold.

Weber, Thomas. 1988. *Hugging the Trees: The Story of the Chipko Movement.* New Delhi: Viking.

Weiner, Myron. 1989. *The Indian Paradox: Essays in Indian Politics.* New Delhi: Sage Publications.

Wells, Roger. 1994. "E.P. Thompson, customs in common, and moral economy," *Journal of Peasant Studies* 21, 2: 263–307.

Williams, G.R.C. 1874. *Historical and Statistical Memoir of Dehra Doon*, reprinted 1985. Dehra Dun: Natraj Publishers.

Williams, Raymond. 1983. *Keywords: A Vocabulary of Culture and Society*, revised edn. New York: Oxford University Press.

Woodman, Dorothy. 1969. *Himalayan Frontiers: A Political Review of British, Chinese, Indian and Russian Rivalries.* London: Barrie & Rockliffe.

Yapa, Lakshman. 1993. "What are improved seeds? An epistemology of the Green Revolution," *Economic Geography* 69, 3: 254–73.

11

THE POLITICAL ECOLOGY OF AGRARIAN REFORM IN CHINA

The case of Heilongjiang Province

Joshua S.S. Muldavin

The struggle to control resources is a key factor in the articulation between local societies and larger entities. In particular, the extraction of surplus from rural areas by the state and transnational capital is contingent upon penetrating, dismantling, and transforming structures capable of maintaining local natural resources. Based on fieldwork over a twelve-year period, this chapter uses village-based case studies to show how privatization and deregulation in China have permitted the exploitation of labor and resources. The apparent success of the contemporary rural China model relies on drawing down communal capital. The severe ecological consequences of this practice challenge the Chinese version of the triumphalist argument outlined in Peet and Watts's introduction, and raises questions about post-socialist transitions, market reforms, and their sustainability. It also demonstrates the power of a political ecological analysis in a socialist context: changes in political economy and patterns of resource use explain increases in environmental degradation (Blaikie 1985).

Some writers herald the reforms of 1978 as the proper pathway for China to follow, given the failures of the communal period, and the obvious success of market reforms throughout the world (Byrd 1990; Dorn and Wang 1989; Friedman 1990; Lin 1993). Other accounts portray the reforms as a necessary step in the right direction, but with difficulties in the transition to a market economy (Croll 1988; Deng *et al.* 1992; Feuchtwang *et al.* 1988). Still another group finds fundamental problems with certain aspects of the reforms which undermine their supposed success in solving weaknesses of the communal period (Chai and Leung 1987; Chossudovsky 1986; Davin 1988; Delman 1989; Hinton 1990; Muldavin 1986, 1992). This chapter situates the Chinese reforms in theoretical relationship to political ecology and the new industrial geography, exploring the social impacts and ecological consequences of reform, and subsequent state attempts to maintain legitimacy in the face of growing resistance to them.

Peet and Watts's "great paradox of the 1990s" (see Chapter 1) has a counterpart in China's transformation over the last quarter of the twentieth century. In the context of astonishing economic growth rates, the World Bank trumpets China as its number one success story for post-socialist transition, in contrast to the difficulties experienced by Russia. I challenge this notion of "success" by focusing on its environmental costs and, more generally, on the whole question of sustainability, which the World Bank defines strictly using short-term *economic* indicators of growth. Additionally, the contradictions of decollectivization accompanied by market reforms and increased privatization suggest that the current "crisis" is not transitional; it stems from three contradictions inherent in the reforms: the potential for internal divisions based on ethnicity and territoriality and fueled by nationalism and racism (Gladney 1994); environmental problems resulting from the very "success" of the reforms; and the appalling spectacle of deepening social polarization, as the gulf between the haves and have-nots in China becomes a chasm. This chapter focuses primarily on the second and third contradictions, using two rural areas in Heilongjiang Province, one primarily grassland, and one primarily cropland, as case studies.

Resistance to the new order has emerged among workers and peasants. Whether the underlying contradictions will become an openly active politics, or surface through everyday resistance, depends to a large degree on the development of a social (civil) space for dissent and discourse. At this time, there is limited potential for organized social movements on a broad scale in China, demonstrating a transfer of authoritarian politics to the new hybrid system emerging there. As the Chinese leadership has made abundantly clear (Tian'anmen being just one example), it will not make the 'mistakes' of the former USSR by freeing politics along with economy.

DERIVATIVE SOCIALISM, DERIVATIVE CAPITALISM: A HYBRID SYSTEM

China is all things to all people: it is "capital's salvation" – a vast new market; a place to invest East Asia's surplus capital; a huge labor pool to discipline workers "back home"; a vast resource base; a powerful industrial system; a diversified agricultural and industrial economy. The World Bank's proclaimed greatest development success is still a Third World economy, though rapidly merging into the First World and increasingly identified as a newly industrializing country (NIC). It is a powerful and potentially destabilizing military force. It is a political enigma.

Besides being huge by all measures, China is extremely important as a model for the Third World. A new system has emerged – derived from a synthesis of command socialism and advanced capitalism – which seems to combine the worst of both systems, particularly in terms of environmental impacts and long-term sustainability. Socially, an eruption of new poverty has accompanied the creation of concentrated personal wealth. It is speculative in capitalist investment terms – despite claims of pragmatism; speculative in social terms – particularly in the state's ongoing attempts to legitimize the current transformation; and speculative

228

in its relations with nature – as I shall shortly demonstrate. In China's hybrid system, the strong one-party state has retained its complete intolerance of dissent. The exact nature of this hybrid system is difficult to pinpoint. It is simultaneously a cross between an NIC, a Soviet-style industrialization model, an export-led growth model, an import-substitution model, and a sweat shop/subcontractor to the world's corporations. Even the notion "hybrid" is a restricted vision when it comes to the Chinese system. Actually we may be witnessing the evolution of an entirely new phenomenon, one historically specific, difficult to repeat, fitting like a missing piece into the global jigsaw puzzle of the new international order.

There is no one "market socialism" in China. There are pockets of pure capitalism (frontier style, speculative to a large degree), and pockets of Stalinesque socialist growth economies (highly controlled, rigid, and bureaucratic). China defies generalization and simplified typologies. Trends exist, but so do the wide range of conditions which produce these trends, with an immense heterogeneity in outcomes. Under such circumstances space intervenes as the canvas on which socioeconomic heterogeneity is translated as uneven development or geographic differentiation in the intensity with which the new "model" appears. Hence, it is almost impossible to talk of China as a whole. More realistic, though less clear cut, is the concept of "multiple Chinas" – multiple development models and paths occurring simultaneously (Muldavin 1992). Yet all share a reliance on the exploitation of labor, nature, and communal capital resources. Generally, there is the development of underdevelopment as wealthier regions (particularly in the eastern portion of the country) utilize the hinterlands in ways which structurally limit those areas' potential for meeting the real needs of the majority of their own populations. More work desperately needs doing on how the new regimes of accumulation of a rapidly changing world order increase environmental degradation, via a production mosaic utilizing nature and labor in new combinations. Much of the recent rapid economic growth has been achieved through mining ecological capital. The ongoing privatization of natural resources is accompanied by the shedding of risk to the lowest levels, forcing decision making towards ever-shorter time horizons. This privatization of the social (welfare, risk, communal capital), this personalization of risk and welfare needs, has profound effects on nature.

As a latecomer into global capitalism, China propels itself into the international arena of capital through the utilization of communal capital built up over the preceding thirty years. This is not a sustainable method of "developing" China, nor is it equitable or desirable for the great mass of China's poor peasants and workers. This new regime de-emphasizes the importance of the questions: development for whom, by whom, how, and towards what goals (Carr 1985)? The assumption is that growth in almost any form is good. In this way it shares the mistakes of state socialism and capitalism by focusing almost exclusively on growth as a means of defining success, mystifying who benefits and what is harmed, in this case nature. China's production transformation is a subset of the broader restructuring taking place in the United States, Europe, and Japan. There are global environmental implications of this restructuring process; none of these is positive.

The hybrid system is tied to global restructuring: first, through new sub-contracting forms between state and peasants (risk transfer); second, through mining communal capital for rapid growth with severe ecological consequences; and third, through new forms of sharecropping in the rural economy, and sweat shopping in the urban and suburban economy. If we view contemporary China as an evolving new regime of accumulation – accumulation for China's elites and international capital – we are led to ask different questions about how the reforms are affecting the environment. First, we might focus on the ways global sub-contracting networks, as they penetrate China, utilize geographic unevenness to increase transnational companies' (TNCs) leverage against organized labor and environmental movements in *other* parts of the world. Second, we might be able to get beyond the rhetorical and ideological battles over "market socialism" in China (Bardhan and Roemer 1993), to see new and more efficient forms of surplus extraction, through rapid increases in exploitation of labor and nature. What are the destinations of this surplus? Who controls it? How is the system of extraction and accumulation of that surplus now organized? And how does this affect nature? In this chapter I challenge the sustainability of market socialism (Nee and Stark 1989) using political ecology as critical analytic.

A POLITICAL ECONOMY OF RURAL CHINA: THE REFORMS OF 1978–95

The Maoist development model of revolutionary China was founded on a strategy of self-reliance. Joint development of agriculture and rural industry, supplemented by improved infrastructure, would bring long-term productivity gains benefiting the entire community. The model emphasized collective labor organization and an egalitarian distribution of surpluses, with individual incentives deriving from social and political ideals of the collective good. Much of the legitimacy of the Maoist period was gained from the preceding "feudal" history of oppressive landlordism.

The Maoist model differed from the former USSR's collective agriculture. It was composed of a spectrum of organizational forms, from state farms to communes, and ranged from large-scale mechanized capital-intensive production on flat open plains, to small-scale non-mechanized labor-intensive production on narrow terraces. Communes and state farms were part of a national system of planning; co-ordination was via top-down, and bottom-up, communication and negotiation (Gurley 1976). The commune was hierarchically divided into brigades organized around traditional villages. Within the commune the greatest responsibilities, cohesiveness, and sharing of risk rested with teams of 10–20 households. At all three levels, collective labor was applied to social and physical infrastructure to enhance productivity and quality of life. Socially oriented infra-structure consists of housing, clinics, schools, and administrative buildings.

Bio-physical infrastructure is directly related to production – dams, levees, canals, roads, reforestation projects, terracing, and improvements in soil fertility, grassland, and cropland quality. Collectives purchased machinery and operated small industrial facilities processing agricultural products, producing construction materials, and supporting productive efforts. I refer to this combination of social and physical infrastructure as "communal capital" – built up under a communal system of production organization with the expectation of long-term benefits for all members of the commune.

China experimented with collective organization of production between 1948 and 1978. With the post-Mao "pragmatic" period, beginning in 1978, attention shifted to decollectivization, the dismantling of the commune system in rural areas, and a move towards a market-oriented economy. Collective production was seen by many as stagnant and economically restrictive; a return to household- and individually based systems was thought the best hope for countering these weaknesses. Yet the results of reform are complex. While impacts are regionally and locally specific, major structural shifts in China's development strategy, presented in Figure 11.1, provide a general framework for addressing my central question: what effects do rural reforms have on the long-term sustainability of production – that is, what effects do they have on environment?

This question can be clarified by examining three arenas of change: in the predominant mode of economic co-ordination; in the base unit of production; and in accompanying decision-making strategies and goals. The most significant shift was from a predominantly planned economy to the decisiveness of market forces. Although planned resource allocation still exists, it functions now within an increasingly competitive framework, individually as well as institutionally. That competitive framework, in turn, makes planning problematic. Simultaneously, there has been a shift from collective production to the household and individual, an immense and historically important transfer of control over resources and means of production. The transfer corresponds to a shift in control at the highest levels of government. Finally, a shift in decision-making strategies and goals supplants long-term egalitarian collective gains with short-term competitive individual gains, the trend being towards short-term decision making, combined with increased social inequalities; a shift from long- to short-term outlooks results from the transfer of risk in production from collectives to households and individuals. The result of this complex of changes is increased insecurity and instability throughout the economy, with immediate and ongoing material returns necessary for providing political legitimation for the state and a strategic base for further reforms.

Some of the potentially negative results of the rapid agrarian change since 1978 are: generalized decline in the role and power of the collective in organizing production for long-term communal goals; changes in cropping patterns to more intensified and soil-taxing regimes; increased vulnerability and risk for individual households as a result of a shift to cash crops and increased dependence on world

	Pre-reform (up to 1978)		Post-1984 reform period
Predominant mode of economic coordination	**Plan**		**Market (increasing)**
Commodity allocation and resource control	via state and collectives	*Transition period 1978 - 1984*	via state/collectives and increasing proportion via markets through independent households and new-style collectives/partnerships
Base Unit	**Collective**		**Household/Individual**
Decision making Strategies & Goals	**Long term; egalitarian social gains**		**Short term; competitive/individual household gains**
National policy goal	Self-reliance		Global interdependence
Local policy goal	Compliance with plan via negotiation		Coordination of markets and contracts
Incentives & risk	Socially & politically based incentives with the state and collective carrying most risk		Economically based individual/household incentives via increased competition and risk

Figure 11.1 Shifts in China's rural political economy
Source: Muldavin (1994b)

markets; short-term decision-making parameters enforced through market mechanisms as well as the contract system; and more room for corruption (Cheng 1991).

Reforms, 1978–95

The political openings for the reforms were made by the failure of certain commandist structures. The communes of the Maoist period proved a mixed success, with one-third failing, one-third holding their own, and one-third running quite efficiently by 1978 (Chossodovsky 1986; Hinton 1990; Muldavin 1986). More generally, the "reforms" (in particular the initial rural ones) are connected with a long history of class struggle surrounding the extraction of surplus from a large peasant population while maintaining state legitimacy. Deng's brilliance lay in playing on problems of collective agriculture (as he had done earlier following the Great Leap Forward of the early 1960s) as a means of catapulting himself into power. Deng united the Chinese peasantry's histori-cally rooted desire for land with its strong patriarchal family structures. Disillusionment with command production in the socialist period, combined with disillusionment (particularly of the urban elite, but also some peasants) as a result of the Cultural Revolution, led to a general willingness for abandoning the political and the long term for the material and the short term. Over a period of six years, beginning in 1978, Deng was able to dismantle much of the collective control of land through the introduction of subcontracting and a land division which parceled out per capita shares of equivalent pieces of land to every family – the Household Responsibility System. Peasants are required to provide the state with an annual quota of grain at below-market prices in exchange for long-term use of the land – a form of rent or sharecropping with the state as "owner." The result is a devolution in farming practices from large to small scale. Combined with rapidly increasing prices for grain via heavy subsidization by the state, and an equally rapid increase in fertilizer availability and use, as factories built in the communal period came on line, and a mining of communal capital, the production of grain increased rapidly, legitimizing Deng's calls for the complete decollectivization of agriculture and the transformation of other areas of the society. By 1984, 99 percent of all communes had decollectivized agriculture and implemented the reforms.

Parallel to this was the restructuring of collective industries. As in Russia, this can be interpreted as a transfer of collective assets to the elite, or the transformation of political power into material inheritable wealth (Cooper 1993). This process was more typical of urban areas dominated by the new "princelings" – the children of China's aging leadership (Malhotra and Studwell 1995). The resulting class stratification is a national, as well as a highly localized and intrafamilial, phenomenon. In rural areas, collective industries existed at a number of levels – team, brigade, commune, and county. National reform polices were interpreted locally in heterogeneous ways and this produced a varied landscape of industrial change. At the team and brigade (village) level, most collective industrial works and sideline industries were contracted out to individuals, families, or small groups of families; at the commune level a combination of individuals and hired managers was utilized; at the county level, most industries remained in the hands of the local state. Since industrial assets were often limited (tofu factories, flour mills, machine shops, etc.), the most unequal distribution occurred at the village level. This led to rapid social stratification in many rural areas, as a few lucky peasant households gained control of collective assets and were transformed into a new class of small industrial and commercial entrepreneurs. By comparison, land distribution was carried out (in most places) in a roughly egalitarian manner. Communes also gave up much of their regulatory power, while distancing themselves from the risks associated with production. Privatization undermines important revenue sources to the state, while leaving high expectations for delivery of social welfare services and infrastructure maintenance. The resulting gap between expectations and service provision undermines state authority and legitimacy.

In sum, the transfer of production risk to households within a deregulatory atmosphere, combined with decreasing social welfare, and increasingly unstable local markets, shifted production practices towards those providing short-term

233

returns, often through rapid (and often degrading) exploitation of natural resources and labor (household and hired). Much of the resulting exploitation came from mining communal capital (Muldavin 1986). Granted, there had been numerous examples of environmentally unsound decisions prior to the reform period (Howard 1993; Ross 1988; Smil 1984, 1993), but the reforms introduced new negative elements which exacerbated environmental problems and increased unsustainability in the new hybrid system of China (Edmonds 1994; Muldavin 1992).

Effects on environment

Intensively using land for crop production over thousands of years, China necessarily developed sustainable organizational forms in farming (King 1911). *Ch'i Min Yao Shu: An Agricultural Encyclopedia of the 6th Century*, carries detailed instructions for proper maintenance and improvement of cultivated land – crop succession and green manure practices to improve soil fertility, field infrastructure to prevent erosion and improve yields, and various methods for insect protection – as well as methods for crop storage and food processing. Agricultural practices 2,000 years ago involved livestock, fish, crops, and humans in a complex and highly evolved system of production (Shih 1962). King (1911) describes many of these systems still in effect during his travels in early twentieth-century China. Yet despite this vast knowledge, environmental degradation also occurred largely as a result of the combination of impoverishment and heavy surplus extraction by the state and local landlords. Historically, war and rebellion have also taken their toll, devastating large areas, for example in the early Qing dynasty, followed by land reclamation and repopulation (Purdue 1987). Only through monumental exertion of human labor was degradation slowed or occasionally reversed (Cressy 1934). Most agricultural land producing crops today is in a degraded state. Massive inputs of nutrients and labor make crop production possible, but at yields far below potential. The post-revolutionary period is just a snapshot in this long evolution of land use in China. But it is one in which the history of sustainability is being rapidly forgotten.

The People's Republic of China set out in 1949 to reclaim and repopulate lands lost to production during Japanese occupation, the Second World War, and the subsequent civil war. The thirty years following liberation are replete with heroic accounts of turning back the desert, converting wastelands into productive croplands, reforesting denuded hillsides, controlling rivers and bringing water to arid lands. Indeed, an entire discursive formation was built on the notion of the socialist transformation of nature in China. It must be stressed that the large-scale organization of labor by collectives focused on building the necessary bio-physical infrastructure was very much responsible for the long-term gains in production that were achieved.

Yet between 1984 and 1989 total grain production fluctuated below the 1984 peak, while unit yields stagnated (see Figures 11.2a and b). Causes of the

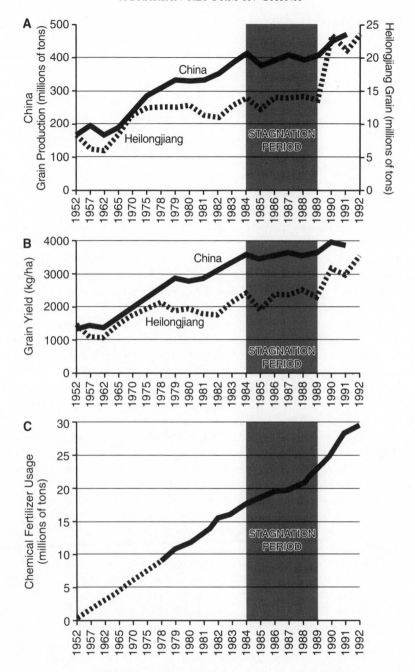

Figure 11.2 Grain production, yields, and fertilizer use (1984–9 focus)
Source: *ZTNJ* (1992: 222, 312, 323–4; 1993: 332, 348, 349); *HSY* (1992: 347, 351, 354); Liu (1992); *ZNTN* (1992: 272)

235

stagnation in grain production include decreasing profitability, a subsequent shift to cash crops, and unsustainable production practices. Much of the gain during the boom period (1978–84) can be attributed to technical factors of production, such as fertilizer application, rather than social reorganization of production (Stone 1988). Thus problems associated with these technical factors loom large in the subsequent period of stagnation. When the state monopoly on agricultural inputs was lifted in the mid-1980s, the potential for huge profits in the black market undermined [many] of the expected benefits of freer input flows. In part due to the long chain of middlemen, prices were so high that the already minimal profitability of grain production simply vanished.

As total grain production area has fallen nationally, there has been an expansion in cultivated land devoted to cash crops, such as oil seeds and tobacco; intensification has occurred in areas still devoted to grain. Higher grain yields, achieved through increased use of chemical fertilizers and other changes in agricultural practices, could not be sustained. Increased cultivation of cash crops has also been accompanied by increased use of chemical pesticides and fertilizers. Intensified crop production, in general, has led to a growing dependency on costly agricultural chemicals, as farm households jumped on to the increased-production treadmill. Over-use of fertilizers is causing immense production and environmental problems through soil degradation, organic matter decline, and water pollution. In the summer of 1994 there was an outcry in China over the immense fish kills in major rivers as a result of fertilizer and pesticide residues entering the waterways through agricultural runoff.[1]

With the introduction of modern inputs to agriculture, the world has experienced massive increases in grain production often at the expense of long-term productivity, nutritional quality, and with additional problems of high dependency on vulnerable 'improved' and hybrid species (Perelman 1977). This vulnerability is well documented and is not unique to China. Worthy of note in China however is the recent trend away from historically more sustainable methods, compounded by economic changes favoring short-term decision making and rapid utilization of existing resources. Without a concerted national effort to develop alternative practices, it appears that the sustainability of agriculture in China will end up depending upon expensive non-renewable resources which in themselves have destructive long-term effects (for example, salinization and declining structural integrity of soils through long-term fertilizer use).

Since the introduction of reforms in 1978, and with the intensification of production, the use of land in China has become a short-sighted affair, with rare exceptions. Loss of collective control over much of the resource base results in a run on the bank, so to speak, with a mining of the land and communal capital improvements made through hard struggle between 1949 and the early 1980s. Resource degradation takes many forms: increased use of marginal lands; more intensive cropping patterns; increased and uncontrolled use of destructive chemical inputs; a decline in agricultural infrastructure investments; rapid exploitation of existing assets through industrialization and accompanying pollution; land conversion; and the rampant destruction of ecosystems through industrial mining and the search for construction materials.

POLITICAL ECOLOGY AND THE NEW INDUSTRIAL GEOGRAPHY

Analyzing the effects on the environment of this hybrid derivative of socialism and capitalism requires a similarly hybrid framework, situated in relation to two bodies of theoretical work: combining aspects of political ecology and the new industrial geography allows a comprehensive multi-level analysis of environmental degradation. Despite the broad sweep of their theoretical introduction, Peet and Watts do not explicitly mention work on capitalist dynamics of production (see Harvey 1982; Massey 1984; Page and Walker, forthcoming; Sayer and Walker 1992; Scott 1988; Storper and Walker 1989).

Political ecology emphasizes the importance of political economy in the understanding of environmental degradation – an historically informed attempt to understand the role of the state, the social relations within which land users are entwined, and resulting environmental changes. Much work has been done on the transformative effects of the penetration of capital into different aspects of the production process, the commodification of production and reproduction, and subsequent changes in land use towards more degrading practices (Blaikie 1985; Blaikie and Brookfield 1987; Carney 1993; Neumann 1992). Little work in political ecology has unraveled specific aspects of the production process, and thus the means by which capital penetration actually occurs. The environmental effects of changes in the organization of production is another under-theorized area. Where studied, emphasis has lain on interactions between peasant/ indigenous and capitalist forms of production organization. The new industrial geography literature helps link together various scales of restructuring, bringing increased sensitivity to changing production relations and organization. This sensitivity stems from an analysis of contracting, flexibility, and risk; and the particular institutional arrangements of emergent regimes of accumulation. Recent insights derived from political economy decisively alter the way in which the process of industrial location is conceptualized. In contrast to a focus on allocative market functions or corporate hierarchies, the new industrial geography emphasizes the propulsive force of capitalist production as the prime determinant in the trajectory of industries (Harvey 1982; Massey 1984; Scott 1988; Storper and Walker 1989). The focus is on the developmental dynamics of capitalist growth and industrialization characterized by an ever-changing pattern in which new industries are periodically generated while old ones are restructured. In this process of economic expansion, new sites emerge and grow with the rising or renewed industries of their time, while others decline. In this view, industries create and re-create places by re-investing capital, expanding commodity output, improving production methods, multiplying the division of labor and competing vigorously. Industrialization is then highly differentiated, with individual industries exhibiting distinctive paths of spatial development based upon critical differences in material base, labor, technology, and organization, as well as upon the special qualities of people and places that enable or constrain industrial growth. The new industrial geography is concerned with cycles of investment, technical and organizational innovation, competition, buyer–supplier linkages and place-bound labor relations (Page 1993).

By comparison, agriculture has been comparatively under-theorized within geography as an arena of capitalist development. Indeed, agriculture serves most often as mere backdrop to analyses of industrialization. Yet it is critical to recognize the ways in which farming and manufacturing are bound together in the process of industrial development (Page and Walker 1991). Recent developments in the new industrial geography, particularly helpful for political ecology, focus on urban–rural linkages and regional production complexes (Page and Walker, forthcoming). My main point is not that industry and agriculture are linked, but that agriculture has to be seen as an economic activity subject to the same kinds of restructuring processes studied by the new industrial geography. The primary problem of the new industrial geography is that it fails to seriously engage production's impact on nature – hence the need for an integration with political ecology.

By taking the best of both of these areas of theoretical inquiry we can begin conceptualizing a new theoretical framework. At the macro-level, broad changes in the capitalist production system – what might generally be termed global restructuring – provide more efficient means for transnational capital to penetrate societies and, in the process, transform nature in new and often disastrous ways. Taking advantage of geographical unevenness, TNCs restructure the organization of production with significant impacts locally on capital–labor relations, natural resource extraction, and the environment. The new social divisions of labor ratchet down regulatory restrictions on environmental pollution and occupational hazards and contribute to lower wages by undermining organized labor (Epstein *et al.* 1993). The evolving mezzo-level patterns, which directly affect the environment, can be clearly identified using the theoretical work in the new industrial geography on urban–rural linkages, subcontracting networks, alternative methods of organizing production, and new regimes of accumulation (Page and Walker 1991; Sayer and Walker 1992; Scott and Storper 1986; Storper and Walker 1989). Non-state actors and institutions transform the production process *despite* the state's attempts to control them. At the micro-level, changes in the organization of production, and in marketing and consumption patterns, result in an intensification of natural resource use increasing environmental degradation. Thus, theorists of the new industrial geography analyzing processes of First World industrial capitalism engage similar issues to those raised by political ecologists analyzing agriculture in Third World rural contexts. These similarities provide analytical tools for identifying the means by which local processes of various kinds are embedded in, and structured by, global processes of capitalist expansion. In sum, exploitation of uneven development and flexible regimes of accumulation (in particular subcontracting networks) increase ongoing concentration of control in the struggle over assets (capital, labor, and natural resources). This leads to fiscal crisis for the state, providing the context for multilateral intervention and "adjustment" policies which often result in expanded deregulation of capital, labor, and resource use and thus increases in environmental degradation.

Drawing on this framework I analyze the interaction between changes in production and state policy in China's reform period. The rapid transition from collective to household forms of production organization in contemporary rural China provides a unique analytical opportunity for political ecology. By comparing collective with household forms we can identify how production organization affects land-use practices, for example through shifts in decision-making strategies involving changing perceptions of risk, social security, and stability. I extend political ecology in the Chinese context with a special emphasis on organizational forms and the mining of communal capital built up under the previous socialist economy. The use of political ecology in China, focusing on long-term environmental effects, forces a re-evaluation of the impacts of short-term coping strategies under shifting regimes of accumulation – from plan to market, collective to individual, long-term to short-term. It also provides a framework for analyzing indirect forms of resistance, such as claims by peasants on the state via 'natural' disasters (Blaikie et al. 1994).

In China, it is the rural industrial complexes which now provide the greatest dynamism in the economy, while simultaneously affecting nature in unprecedented ways. We can better illuminate the rapid change taking place by conceptualizing rural industrialization as part of a regional transformation (Page and Walker 1991). As "flexibility" becomes the watchword of global capitalism (Sayer and Walker 1992), global restructuring processes are paralleled by shifts in state policies in China since 1978 towards privatization, subcontracting, and deregulation. This transfer of risk to individuals and decreasing responsibility of the state and large corporations/collectives is accompanied by international artic-ulation – the process by which the dynamics of global capitalism disarticulate Third World economies (de Janvry 1981). Functional dualism, in the Chinese context, facilitates labor migration undermining local conservation efforts. In response to competitive pressure to attract international capital, and in order to grab market share, China has become increasingly enmeshed in the global economy, adding to the overall ratcheting down effect. Current rates of envi-ronmental degradation are difficult to tally for the whole country. Through anecdotal accounts[2] we can see waves of environmental problems rolling through Heilongjiang Province, including erosion and soil degradation, air and water pollution, forest decline, and grassland desertification.

THE CASE OF HEILONGJIANG PROVINCE: 1978-95

My case studies are drawn from two counties, Zhaozhou and Bayan, in Heilongjiang Province (Figure 11.3). With a total area of 454,000 square kilometers, and more than 35 million people (*HSY* 1992), Heilongjiang Province is a major grain base with important exports of agricultural commodities to the rest of China. In addition, Heilongjiang is a major industrial center and source of raw materials like oil and lumber. One-twelfth of the total cropland and one-sixth of the total commercial grain production of China are concentrated in this frontier province (Zhao 1981). In addition, vast areas are devoted to grazing

Figure 11.3 Heilongjiang Province
Source: Muldavin (1994b)

240

cattle, sheep, and horses, as well as exotic species, such as deer. There are also important fisheries in the province. Along the Heilongjiang River, on the northern provincial border, sturgeon factories process caviar for the world market.

Heilongjiang was transformed from wilderness frontier into "the great northern granary" in less than a century. With particularly harsh and long winters the region is limited to single-season crop production. This prevents the intensification of production through multiple cropping prevalent in other parts of China. However, in many other respects, Heilongjiang serves as an important indicator of the wider changes. A striking example is a rapid expansion in livestock, a reflection of national development policy goals for the region.

Decollectivization and privatization have important consequences for nature. Using a revised political ecology framework to analyze the reforms of the emergent new hybrid system, we can look at the impacts in rural Heilongjiang Province of the following: (1) land-use intensification; (2) agro-industrial pollution; (3) declining social/communal capital.

Intensification of land use

Changes in agricultural production practices stem from a series of interrelated "choices" imposed on peasant households by the necessity for intensifying production under a situation of increasing risk. Intensification of land use in Zhaozhou and Bayan counties in the early 1980s led to a rapid decline in overall soil fertility. Organic matter declined rapidly in the years immediately following the reforms, signifying a departure from sustainable agronomic practices (Muldavin 1986, 1992). Increased use of chemical fertilizers, a continuing decline in the use of organic manure, and a 50 percent decline in green manure area intensified organic matter decline and decreased soil fertility and water quality overall (field notes 1989; Liang 1988). For example, all green manure cropping ended one year after the implementation of reforms in Zhaozhou County. Likewise, manure delivery to fields for composting and spreading declined rapidly in Bayan County; these were reinstated only through an enforced system of compost quotas in the late 1980s. Crop rotation was simply abandoned throughout Heilongjiang in favor of monoculture – the choice of crop (especially corn) depending upon short-term profitability. The impact on soil fertility was partly countered through increasing chemical fertilizer use, from 10 million metric tons in 1978 to over 30 million metric tons by 1994 (see Figure 11.2c). Indicative of a growing dependence on a complex of unsustainable practices, this led to a series of problems difficult to resolve – salinization, groundwater pollution, and micro-nutrient deficiencies (Muldavin 1983, 1986, 1992). As a result, yields stagnated and then declined. Peasant farmers complain about the "soil burning" of long-term fertilizer use. Because of a loss of structure and decline in overall quality, the soils become harder, less friable, with available nutrients actually diminishing despite large additions of chemical fertilizer. A combination of insect resistance to pesticides and repetitive monocropping intensifies disease and pest problems. With a lack of investment in the crumbling irrigation infrastructure, and declining availability of water, agricultural production has become more unstable as it has intensified.

The most common example of marginal land invasion in Zhaozhou County was the transformation of tall prairie grasslands into field crop or intensively grazed pasture areas. The movement into marginal lands and the shift of land use from forests, marshes, and grasslands to more intensively utilized cultivated land (extensification of agriculture) were other means of increasing overall production. These shifts often bring short-term positive results, but also result in serious damage to areas unsuitable for agriculture. Whether the choice was to rapidly increase grazing, or plow-under the grasslands, the end result is the same – massive problems in production within a short time. As cropland is expanded into marginal grasslands in the northeast China plain area, such lands can only support crop production for a limited amount of time. The results have been disastrous – rapid expansion in the area of barren sodic alkali land (Muldavin 1986). It takes large amounts of water and expensive tile drainage systems to flush out the salt accumulations. This is not economically viable for large areas of extensive grain production such as the northeast China plain.

A more intensified use of grazing lands in Zhaozhou County in 1983–5 resulted from decollectivization and subsequent decontrol over livestock numbers and grazing area rights. Peasant herders rapidly increased their herd sizes after the reforms, as a means of improving security and laying claim to previously communal grasslands (see Figure 11.4b). Herders moved their expanding flocks further away as pastures near villages were either degraded or transferred to crop production. Temporary housing was set up to guard investments and help stake claims to land. But the effective deregulation of grasslands meant that the rapid decline in land quality was difficult to monitor, let alone reverse. In some areas no contracts (à la the household responsibility system) have been made between herders and the village government, which might provide at least minimal land-use regulation, and therefore lands are used by whoever can seize control. Given the increased risk borne by individual households, there are strong incentives for increasing herd size. Hinton (1990) documents the same process in nearby Inner Mongolia. Enclosures of grassland by wealthier peasants forces larger numbers of livestock into an area of diminishing size (field notes and interviews 1984, 1988, 1989). As large areas go out of productive use through rapid degradation, grazing pressure is increased on the remaining lands.

The process of accelerating internal desertification in Heilongjiang Province is tied closely to the rapid increase in the use of marginal lands coming with the decline in collective control (Muldavin 1986). A direct correlation between decollectivization and increased use of marginal lands is a result of the fact that many of the most marginal lands are part of the collective holdings. Officially designated desert area in China increased from 1.3 to 1.5 million square kilometers between the late 1970s and 1990 (Liang 1988). As land use

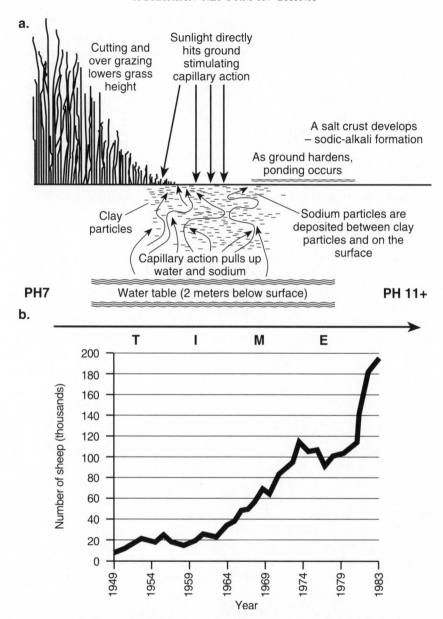

Figure 11.4 (a) The process of sodic alkalinization in grasslands
(b) Sheep population 1949–83, Zhaozhou County, Heilongjiang Province
Source: Muldavin (1986, 1992)

in general is intensified, production increases are achieved through environmentally destructive practices, often resulting in desertification of vulnerable regions.

243

With a decline in soil cover due to intensified grazing, a process of sodic-alkalinization begins, raising the pH in the upper horizons of the soil from 7 or 8 to 10 or 11. At the higher pH levels, what is left of the grasses dies off, leaving only clumps of high-pH tolerant weeds with low nutritional value (Muldavin 1986). The physical process of degradation is as follows (Figure 11.4a). There is decreased grass cover on land with only slight elevation variations. Following significant precipitation (particularly in the summer rainy season) water ponds on the soil surface. Exposed to sunlight, much of the water evaporates before percolating down into the soil, drawing salts to the surface through capillary action and forming crusts of white particles. The soil porosity is reduced further by the lodging of sodium (salt) particles between clay particles, thus further restricting downward percolation of water and speeding up the process of alka-linization. The result is an expansion of barren areas with extremely high pH soils covered with a salty crust. Erosion increases, exposing even less fertile sub-horizons, further complicating rehabilitation.

Lands lost to animal and crop production via improper and intensified utiliza-tion are often used as *adobe* sources. Since the high pH and sodic-alkali nature of these soils acts as a water repellent, the topsoil makes bricks with good endurance qualities. Soil for bricks is only taken from the top meter, often in areas that still have a short grass cover. In Zhaozhou County thousands of hectares have been pockmarked in this fashion. This practice has increased rapidly due to the rural and urban construction boom of the last fifteen years, compounded by the building of larger homes. There is a grave problem of increased standing water in this minefield of holes, supplying a perfect breeding ground for mosquitoes and other disease vectors. In this area the mosquito carries Japanese encephalitis, a very serious disease. Yet, one of the (conveniently) less-noted results of decollec-tivization is a breakdown of large-scale pest control programs. Also there is increased exposure to harmful pesticides, particularly for children and the elderly. In Zhaozhou County the collective's former method of running sheep through a cement-lined pesticide dip to control scabies is no longer practiced. Responsibility for control of such diseases falls on the household, usually untrained female children, with little or no protective measures taken against exposure to toxic chemicals.

The construction boom increased the demand for wood, cement, iron products, and other housing materials. Rampant cutting of roadside windbreaks, nurtured collectively over thirty years, became unstoppable. It is now possible to drive for miles among stumps of trees felled in the last two or three years. Peasants call this "nibbling" describing, in interviews, how every night a few more trees disappear. Despite fines and increased policing the practice only accelerates (Muldavin 1989c). Combined with intensified grazing pressures, cutting forest for wood products and fuel increases soil erosion by breaking through the protective cover (Hinton 1985, 1990; Muldavin 1986, 1992). Critical reassessments of recent reforestation programs show large areas of poor quality stands (Delman 1989: 59).

Changes in production practices were accompanied also by a shift in scale, from large unified plots to a mosaic of small holdings reminiscent of pre-revolutionary China. Large tractors and harvesters were abandoned after the reforms, for it was no longer possible to utilize them efficiently on the myriad individual plots (Muldavin 1983, 1986, 1992). The return to labor-intensive methods in agriculture has been dubbed the "feminization of agriculture," as increased labor demands are borne primarily by women and children. In turn, there has been a realignment of power relations within the household, with a corresponding reinforcement of patriarchal domination in decision making, and a cultural devaluation of women's labor despite its increased importance.

These changes in production practices are indicative of a shift from long- to short-term goals in rural decision making. The intensification and transformation of arable land through rapid and unregulated rural industrialization, expanded home building, and population increase is an immensely difficult problem. Intensification resulting from increasing population pressure tends to be a long-term trend. On the other hand, the current shift in production practices directly followed decollectivization. Deregulation and decentralization of agriculture and the increasing control at local and household levels led to production decision making in response to perceived market signals. Short-term goals are increasingly emphasized in a riskier production environment combined with associated social welfare declines. The consequences for the environment are disastrous.

Agro-industrial pollution

Industrial pollution occurs on such a large scale in both rural and urban areas, and is so completely uncontrolled, that China faces an imminent ecological disaster. With over 80 percent of industrial waste water re-entering waterways untreated it is no wonder that almost 50 percent of inland water is too polluted to drink or support aquatic life (Liang 1988). Unregulated township and village enterprises (TVEs) pose serious health threats at the village level through groundwater contamination and other forms of pollution. TVEs expanded rapidly in the new hybrid economy, many as subcontractors to national and international firms. Their rapid growth is due to a combination of flexibility and low-cost production: in other words, their ability to hire cheap labor, utilize older highly polluting methods and machinery, and ignore all environmental and occupational health and safety regulations. Often built on the limited arable land (declining by 1 percent each year, nationally), and being geographically far-flung, they are virtually impossible to monitor or regulate, making them potentially more hazardous than spatially concentrated urban industries.

In the three villages where I worked in Heilongjiang, industrial and agricultural pollution were largely undetected contributors to public health problems. These problems are compounded further by the declining quality of health care over the last fifteen years as collective services disappear through lack of funding and organization. The potato starch noodle mills in Hesheng village simply empty their effluents into dirt-lined ditches lining the roads. A purple-colored chemical is used to process potatoes into starch; thus it was easy to identify the contamination of local groundwater, as families began drawing violet water from their household wells. In another village in Bayan, all the fish raised by two households were killed by chemically contaminated water from a number of small village industries. Fishermen along the Songhua River reported rapid declines in fish yields as chemical runoff from TVEs and agriculture increased dramatically.

Brick-works in many villages burn huge quantities of low-quality coal. By the end of the day, a low, fog-like inversion spreads across the villages. Combined with widespread use of small fires for cooking meals this makes "rural" air quality abysmal. The high level of particulate matter in the air leads to rising rates of associated illness and disease, according to local health workers (Muldavin 1989a, 1989b). It is not only air quality which is affected by the widespread use of high-sulphur coal in TVEs. Despite producing a third of China's industrial output, TVEs receive only 20 percent of state-supplied energy inputs for industry (FBIS 1992). The gap in supply is filled though production from largely unregulated small-scale collective and individually run coal mines – 60,000 of which are scattered throughout the countryside. These small mines are responsible for widespread destruction of adjacent forests and fields. Mine tailings are a further health hazard, polluting land and water alike.

Declining social/communal capital

Neo-classical (Lin 1993) and Maoist (Hinton 1990) analysts recognize that state investment in agriculture and infrastructure has fallen significantly since the reforms were implemented, with resulting negative impacts on production. Less agreed on are the fundamental causes of this situation and its solution. By 1984, the collective ceased to exist as allocator of capital and labor. There are two major interconnected aspects of the resulting situation: a decline in capital and labor investment in large-scale infrastructure; and the redirection of capital and labor into short-term projects and investments.

Villagers have been hard put to organize the necessary labor and capital investment for agricultural infrastructure. Reservoirs, dikes, irrigation canals, tube wells, erosion control, tree planting, all critical to sustaining and increasing production, receive little investment for maintenance, let alone improvement or expansion, and are in a state of serious disrepair (field notes 1994; see also Hinton 1990; Muldavin 1992; *Peasant Daily* 1989). Furthermore, loss of control over the utilization of resources led to severe overdraft of water resources in many areas (Hinton 1990). Although collectively owned equipment and facilities were often distributed along with land (*Peoples' Daily* 1982), water control infrastructure remained under collective management. Yet, such collective structures experienced a massive decline in investment capital while simultaneously being stripped of authority and assets (*ZNTN* 1992). A rapid acceleration in local-level "natural" disasters in the last few years is attributable largely to the delayed effects of this decrease in capital investment. The falling into disrepair of irrigation systems leaves villages without any back-up to the unpredictable rains. Villagers once again pray to ancestors and other deities as a last resort. Village informants describe in detail the contradiction between their attempts to meet subsistence needs and the demands of local state authorities for higher taxes despite a precipitous decline in local infrastructure investment. This engenders widespread anger towards local representatives of a delegitimized state, an increasing trend throughout much of rural China (field notes 1987, 1988, 1989, 1994; Muldavin 1994a).

In Zhaozhou County in 1983 and 1984 I witnessed the abandonment of tree-planting programs, as well as the subsequent rapid harvest when forest plots were contracted for management to individual households. Rapid destruction of tree-based erosion control systems was further along in Bayan County in 1988 and 1989 (field notes 1989). Both counties are unable to organize sufficient labor and funds to replant denuded hillsides or windbreaks; this is representative of the recent decline of investment in sustainable land use. In merely four years, between 1985 and 1989, there was a 48 percent decline in area covered by windbreaks nationwide (ZNTN 1992).

Expenditure on agriculture as a whole, as a percentage of total (national) state expenditure, has been in continual decline since the late 1970s. Over 13 percent of the state budget at the time of Deng's ascent, state expenditures had fallen to less then 9 percent by 1981, and were 8 percent of the national budget in 1986–90 (ZTNJ 1991, 1992, 1993). This rapid decline in state capital investment in agriculture prefigured the agricultural stagnation between 1985 and 1989. It remains a major barrier to the improvement of land and labor productivity in agriculture (Lardy 1984; Muldavin 1986; Stone 1985; Walker 1984).

Repeated calls have been made in the last few years for diverting state revenues to agriculture. In 1989, the Vice Minister of Agriculture said that large capital construction projects in agriculture were indeed being successfully implemented,[3] with a new land conversion tax as major source of funds. Yet problems in collection of this and other agricultural taxes hinder proposed increases in investment. Another difficulty is the translation of top directives and transfer of monies down through the institutional maze into productive investment, without most getting "lost" along the way. Total projected investment needs of rural China over the last fifteen years of this century exceed 1.5 trillion RMB (Renminhi, People's money), with the state expected to contribute less than one-fifth and the rest expected to come from peasant savings, the savings of TVEs, and the weakened collectives (Delfs 1984; Du 1985; Riskin 1987). With institutional structures at a local level severely weakened by reforms, it is difficult to imagine the funding and smooth implementation of this project.

The attempt to build a levee in Hesheng village is a case in point. Work has been organized annually using contracted labor-duty days to construct a levee protecting the village from floods. Yet the village is unable to complete construction. Failure to collectively organize, and invest the requisite capital and labor, necessitates starting the project over from scratch, as the product of each year's labor is washed away. The villagers are so demoralized by the Sisyphean nature of this task, that it is no longer possible to rally their efforts around the project. The result is the progressive impoverishment of Hesheng village. For a number of years the village could not deliver its quota of grain to the township because the entire harvest was destroyed and the village completely dependent on state relief. Families trying to survive have moved out, further complicating the attempt to complete the levee. These migrants join the millions of peasants wandering across China in search of any kind of work (Muldavin 1989a). In late 1994 it was impossible to reach the village for a follow-up research visit because it was under water (Muldavin 1994a).

Capital previously used for large-scale infrastructure and collective projects has been diverted to short-term investments, such as increased chemical fertilizer application, plastic sheeting, small pumps for water control, and other agro-chemicals (field notes 1990; *ZNTN* 1992). Credit available from the state tends to favor such short-term investments. Local capital investment by farmers has also been of the short-term type since the reform. Most accumulated funds are spent on house building, consumer durables, and traditional ceremonies (field notes 1983; *BR* 1989; Zhang 1993).[4]

Similarly, investment in TVEs comes at the expense of other needed investments in infrastructure. Here too there is little reinvestment in production for long-term gains. Given the fact that many unregulated TVEs contribute to increased local air, water, and soil pollution, the much-touted change in locus of control over capital investment and planning, from the center to local authorities, has not brought the expected efficiencies when these are defined in terms of long-term sustainability of production.

A major impact of the mining of communal capital is its delayed effect on the frequency of "natural" disasters. Such disasters seem to be occurring in greater numbers, not all of which can be attributed to climatic variation. Overgrazing and clearcutting, as well as cultivation of marginal lands, rapidly increases soil erosion while further limiting water absorption and increasing runoff. Raised silt loads and deposits in lower reaches of the major rivers threaten dike systems weakened by lack of repairs and investment. The effect is felt through greater volatility of river flow and higher crests in flood levels. Reservoirs are rapidly silting up, undermining water conservation efforts, decreasing flood control and the electricity-generating potential of the large-scale hydroelectric projects on China's major rivers. In China there was a 25 percent increase in the area covered and affected by natural disasters between 1985 and 1991 (*ZNTN* 1992: 284). Hinton (1990: 132) discusses the collapsed terraces, groundwater overdraft, and other aspects of declining capital investment as well as resource exploitation run amok. My own fieldwork in Heilongjiang Province shows rapid increases in supposedly "natural" disasters resulting directly from declining investment in agricultural and conservation infrastructure. Combined with the intensified use of forests and grasslands, many of these natural disasters can be traced to distinctively human-induced causes.

STATE LEGITIMACY AND RESISTANCE

What forms does resistance take in this hybrid society? Is there a tendency for organized social movements? The traditional alliance between the Communist Party and the peasantry is threatened by the new hybrid regime and appears to be unraveling in many areas of the countryside. A decline in state subsidies aggravates the problems caused by the reforms and concentrates the effects of contradiction on the state. Subsidies are one interesting point of contradiction in this new system – a leftover from "communism" as well as an integral part of all "market" systems, at least for agriculture, they are certainly an ignition point for China! Chinese agriculture, as it becomes further enmeshed in national and international markets, shares the rural problems experienced all over the world. The state is needed to maintain stable markets for inputs and outputs via subsidies, which results in periodic fiscal crises for the state.

Subsidies

The early reform period saw high levels of state subsidies to agriculture. A resulting fiscal crisis now gives the state few options other than cutting back on many subsidies, bringing about further intensification of land use and environmental degradation. As a direct result of the reforms and the reorganization of production and exchange relations, peasant producers in the new system have been forced onto production treadmills similar to those found in capitalist agriculture (Muldavin 1994b; Perelman 1977). In both cases the state intervenes to mitigate the worst consequences, but in neither case are the fundamental structural conditions addressed. The rise in agricultural production costs takes place within a context of rising levels of surplus extraction from rural areas – causing farmers to redouble their productive efforts (Bernstein 1977).

Subsidies in the Chinese system range from price subsidies for inputs or commodities, to inflation-proof savings for urban and rural residents, where the state provides subsidies for interest on savings guaranteeing an acceptable return above the rate of inflation. The total percentage of state revenue used for subsidies to rural producers and urban consumers was roughly 38 percent by 1989. Necessary state investment in long-term production, both agricultural and industrial, has been redirected primarily to price subsidies in order to "keep the peace" with peasants and workers. The resulting fiscal squeeze forced the proportion of total government expenditures allocated to agriculture to drop precipitously (CSY 1991). Declining input subsidies place peasant farmers in an increasingly competitive environment. This forces a variety of methods to expand overall production, including soil mining (i.e. relying on pre-existing fertility of the soil without maintaining organic and inorganic fertilizer input levels) and increased use of marginal lands. These methods failed to halt the overall decline and stagnation in agricultural production in the late 1980s. Social tensions rose, within villages and between peasants and representatives of the state (Muldavin 1992). Increasingly in the early 1990s, input prices again rose rapidly, eroding the profits of peasant agricultural producers, which in turn led to more frequent incidents of militant peasant resistance (field notes 1992, 1993). These trends were mirrored in the urban sector, where there has been a decline in overall subsidies, adjusted for inflation, since 1984 (*CSY* 1992), angering urban residents on relatively fixed incomes.

The declining agricultural subsidies of the post-Mao period are representative of a larger crisis of the Chinese state, which essentially involves mortgaging the future for the immediate satisfaction of part of the current generation. This situation has been dictated by political necessity, as a means of legitimating a set of reforms that harm a wide array of China's people. This requires a rapid increase in international borrowing. By 1988 one-third of state investment in agriculture was financed by foreign sources, principally the World Bank.[5] China's indebtedness in 1990 topped US $44 billion, with some figures showing it to be US $52 billion (*BR* 1993a), the entire debt having been incurred during the 1980s (World Bank 1990). By 1993 the debt was approximately US $70 billion, with repayment beginning on earlier loans (*ZTNJ* 1993). The debt is symbolic of a rapid change in China's international economic position, and its ability to set the terms of its own articulation into the global economy.

Also, during the 1980s inflation rose from less than 3 percent a year in the early 1980s, to a peak of an annual rate of 27 percent in the first quarter of 1989 (*CSY* 1992; *BR* 1993b). With the contraction of the economy after Tian'anmen in June of 1989, inflation declined dramatically. But by 1994 it had risen again to 21.7 percent a year, the highest official rate since reforms began (*BR* 1995).[6] Inflation worsens the reproductive squeeze, thus intensifying land use. Real growth in GDP has been relatively high but variable, with inflation severely undermining its potential impact (*BR* 1993b, 1995; *USNWR* 1990).[7] National growth figures hide the effects of the production squeeze on land intensification and the environment. Immense regional variations in growth exist in China. Most growth is focused in a narrow band of primarily urban and suburban areas in the eastern portion of the country. Even within this area there is great spatial and social heterogeneity. Contrary to popular belief, most of the benefits from economic growth in China are garnered by a small portion of the population. The vast majority have not gained in the ways that generalized figures of GDP seem to suggest. A significant number of China's peasants, and to a lesser degree urban residents, experience decline in living standards (Muldavin 1992).[8] In 1993, while 4.3 million people saw their incomes skyrocket to thirty-two times the rural average, the incomes of 400 million people declined (Schell 1995). Social stratification and unequal distribution of benefits and costs of the reform process is a fundamental cause of rural and urban unrest. This has strained the ability of the state to maintain its reform program and undermined its legitimacy in the eyes of the majority.

The importance of the inflation and growth figures is that, combined with the subsidy figures, a very different picture of China's prosperous period of the 1980s and 1990s begins to emerge. Inflation in the latter 1980s delegitimized the state to the point where urban and rural interests allied in a series of protests, Tian'anmen being only the urban and most obvious manifestation of what was a widespread and also rural phenomenon. For ordinary workers and peasants inflation meant a real decline in wages and living standards, undermining many of the gains of the past fifteen years, and imposing a work speed-up without the anticipated rewards. Despite worker and peasant protests being potentially much more important, only the student protests were picked up by the Western press. Widespread dissatisfaction provided a potential foundation to transform an essentially silent social movement (at least in Western-media terms) into a vocal coalition representing the majority of workers and peasants. Unfortunately, this was not the goal of the student leadership in 1989, nor was it the outcome of their ill-fated movement.

The structural problems of the hybrid system are manifested in rural China in numerous ways: stagnant output of grain; intensification of land use; agro-industrial development and pollution; changes in capital investment and in the locus of control and risk; subsidies and the state revenue crisis; and inflation. These problems are important components of the subsequent delegitimization of the state. The effects on the environment are often underestimated, with grave consequences for the future. Although some difficulties may be transitory, the vast majority are fundamental, giving rise to long-term problems in social reproduction. They call into question the long-term viability of this derivative economic system.

Rural resistance

Rural resistance exists though it is not much publicized. Indeed resistance periodically flares into local, but nonetheless important, incidents of rebellion. Inflation plays a role, as does the fiscal crisis of the state. Following reform, rising real costs of agricultural and agro-industrial inputs caused unrest and demonstrations among the peasantry by 1987. In 1992 there were dozens of reported incidents of rural unrest due to the reproduction squeeze. A recent peasant revolt in Renshou in Sichuan Province highlights the difficulties of a state trying to impose levies on rural peasants when peasant income is stagnating or declining (Muldavin 1992).

Poor people in areas without "poor" status bear the greatest burden of an immensely regressive taxation system (Riskin 1995). When peasants refuse to pay taxes in rural Heilongjiang, local police (whose wages are dependent on tax income) go with local cadres to peasant homes to demand payment. If the peasants don't pay, the police take household goods of equivalent value. In cases where peasants accuse the local leaders and police of unfairness in treatment, they are punished through the cancellation of the state contract for their fields. They are left with a small quantity of ration (subsistence) land (field notes 1990, 1993).

Instead, peasants develop indirect forms of resistance (cf. Scott 1985; also Peet and Watts, Chapter 1 in this volume). In a state which declares that each person contributes according to his/her ability, disasters are legitimate calls on the state's forbearance with regard to taxation – the state has little moral choice but to come to the aid of the stricken communities. Despite sustained favorable weather during the mid-1980s there was a "strange" increase in peasant declarations of natural disasters as village cadres mastered the art of "poor-mouthing" to strengthen their claims (Muldavin 1989a, 1989c). Thus between 1981 and 1989 natural disaster claims were made for seven of nine years on behalf of villages in Songhuajiang township (Table 11.1) to justify lower annual quota requirements, disaster relief, development aid, cheaper and more abundant inputs, and cheap or free credit. This method of garnering state funds, or refusing to relinquish local surplus to the state, provides an avenue for legitimate and illegitimate claims on state assets (Swift 1989), and is an indirect method of peasant resistance. In Hesheng village the good repayment record for credit from the township credit co-operative declines in years of natural disasters, when loans are either pushed forward or written off. This adds to local officials' and peasants' temptations for having their area declared eligible for disaster relief. Such costs accrue to the provincial and central governments, which must replenish credit sources each year.

Peasants can lever assets from the state based on historical agreements between the Communist Party and the peasantry. In a sense, the "party of the peasants" has to provide a morally correct patronage system of asset distribution to its

Table 11.1 Natural disaster claims for Songhuajiang township, 1981–9

Year	Actual conditions	What was claimed
1981	Climate normal	Minimal claims
1982	Climate normal	Serious pest infestation affected a large area
1983	Climate normal	Minimal claims
1984	Climate normal	Spring drought and flood
1985	Severe flooding (village under water)	Flood relief necessary
1986	Severe flooding (village under water)	Flood relief necessary
1987	Climate normal	Late freeze and unseasonably low temperatures late into spring, hail storm (8/22/87), flooding
1988	Climate normal	Flooding (7/7/88–173 mm of rain in two days)
1989	Climate normal	Spring and summer drought

supporters. If it were to lose its rural support base, it would be further weakened in relation to pressing urban demands. In this derivative system problems accumulate around the state presenting it with an intractable legitimation crisis.

CONCLUSION

It is a commonly held view that reform of collectively based planned economies requires decollectivization to occur simultaneously with the promotion of market mechanisms. In fact, the two need not be tied together. A shift from collective to household economy is not a necessary corollary to the transition from command to market economies (Bardhan and Roemer 1993; Nove 1983). Within the local and regional heterogeneity of China's political economy reforms take diverse paths. Alternate pathways are emerging, for example a shift to markets within a predominantly collective economy allowing peasants to employ long-term decision-making strategies in production alongside short-term market ones. This arrangement can spread the increased risk of market mechanisms over larger groups (Bardhan and Roemer 1993). Such an institutionalized competition within a collective market economy could reduce the mining of communal capital. In Heilongjiang Province re-collectivization by poor peasant households into small groups to spread risk, and increase efficiency and power, is one form of coping strategy seemingly contrary to the overall move towards highly competitive individualized alienated households within an unstable economy. But this kind of collective market economy may soon predominate, according to the deputy director of agriculture for the province, and is being closely watched in Beijing as a strategy to overcome some of the contradictions of the more general, hybrid system.[9] What is critical in any assessment is to learn in what ways small-scale reunification is truly a voluntary collective or, in contrast, an imposed and more efficient means of surplus extraction by the state through collective organization. What may in fact be emerging is a new form of "micro-feudalism," with the local neo-gentry providing protection, as well as supervision of production practices, while guaranteeing a certain level of surplus extraction for the state.

Another alternate path is to have individual decision making within a planned economy with certain state guarantees left in place (individual planned economy). This path could provide adequate social welfare and stability of markets for individuals to make long-term decisions in regard to production. In fact, intervention of this sort (though short-term and reactive) has been the method most often used to deal with crises arising from the reforms.

Both alternatives already exist in certain of China's regions and localities. Thus, a wide range of options is already in place to deal with the problems facing rural China – all of them better than a complete turn to a market economy composed of millions of small producers. The resulting picture of agrarian and ecological change, emerging from an array of often contradictory forces – from state policy to local resistance – has evolved towards a mosaic of mixed market and socialist forms, some with government sanction, some without. These forms range from individual to collective, and in recent years in Heilongjiang Province at least buck the trend by increasingly relying on government intervention in the local economy for long-term market planning and stability. All forms share a common goal of growth, but with differing emphases placed on quality of life issues, such as a sense of security and shared risk.

China's restructuring, coming from its experiments with socialism and its articulation with the global economy, significantly increases its dependency on international markets, as outlet for products, and as source of capital, technology, and imported commodities. Therefore China's problems increasingly resemble the restructuring taking place in the capitalist economies (Sayer and Walker 1992). Both emerge from the need to increase the flexibility of surplus extraction on the part of dominant economic structures, institutions, classes, and actors. With articulation into the international economy global 'ratcheting down' effects occur domestically. China's reforms involve new kinds of institutionalized regimes of accumulation, part state, part market, relying on drawing down communal capital and increasing exploitation to enable rapid increases in total production. These new regimes have geographically far-flung subcontracting networks in which risk shed by larger firms and institutions is transferred to small firms and peasant households. This chapter shows that costs and risks in these new regimes are also transferred to nature, resulting in rapid environmental degradation. Combined with unstable and changing tenure relations, short-term practices are emphasized over long-term sustainable ones. Thus, the problems of market triumphalism, discussed by Peet and Watts (Chapter 1 in this volume), are mirrored by quite similar problems of socialist-market triumphalism in China. Rapid social stratification has occurred, in line with global trends, as control of productive assets, particularly non-agricultural, has been concentrated in the hands of a new wealthy elite (Odgaard 1992). The failure of the commandist state – or the magic of trickle down – to redistribute the wealth of the rapidly developing coastal and southern regions to the poorer, agriculturally dependent hinterlands amplifies the sense of multiple Chinas. Crises ranging from overproduction and declining/unstable returns on investment in agriculture, to stagnation and declines in productivity, amplify contradictions in population policy. A lack of social security and increasing economic instability promotes a rise in human fertility, while the state's population policies penalize the peasant households for this rational response (Davin 1988).

The abrupt dismantling of collectives, and their replacement by individual/household production units, in concert with the shift from plan to market, causes a complex of significant problems, many of which have severe environmental consequences and raise fundamental issues of sustainability. Within the new economic context provided by the hybrid system there are deepening *contradictions* – lowering the productivity of the resource base and bringing a decline in the long-term development potential in many areas of rural China. Taking into account the immense heterogeneity of China, locally based collective action (potentially finding its strength and direction in the growing rural resistance) and long-term production strategies must be promoted if the difficult problems of sustainability are to be resolved. The state's ability to maintain legitimacy through its paraphernalia of subsidies and debt is structurally limited by fiscal crisis. Further, the inequitable growth of the past fifteen years has been achieved through mining of communal capital. Thus, China's emergent derivative hybrid of socialism and capitalism – authoritarian, corrupt, speculative, exploitative of labor and nature – is based on fundamental contradictions that will not simply go away with the completion of "transition" to a market economy.

NOTES

I would like to acknowledge the Fulbright-Hays Doctoral Dissertation Research Fellowship, and the IEE/ITT Fulbright Fellowship for their generous support during a period of research which forms a substantial basis for this chapter. I would also like to thank the following people for extremely useful close readings and/or critical comments on drafts along the way – Piers Blaikie, Claudia Carr, Alex Clapp, Carmen Diana Deere, Donald Gauthier, Lucy Jarosz, George Leddy, Brian Page, Richard Peet, Teodor Shanin, Ivan Szelenyi, Craig Thorburn, and Michael Watts. Responsibility for the ideas presented is completely my own.

1 Conversation between author and environmental program offices at UNDP, Beijing, August 1994.
2 Based on an untitled documentary video shown by environmental geographers from the Institute of Geography at Beijing, at the IGU meeting in Beijing, August 1990.
3 Interview with Wang Lianzheng, Vice Minister, Ministry of Agriculture and President of the Chinese Academy of Agricultural Sciences, April 1989.
4 Information also derived from collected field notes from work in Heilongjiang (1983–9), Jilin (1984), Xinjiang (1984–5), Sichuan (1985), Anhui (1985), Yunnan (1989), Guangxi (1988), Guangdong (1987), and Henan (1989) provinces.
5 This is the World Bank's own estimate. Taken from notes of discussions with members of the staff at the World Bank office in Beijing, 1989, 1990, 1991, and 1992.
6 Also based on discussions between the author and economists at the Chinese Academy of Agricultural Sciences, 1993. Inflation estimates by World Bank officials in Beijing.
7 Also based on discussion with the vice-head of the World Bank office in Beijing, April 1993.
8 Also based on discussion with William Hinton, Beijing, April and July 1993, and with the vice-head of the World Bank office in Beijing, April 1993.
9 Conversation with Sun Jia, Deputy Director of Agriculture, Heilongjiang Province, 1990.

REFERENCES

Bardhan, P. and J. Roemer. 1993. *Market Socialism: The Current Debate.* Oxford: Oxford University Press.
Bernstein, H. 1977. "Notes on capital and peasantry," *Review of African Political Economy* 10: 60–73.
Blaikie, P. 1985. *The Political Economy of Soil Erosion in Developing Countries.* Essex: Longman.

Blaikie, P. and H. Brookfield. (eds.) 1987. *Land Degradation and Society*. London: Methuen.

Blaikie, P., T. Cannon, I. Davis, and B. Wisner. 1994. *At Risk: Natural Hazards, People's Vulnerability, and Disasters*. London: Routledge.

BR. 1989. *Beijing Review* 32, 11, 1–7 May.

—— 1993a. *Beijing Review* 36, 4–5, 3–9 May.

—— 1993b. *Beijing Review* 36, 5, 11–17 Jan.

—— 1995. *Beijing Review* 38, 5, 23–9 Jan.

Byrd, W.A. (ed.). 1990. *China's Rural Industry: Structure, Development, and Reform*. Oxford: Oxford University Press.

Cannon, T. and A. Jenkins. 1990. *The Geography of Contemporary China: The Impacts of Deng Xiaoping's Decade*. London: Routledge.

Carney, J. 1993. "Converting the wetlands, engendering the environment: the intersection of gender with agrarian change in The Gambia," *Economic Geography* 69, 4: 329–48.

Carr, C.J. 1985. "The ideology of development," working paper, Department of Conservation and Resource Studies, University of California, Berkeley.

Chai, J.C.H. and C. Leung (eds.). 1987. *China's Economic Reforms*. Hong Kong: University of Hong Kong.

Cheng, H. 1991. *Federal Reserve Bank of San Francisco Weekly Newsletter*. February 22.

China Statistical Yearbook (CSY). 1991. Beijing: China Statistical Publishing House.

—— 1992. Beijing: China Statistical Publishing House.

Chossudovsky, M. 1986. *Towards Capitalist Restoration? Chinese Socialism after Mao*. New York: St. Martin's Press.

Cooper, M. 1993. "Transformation in contemporary Russia," lecture given at UCLA, 28 March.

Cressy, G.B. 1934. *China's Geographic Foundations: A Survey of the Land and its People*. New York: McGraw Hill.

Croll, E.J. 1988. "The new peasant economy in China," in S. Feuchtwang, A. Hussain, and T. Pairault (eds.) *Transforming China's Economy in the Eighties: The Rural Sector, Welfare, and Employment*. Boulder, CO: Westview Press.

Davin, D. 1988. "The implication of contract agriculture for the employment and status of Chinese peasant women," in S. Feuchtwang, A. Hussain, and T. Pairault (eds.) *Transforming China's Economy in the Eighties: The Rural Sector, Welfare, and Employment*. Boulder, CO: Westview Press.

de Janvry, A. 1981. *The Agrarian Question and Reformism in Latin America*. Baltimore and London: Johns Hopkins University Press.

Delfs, R. 1984. "Agricultural yields rise, but the boom cannot last," *Far East Economic Review* Dec.: 68.

Delman, J. 1989. "Current peasant discontent in China: background and political implications," *China Information* 4, 2: 49.

Deng, Y., J. Chen, Y. Xue, and J. Liu. 1992. *Chinese Rural Reform and Development: Looking Back and Looking Forward* (Modern Socialism Series – in Chinese). Guangdong, China: Guangdong Education College Publishing House.

Dorn, J.A. and X. Wang (eds.). 1990. *Economic Reform in China: Problems and Prospects*. Chicago: University of Chicago Press.

Du, R. 1985. "Lianchan chengbaozhi he nongcun hezuo jingjide xin fazhan" (New developments in the contracting system of united production and the cooperative economy in the countryside), *Renmin ribao* 7 March: 2.

Edmonds, R.L. 1994. "China's environment," in William A. Joseph (ed.) *China Briefing*. Boulder, CO: Westview Press.

Epstein, G., J. Graham, and J. Nembhard (eds.). 1993. *Creating a New World Economy*. Philadelphia: Temple University Press.

FBIS (Foreign Broadcast Information Service). 1992. Broadcast on "Rural areas suffering energy shortages," (HK 230 1043992) January 31: 50–2. English article by Zhai Feng in *China Daily* January 23, 1994, p.4.

Feuchtwang, S., A. Hussain, and T. Pairault (eds.). 1988. *Transforming China's Economy in the Eighties.* Boulder, CO: Westview Press.

Friedman, M. 1990. "Using the market for social development," in J.A. Dorn, and X. Wang (eds.) *Economic Reform in China: Problems and Prospects.* Chicago: University of Chicago Press.

Gladney, D. 1994. "Ethnic identity in China: the new politics of difference," in William A. Joseph (ed.) *China Briefing.* Boulder, CO: Westview Press.

Gurley, J.G. 1976. *China's Economy and the Maoist Strategy.* New York: Monthly Review Press.

Harvey, D. 1982. *The Limits to Capital.* Oxford: Basil Blackwell.

HSY 1988. *Heilongjiang jingji Tongji Nianjian* (Heilongjiang Statistical Yearbook). Beijing, China: Heilongjiang Province Statistical Bureau, China State Statistical Bureau Publishing House.

—— 1989. *Heilongjiang jingji Tongji Nianjian* (Heilongjiang Statistical Yearbook). Beijing, China: Heilongjiang Province Statistical Bureau, China State Statistical Bureau Publishing House.

——1992. *Heilongjiang jingji Tongji Nianjian* (Heilongjiang Statistical Yearbook). Beijing, China: Heilongjiang Province Statistical Bureau, China State Statistical Bureau Publishing House.

Hinton, William. 1985. "The situation in the grasslands," *Inner Mongolian Journal of Social Science* August. Reprinted in *The Great Reversal,* by Hinton, 1990. New York: Monthly Review Press.

—— 1990. *The Great Reversal: The Privatization of China, 1978–1989.* New York: Monthly Review Press.

Howard, Michael (ed.). 1993. *Asia's Environmental Crisis.* Boulder, CO: Westview Press.

King, F.H. 1911. *Farmers of Forty Centuries, or Permanent Agriculture in China, Korea and Japan.* Madison, WI: Mrs F.H. King.

Lardy, N.R. 1984. "Consumption and living standards in China, 1978–83." *The China Quarterly* 100: 849–65.

Liang, M. 1988. *Jingji Yanjiu Cankao Ziliao* (*Economic Research Reference Materials*). 34: 32–40. (Beijing Chinese Academy of Social Sciences.)

Lin, J. 1993. "Chinese agriculture: institutional changes and performance," paper presented at the Conference on Rural Transformation and Comparative Socialist Transition, Shanghai, September.

Lin, Xiaoran. 1992. "Setting up a grain reserve system to strengthen the state macro-regulation capacity," *Problems of Agricultural Economy* 9: 47–50.

Malhotra, A. and J. Studwell. 1995. "Revolution's children," *Asia, Inc.* 4, 1: 28–33.

Massey, D. 1984. *Spatial Divisions of Labour: Social Structures and the Geography of Production.* London: Macmillan.

Muldavin, J. 1983. *Fifteen-Year Plan for the Agricultural and Agro-Industrial Development of Zhaozhou County.* Beijing: Ministry of Agriculture.

—— 1986. "Mining the Chinese earth," MA thesis, Department of Geography, University of California at Berkeley.

—— 1989a. "Hesheng Village Survey (HSVS)," Bayan County, Heilongjiang Province.

—— 1989b. "Bayan County Survey I (BCSI)," Bayan County, Heilongjiang Province.

—— 1989c. "Fendou Village Survey (FDVS) and Fuxiang Village Survey (FXVS)," Bayan County, Heilongjiang Province.

—— 1992. "China's decade of rural reforms: the impact of agrarian change on sustainable development," Ph.D. dissertation, Department of Geography, University of California at Berkeley.

—— 1994a. "Bayan County Survey II" (BCSII), Heilongjiang Province.

—— 1994b. "Production treadmills in socialist transition," manuscript.

Nee, V. and D. Stark. 1989. *Remaking the Economic Institutions of Socialism: China and Eastern Europe*. Stanford, CA: Stanford University Press.

Neumann, R. 1992. "Political ecology of wildlife conservation in the Mt. Meru area of northeast Tanzania," *Land Degradation and Society* 3: 85–98.

Nolan, P. 1988. *The Political Economy of Collective Farms*. Boulder, CO: Westview Press.

Nove, A. 1983. *The Economics of Feasible Socialism*. London and Boston: Allen & Unwin.

Odgaard, O. 1992. *Private Enterprises in Rural China: Impact on Agriculture and Social Stratification*. Hants: Avebury.

Page, B. 1993. "Agro-industrialization and regional transformation: the restructuring of Midwestern meat production." Unpublished Ph.D. dissertation, Department of Geography, University of California, Berkeley.

Page, B. and R. Walker. 1991. "From settlement to Fordism: the agro-industrial revolution in the American Midwest," *Economic Geography* 67: 281–315.

—— (forthcoming). "Staple lessons: Harold Innis, agriculture and industrial geography," in M. Gertler and T. Barnes (eds.) *Regions, Institutions and Technology: A Centennial Celebration of Harold Innis*. London: Routledge.

Perelman, M. 1977. *Farming for Profit in a Hungry World: Capital and the Crisis in Agriculture*. Montclair: Allanheld, Osmun & Co.

Purdue, P. 1987. *Exhausting the Earth: State and Peasant in Hunan, 1500–1850*. Cambridge, MA: Harvard University Press.

Riskin, C. 1987. *China's Political Economy: The Quest for Development since 1949*. Oxford: Oxford University Press.

—— 1995. "Non-regional poor in China," paper presented at the Association of Asian Studies Annual Meetings in Washington, DC. 3 April.

Ross, L. 1988. *Environmental Policy in China*. Bloomington: Indiana University Press.

Sayer, A. and R. Walker. 1992. *The New Social Economy: Re-working the Division of Labor*. Oxford: Basil Blackwell.

Schell, O. 1995. "Twilight of the titan: China – the end of an era," *The Nation* 17/24 July, 61, 3: 92.

Scott, A. 1988. *Metropolis: From the Division of Labor to Urban Form*. Berkeley and Los Angeles: University of California Press.

Scott, A. and M. Storper. 1986. *Production, Work, Territory: The Geographical Anatomy of Industrial Capitalism*. Boston: Allen & Unwin.

Scott, J.C. 1985. *Weapons of the Weak: Everyday Forms of Peasant Resistance*. New Haven and London: Yale University Press.

Shih, S. 1962. *A Preliminary Survey of the Book "Ch'i Min Yao Shu": An Agricultural Encyclopaedia of the 6th Century*. Peking, China: Science Press.

Smil, V. 1984. *The Bad Earth: Environmental Degradation in China*. New York: M.E. Sharpe.

—— 1993. *China's Environmental Crisis: An Inquiry into the Limits of National Development*. New York: M.E. Sharpe.

Stone, B. 1985. "The basis for Chinese agricultural growth in the 1980s and 1990s: a comment on Document No. 1, 1984," *China Quarterly* 101.

—— 1988. "Developments in agricultural technology," *China Quarterly* 116: 767–822.

Storper, M. and R. Walker. 1989. *The Capitalist Imperative: Territory, Technology and Industrial Growth*. Oxford: Basil Blackwell.

Swift, J. 1989. "Why are rural people vulnerable to famine?" *IDS Bulletin* 20, 2: 27.

USNWR (U.S. News & World Report). 1990. 12 March: 44.

Walker, K. 1984. "China's agriculture during the period of readjustment, 1978–83," *China Quarterly* 100: 783–812.

World Bank. 1990. *China: Managing an Agricultural Transformation – I* (Vol. II: Working Papers 4–8), World Bank, China Department, Agricultural Operations Division, Asia Regional Office.

Zhang, Bingwu. 1993. "Yingjie xiandaihua tiaozhan de dangdai nongmin" (article on a household study in Baomiancheng zhen, Changtu xian, in Liaoning Province, entitled "Modern peasants facing the challenge from modernization"), *Shehui kexue jikan* (Social Science Periodical) 1: 32–9.

Zhao, Songqiao. 1981. "Transforming wilderness into farmland," *China Geographer* 11 (Special Issue on agriculture).

ZNTN. 1992. *Zhongguo nongcun tongji nianjian* (China Rural Economy Statistical Yearbook). Beijing: Beijing Statistical Publishing House, State Statistical Bureau.

ZTNJ. 1991. *Zhongguo tongji nianjian* (A Statistical Survey of China). Beijing: Beijing Statistical Publishing House, State Statistical Bureau.

—— 1992. *Zhongguo tongji nianjian* (A Statistical Survey of China). Beijing: Beijing Statistical Publishing House, State Statistical Bureau.

—— 1993. *Zhongguo tongji nianjian* (A Statistical Survey of China). Beijing: Beijing Statistical Publishing House, State Statistical Bureau.

12

CONCLUSION
Towards a theory of liberation ecology
Michael Watts and Richard Peet

> The most likely center for a possible coalescence of a multitude of new
> social movements into a major social movement . . . is the societally-basic
> relationship, nature–society.
>
> (Oloffson 1988: 15)

Liberation ecology integrates critical approaches to political economy with
notions derived from poststructural philosophy. The quest is to understand the
ways human practice transforms the Earth and the ways in which environmental
practices, institutions, and knowledges might be subverted, contested, and
reformed. In this sense *Liberation Ecologies* speaks to a critical analysis of environmental degradation and rehabilitation framed by something called development,
and also the liberatory potential of struggles and conflicts exactly around these
processes. Liberation ecology starts from Marx's presumption that society–nature
relations are the outcomes of the metabolic activity of the labor process –
by which nature is humanized and humans are socialized – but posits this
metabolism as social, cultural, and discursive as much as it is narrowly
"economic." If nature, to use O'Connor's language, is a production condition in
contradiction with the impulses of a profit-driven commodity economy, the
particular ways in which it is configured and transformed must, in our view, be
linked to the complex ways in which power, knowledge and institutions sustain
particular regimes of accumulation. This so-called "third leg" of the analysis of
modernity – the Foucauldian complement to Weber and Marx – represents a
central part of liberation ecology as a theoretical enterprise.

In this regard, poststructural criticisms of Western rationality and modernity
provide critical vantage points from which to assess Enlightenment claims to
absolute scientific objectivity. Indeed we might add to the poststructural critique
of reason and science the main criticism of modernity from the perspective
of environmentalism. Namely, that the First World which claims to know,
speak, and practice truth through the medium of science is simultaneously
nature's horsemen of the apocalypse: Reason's unreason is displayed in society's
demonstrated ability to destroy its natural conditions of existence. Yet as
Marshall Berman (1982) argued in *All That Is Solid Melts Into Air*, recognizing

260

the contradictory qualities of modernity – its powers of creative destruction – does not necessarily imply rejecting it *tout court*. For us the response to unreason has to be a different kind of reasoning, to guide a different social order of practice on nature, with a knowledge of natural processes, and the effects of human activity on them, consciously integrated into the very relations that make up society. The causes of the "reversal of reason," and the sources of resolution of ecological crises, must both be sought in the social relations which unify thought with material practices. In this regard, theories of ideology still hold great attraction, not in the sense of counterposing a Marxian claim to objective truth, but in the sense that all thought systems are understood as serving the interests of specific social forms of power.

If we call "reasoning" the contested process of constructing logic – in contra-distinction to "rationality" as the discovery of a logic latent in things – then reasoning as an active process means contemplating the consequences of practice a priori; "reasonable behavior" with regard to nature involves practices disciplined by prior knowledge of their effects. The lack within modern life is a *poverty of reasoning* and a lack of disciplined behavior rather than an excess (Thompson 1979). Harms against nature – that is acting on nature in ways which we already know or, given the complexity of the interaction, deeply suspect, to be fundamentally destructive – reflect precisely this surfeit of *un*-reason. But branches of poststructural and ecological thought which abandon reason and science as guides to human action, in our view, run the grave risk of an idealism which throws the baby out with the bath water. Reason must be re-reasoned rather than rejected, science should be changed and used differently, not abandoned. A liberatory ecology thus retains the modernist notion of reasoned actions on nature, accepts much of the Marxian and poststructural critique of capitalist rationality, yet wishes to substitute for it a democratic process of reasoning – a sort of environmental public sphere – in a transformed system of social and natural relations.

As we argued in our introductory chapter, Marxian notions of ideology, post-structural critiques of Enlightenment reason, and the postmodern questioning of the entire process of Western development, open the way for more serious consideration of *alternate* forms of environmental practice and knowledge. The possibility emerges of joining a critique of the West, especially its environmental relations and practices, with a critical appreciation for alternative rationalities, productive relations, and environmental practices in non-capitalist societies. Nonetheless, we would caution against the sort of idealism expressed by Vandana Shiva (1991) when she claims confidently that the universal feature of Indian and other environmental movements is that they create new values, new rationalities, and "a new economics for a new civilization." Southern green movements have all too readily been mystified, glorified, and idealized as "authentic" forms of agency and conservators of traditional green lifestyles (Linkenbach 1994: 70). There is no pure, perfect, or easy solution waiting to be found stored in the non-polluted minds of shamans or retained by all-knowing Third World peasant agro-ecologists. Hence an opening of poststructural materialist thought to the world of environmental experiences is as much an exercise in *critique* as it is an appeal to the virtues of local or subaltern knowledges.

FROM POLAR TO HYBRID POLITICS: LESSONS FROM THE CASE STUDIES

From the chapters of this book it should be clear that the structure, content, and even the direction of the intellectual and political debates over modernity, post-modernity, and the environment are far from closed issues. In Chapter 2, Arturo Escobar argues for a poststructural political ecology which reflects the belief that nature is socially constructed – both known and "produced" through discourses and social practices. He employs this perspective in a discursive-materialist analysis of the contemporary phase of ecological capitalism, that is the movement from modern capitalism, with its exploitation of an external nature, to post-modern capitalism, with its conservation of an internalized "capitalized nature." Both tendencies are associated with complex cultural and discursive articulations. For Escobar, there cannot be a material analysis that is not, at the same time, discursive. In Chapter 3, Lakshman Yapa's account of development's production of modern poverty is cut from the same cloth as Escobar, but employs the analytic of a "nexus of production relations," in other words the notion that production is determined within a web of mutually constitutive technical, social, ecological, cultural, and academic relations. Hence, for example, "improved seeds," part of the Green Revolution, are not simply technical and ecological, but are also social and cultural insofar as they bear the imprint of the hegemonic culture of capitalist modern science. Both chapters elaborate notions of culture and discourse as active agents creating reality, reproducing nature, and framing our knowledge of this process.

This social constructionist argument adds depth to the earlier work of political ecologists such as Blaikie who, nevertheless, pointed to the perceptual and cognitive nature of environmental problems (for example, the wildly disparate and inconsistent evidence for an ecological crisis in Nepal offered by a variety of state, multilateral, and private institutions and researchers). But there are, in our view, grave limitations associated with a strong social constructionist position, that is the position that nature is "constructed" not only in the sense of being "known" through socially conditioned minds, but also "historically produced" by discourse and knowledge. Its idealist tendencies – the notion that ideas exist first and reality, even nature, is their materialization – and its failure to recognize that not everything is socially produced (certainly not socially produced to the same degree) are deeply problematic. For us the term "social construction of nature" overestimates the transformative powers of human practice and, as Benton (1989) notes, underestimates the significance of non-manipulable nature. In this sense, the terms "production" and "construction" of nature may often obscure more than they illuminate.

In terms, then, of a more critical liberation ecology, there is a need to counter-balance the "social construction of nature" with a profound sense of the "natural construction of the social." Each society carries what we refer to as an "environmental imaginary," a way of imagining nature, including visions of those forms of social and individual practice which are ethically proper and morally right with regard to nature. As intimated in our introduction, this imaginary is typically expressed and developed through regional discursive formations, which take as central themes the history of social relations to a particular natural environment. Environmental imaginaries are frequently, indeed usually, expressed in abstract, mystical, and spiritual lexicons. However they contain some degree of the reasoned approaches which display or "work out" the consequences of environmental actions referred to earlier as "prior knowledges." Liberation ecology proposes studying the processes by which environmental imaginaries are formed, contested, and practiced in the course of specific trajectories of political-economic change. It borrows from poststructuralism a fascination with discourse and institutional power, yet remains within that tradition of political ecology which sees imaginaries, discourses, and environmental practices as grounded in the social relations of production and their attendant struggles. The environmental imaginary emerges, therefore, as a primary site of contestation; critical social movements have at their core environmental imaginaries at odds with hegemonic conceptions. An environmental imaginary, then, is a particular sort of situated knowledge, to employ Donna Haraway's language. But perhaps most importantly, through the concept of environmental imaginary, liberation ecology sees nature, environment, and place as *sources* of thinking, reasoning, and imagining: the social is, in this quite specific sense, naturally constructed.

Why do we debate such positions in a liberation ecology which encourages, indeed revels in, contestations over the reasoning process? Liberation ecology shares with Marx the belief that reasoning provides a basis for political action. Liberation ecologies provide a sort of mapping of the social and political relations to nature and guide the strategic interventions which can be derived. In this last sense, and surprisingly perhaps, the authors of what at first appear to be disparate chapters in *Liberation Ecologies* end up with quite similar political accounts. Arturo Escobar concludes with a sympathetic account of Leff's argument on the need for articulating an alternative (ecologically sustainable) production rationality which integrates ecology, culture, and economy and builds on the "ecological cultures" of indigenous peoples. In a similar vein, the practice of indigenous people in Ecuador, sympathetically described by Tony Bebbington in Chapter 4, is a fascinating tale of the selective incorporation of modern techniques into a project of cultural resistance by indigenous people. Likewise, Haripriya Rangan in Chapter 10 argues that there is not one model of eco-development in India's Garhwal district, but (to employ Linkenbach's language [1994: 81]), several models for future life, including one in which villagers have no desire to withdraw totally from modernity, but want to pick and choose and get their share of modern benefits. One lesson from these diverse

263

cases is that a liberation ecology which discourages polarized arguments enriches the hybrid politics which eventually emerge around ecologically sustainable livelihood strategies.

FORMS OF ARTICULATION IN LIBERATION ECOLOGY

Virtually all the chapters in this book take as their starting point the consequences of modern development – whether inspired by industrial or agrarian, socialist or capitalist models – on Third World environments. That is, they examine specific sorts of *societal articulations* with nature seen through the lens of development. The notion of articulation has a lineage traceable to the French-inspired theory of modes of production (see Wolpe 1980), and it remains one of structural Marxism's finer insights. Drawing on the geographies depicted by dependency theory, modern history in this account consists of a series of interactions – articulations – between a capitalist mode of production expanding from its point of origin in Western Europe, and a series of non-capitalist modes located in what became the peripheries of global space (Peet 1991). In spite of its deficiencies – insufficient critical attention to the rationality of capitalist development, a clumsy view of the social relations constituting modes of production, and a lack of case studies of the intricacies of specific interactions between societies – this geographical variant of structuralism offered important insights and a powerful language for thinking and talking about the development experience. A number of themes in the chapters of this book – the series of encounters between development based in Western rationality, with its objectifying, instrumental attitudes towards an externalized nature, and the structures and dynamics, rationalities and imaginaries, of a sequence of Third World societies – resonate still with the lineage of Marxist structuralism and articulation. We remain, in short, in an intellectual environment of post*structuralism* after all.

Yet if the idea of articulation runs across the chapters, it needs to be said that the authors stress quite different aspects and relations than was the case twenty years ago. Let us quickly run through some of these new emphases, which in many ways explore the full range of the meaning of the word "articulation."

Insofar as Western development posits itself as the embodiment of reason then to the same extent must it see its Other as irrational. Poststructuralism's linking of rationality with power enables us to better understand such views as the modalities by which colonial and post-colonial resource use proceeds and multiplies. This is nicely illustrated by Lucy Jarosz's Chapter 7 on shifting cultivation in the tropical forests of Madagascar. There "rational" relations with nature were defined by the colonial (and post-colonial) state's intention with regard to the use of forest resources; rationality is always politically and economically motivated. Peasants' alternative use (shifting cultivation) had to be defined as "irrational" by the state and its academic allies. Jarosz takes this critique in an interesting direction by showing how the banning of shifting cultivation by the colonial state conferred new meanings on the practice –

burning the forest became an act symbolizing defiance of colonial power. Thus environmental struggles in the Malagasy case emerged from contested "readings" of the rationality of resource use. Drawing on a somewhat different intellectual tradition, Chapter 6 by Donald Moore uses Gramscian Marxism to explore how resource actions are mobilized not only by material interests in a direct sense, but by cultural as well as class identities, so that meanings become constitutive forces, shapers of history, space, and nature. And again Karl Zimmerer in Chapter 5 argues that local knowledges, perceptions, and discourses articulated by governmental and non-governmental organizations, peasants and peasant social movements, must be added to the conventional analytical repertoire of political ecology.

These sorts of analyses, which link power, knowledge, and practice, in the context of articulation, are not peculiar to capitalist modernity, but run through the history of actually existing socialisms. The rapidity with which Stalinist models of agro-industrial collectivization were exported to newly liberated Third World socialisms speaks directly to the hegemony of certain "socialist" rationalities. In the same way, only now with the growth and opening up of civil society in parts of the socialist world – including green movements and efforts to rehabilitate pre-socialist environmental practices – is the contestation over different sorts of environmental rationalities seeing the light of day (MacArthur 1995). In this regard, Cuba represents an intriguing case. While Castro and the Communist Party adopted a Soviet-style model of resource use, in which nature became a source of administrative and political use-value through centralized state planning, the impact of the post-1989 collapse of the Soviet Union – the so-called Special Period in Cuba – has compelled the Cuban government to refigure its industrial model of agriculture along agro-ecological lines. The shortage of green revolutionary inputs for state farms and co-operatives – the overall import coefficient for fertilizer was 94 percent in the 1980s – compelled the government to substitute local for imported technologies, a process which is, amidst much debate and contestation, gradually displacing the classical rationality with an alternative model based on IPM (integrated pest management), crop rotations and intercropping, biofertilizers, waste recycling, local knowledge, and popular participation (Rosset and Benjamin 1994). Conversely, Joshua Muldavin's analysis of Chinese socialism in Chapter 11 examines how another sort of articulation – the post-Mao integration of China into global market relations – produces a drawing down of communal ecological capital constructed within the old commune system as decollectivization and privatization extends to the Chinese countryside. Here articulation combines the worst of state socialism and market-induced "rationalities" in generating what threatens to become an environmental catastrophe. In both the Chinese and the Cuban cases, the environment–development nexus turns on a series of unexpected articulations, complex sorts of hybrid associations between nature and society, that emerge from various local sorts of market-driven restructuring.

Another distinctive emphasis within liberation ecology turns on the ways in which social relations are employed to understand the contradictory quality of environment and development. Development initiatives create the potential for new or heightened tensions in social relations to nature but the chapters in this book reveal dramatically how struggles over natural resources are embodied in a panoply of social and cultural forms. In general it appears that custom and tradition provide the symbolic and representational landscape on which struggles over nature are located. But this understanding is, as yet, insufficient for anything approaching a general theory which can predict the social sites of environmental and livelihood struggles which, to put it simply, vary from one place to another. Thus in Judith Carney's Chapter 8, transformation through development and commodification of the use of Gambia's wetlands is played out through domestic, that is to say gender and patriarchal, spheres rather than simply along class lines. This moment of articulation is explored in the active sense of its contention by women, growing in militancy, and the possibility of new forms of collective action. The diversity of sites of contesting articulations is nicely demonstrated in Chapter 9 by Richard Schroeder and Krisnawati Suryanata, which compares agroforestry initiatives in Java, where conflicts emerge along nascent class lines, and in Gambia, where they appear as conjugal conflict and women's solidarity. In some of the other case studies – for example of Bolivia, Ecuador, and Zimbabwe – the idiom of environmental and resource contestation is expressed through generational differences, ethnic identities, lineages, or other aspects of local "tradition" (religion for example), all of which are subject to constant manipulation and reinvention.

Our emphasis on articulations in various guises suggests a number of fundamental starting points for a liberation ecology focused on the developing world. First, at a moment of capitalist triumphalism, the various restructurings and adjustments in the world economy produce neither a simple displacement of the traditional by the modern, the state by the market, or the local by the global but, rather, a complex set of articulations which take the form of hybrid sorts of development and forms of modernity. The challenge for liberation ecology is to root specific environmental practices, outcomes, knowledges, and politics in these hybrid conditions of the late twentieth century. Second, development, in whatever form it appears on the Third World's doorstep, often has the effect of destabilizing the systems of access and control over local resources, and to this extent nature–society relations become objects of struggle, negotiation, and contestation; the idioms in which this occurs tend to be varied and heterogeneous but not infrequently bracketed by some notion of custom or tradition. And third, the various institutions, practices, and discourses through and by which the nature–society dialectic plays itself out under conditions of late twentieth-century capitalist modernization always contains a locally grounded vision of nature itself, an environmental imaginary, typically expressed through a regional discursive formation. It is to this specifically natural discursive formation – the environmental imaginary – that we finally turn.

LIBERATION ECOLOGIES AND ENVIRONMENTAL IMAGINARIES

[T]he Amazonia with which most Americans are familiar is, in large part, an invention, a series of variations upon the guiding theme of El Dorado. By El Dorado I mean both the sixteenth-century explorer's original shining city and a more general complex of ideas involving immense natural wealth, transformative potential and some sort of inaccessibility. My objectives are to suggest the power and longevity of this foundational theme . . . and its multiplicity of forms.

(Slater 1995)

Let us return to that part of our introductory chapter which summarized the work of Cornelius Castoriadis on the social imaginary. Castoriadis (1994) argues that societies create themselves as quasi-totalities held together by institutions, social imaginary significations, and systems of meaning and representation which organize their natural worlds and establish ways people are socialized. While societies are conditioned by natural environments, he says, they are not determined by them. Instead, for Castoriadis, the active moment is the society's creative construction of edifices of signification – that is, their social imaginaries. For us, however, this raises the further issue of *how* societies create these systems of meaning which endow the natural world with significance?

We argue for an active role of the social relations with nature in creating not merely a social, but also an environmental, imaginary. Natural environments, visible still beneath layers of socialization, landscapes which express human use of what remain primarily natural spaces, the places groups of people inhabit, are main sources in the *creation* of their meaning systems, aesthetics, and systems of thought. In particular the "pictures" or images which form the first moments in the creation of thoughts, and which thought constantly employs as material-izations (i.e. visualizations), are representations of specific natural and social environments. Drawing on Castoriadis's (1994: 138) two meanings of the word "imagination" – its connection with visual images and its connotation of inven-tion and creativity – we argue that inhabiting a specific natural environment with, as it were, a given "supply" of image-types, limits yet projects the creative aspect of the imagination. That is, the very act of imaginative creativity builds on, fantasizes about, makes more perfect, a certain world with limited ranges of color, materials, and resource frameworks. Creativity begins with the familiar. The very finest thoughts we have, in which the mind ranges ahead of what is known, can draw only on an aesthetic derived from interactions with a given natural world. The important point here is for social theory to draw out the *creative* potential of necessary relations with the natural world – much in the sense of Sartre's "projects" – which derive aesthetics, innovative thinking, ideals, values, and imaginaries from labor and residence.

While environmental imaginaries stem from material and social practices in natural settings they also guide further practices. There is no implication here

267

that humans contemplate nature and, so-inspired, go out to change the world – a kind of environmental idealism in existential disguise. Instead there is an active interaction between practice and idealization in which imaginaries are constantly rebuilt and refigured, accumulate and change, during practical activities which imagination has previously framed. Yet the word "imaginary" is meant in the full sense of creativity – the projection of thought into the scarcely known – so that it is a vital source of transformational, as well as merely reproductive, dynamics: the imaginary links natural conditions with the construction of new social forms. We should add that (hegemonic) imaginaries accumulated in one region can be carried to others, where they alter perceptions and attitudes often with significant, even calamitous effects. In a dramatic statement, Davis (1995) describes how a "uniformitarian" conception of natural processes, what he calls imaginary norms and averages derived from the European and New England experiences, conditions environmental expectations in the Southern Californian desert, with its episodic bursts of sudden transformation. To this we might add the further complication that images of the California landscape are projected by the visual media into a world of imaginaries as a technicolored stereophonic ideal, altering if not the contents of these imaginaries, then certainly the representational process through which they are formed and displayed. Thus are contradictions and mistakes universalized by an imperialism of the imaginary! There is, then, a complex interplay between natural and social construction with the environmental imaginary as centerpiece, while articulations between societies can be expressed as the interaction between imaginaries armed with different powers and technologies.

Environmental imaginaries are thus prime sites of contestations between normative visions. Unjust property rights and aesthetically offensive uses of nature can spur political opposition to the hegemonic social order. In world systems which destroy broad, even global environments, these have the potential to become widespread social movements – many environmental movements cut across class, gender, and regional divisions. They also can be fundamental movements in that they challenge the very basis of society – how people use nature, how human nature comes about, how imaginations are imagined. At stake in environmental movements is nothing less than the way people understand their humanity. It is almost delightfully naive to assume that the content of the resulting green movements is necessarily progressive. The panoply of green politics in the contemporary United States, for example, reveals political allegiances which cover a multitude of positions and exhibit a diversity of sins. Some deep ecology – indeed the history of ecological thinking itself (Bramwell 1980) – has an affinity with what might legitimately be called eco-fascism. Rather than uncritically endorsing all ecological movements – or all visions of alternative developments – *Liberation Ecologies* hopefully provides a set of critical tools which point up the limitations, intractabilities, and contradictions of various models of development, and registers our support of Edward Thompson (1992: 15) in his admonition that "we shall not ever return to pre-capitalist human nature yet a reminder of its alternative needs, expectations and codes may renew our sense of nature's range of possibilities."

REFERENCES

Benton, Ted. 1989. "Marxism and natural limits: an ecological critique and reconstruction," *New Left Review* 178: 51–86.

Berman, Marshall. (1982) *All That Is Solid Melts Into Air.* New York: Simon and Schuster.

Bramwell, Anna. 1980. *Ecology in the Twentieth Century.* New Haven: Yale University Press.

Castoriadis, C. 1994. "Radical imagination and the social instituting imaginary," in G. Robinson and J. Rundell (eds.) *Rethinking Imagination: Culture and Creativity.* London: Routledge, pp. 136–54.

Davis, M. 1995. "Los Angeles after the storm: the dialectic of ordinary disaster," *Antipode* 27: 221–41.

Linkenbach, Antje. 1994. "Ecological movements and the critique of development," *Thesis Eleven* 39: 63–85.

MacArthur. 1995. MacArthur Foundation Conference on "Collectivization, Decollectivization and the Environmental Record in the Socialist World," Havana, Cuba, 20–6 June.

Oloffson, Gunnar. 1988. "After the working-class movement?" *Acta Sociologia* 31: 15–34.

Peet, Richard. 1991. *Global Capitalism: Theories of Societal Development.* London: Routledge.

Rosset, Peter and Medea Benjamin. 1994. *Two Steps Back, One Step Forward.* London: IIED, Gatekeeper Series #46.

Shiva, Vandana 1991. *Ecology and the Politics of Survival.* New Delhi: Sage.

Slater, Candace. 1994. *Dance of the Dolphin: Transformation and Disenchantment in the Amazonian Imagination.* Chicago, University of Chicago Press.

Thompson, Edward. 1979. *The Poverty of Theory.* London: Merlin.

—— 1992. *Customs in Common.* London: Penguin.

Wolpe, Harold (ed.). 1980. *The Articulation of Modes of Production.* London: Routledge.

INDEX

270